Gold

A Cultural Encyclopedia

SHANNON L. VENABLE

ABC-CLIO

Santa Barbara, California • Denver, Colorado • Oxford, England

Copyright 2011 by ABC-CLIO, LLC

All rights reserved. No part of this publication may be reproduced, stored in a retrieval system, or transmitted, in any form or by any means, electronic, mechanical, photocopying, recording, or otherwise, except for the inclusion of brief quotations in a review, without prior permission in writing from the publisher.

Library of Congress Cataloging-in-Publication Data

Venable, Shannon L.
 Gold : a cultural encyclopedia / Shannon L. Venable.
  p. cm.
 Summary: "This encyclopedia provides detailed information about the historical, cultural, social, religious, economic, and scientific significance of gold, across the globe and throughout history"— Provided by publisher.
 Includes bibliographical references and index.
 ISBN 978-0-313-38430-1 (hardback) — ISBN 978-0-313-38431-8 (ebook)
1. Gold—Social aspects—History—Encyclopedias. 2. Goldwork—History—Encyclopedias. 3. Gold—Metallurgy—History—Encyclopedias. 4. Gold in art—History—Encyclopedias. I. Title.
 GT5170.V46 2011
 669'.2203—dc22   2011000411

ISBN: 978-0-313-38430-1
EISBN: 978-0-313-38431-8

15 14 13 12 11   1 2 3 4 5

This book is also available on the World Wide Web as an eBook.
Visit www.abc-clio.com for details.

ABC-CLIO, LLC
130 Cremona Drive, P.O. Box 1911
Santa Barbara, California 93116-1911

This book is printed on acid-free paper ∞

Manufactured in the United States of America

Contents

List of Entries	vii
Guide to Related Topics	ix
Preface	xiii
Acknowledgments	xv
Introduction	xvii
The Encyclopedia	1
Resource Guide	289
Bibliography	291
Index	303

List of Entries

Aerospace Industry
Alchemy
Allergies
Alloying
Aquinas, Thomas
Archimedes
Argentina
Asante Golden Stool
Assaying
Atahualpa
Au

Balboa, Vasco Nuñez de
Banking and Credit
Baring Crisis
Beauty Products
Bezant
Biblical Magi
Biomedical Research
Bling
Bonding Wire
Bretton Woods System
Briot, Nicholas
Bullion
Bullion Committee of 1810

Ca d'Oro
California Gold Rush
Casting
Catalysis
Cellini, Benvenuto
Chevalier, Michel
Chrysography
Chrysotype
Coin Clipping
Coin Stamping
Coinage Act of 1873
Columbus, Christopher
Cortés, Hernán

Cox, James
Croesus
Crowns, Royal
Cyanide

De Gaulle, Charles
Dental Crowns
Drake, Sir Francis
Ducat

El Dorado
Electroplating
Emergency Banking Relief Act of 1933
Exchange Rates

Fabergé, House of
Federal Reserve System
Ferdinand II of Aragon
Field of the Cloth of Gold
Filigree
Florin
Fuel Cells

Gilded Age
Gilding
Gold Album Certification
Gold Nuggets
Gold Reserves
Gold Standard
Gold Standard Act of 1925
Golden Calf
Golden Fleece
Golden Goose
Golden Rule
Golden Ticket (*Charlie and the Chocolate Factory*)
Goldfinger (film)
Goldschläger
Goldsmithing

LIST OF ENTRIES

"The Goose and the Golden Egg" (Aesop)
Gospel Books of Saint Médard de Soissons
Granulation
Great Recoinage Act of 1696
Guinea

Hagia Sophia
Hamilton, Alexander
Hargraves, Edward Hammond
Hatshepsut
Henry VIII
Holy Grail
Hume, David

Iconoclasm
Illuminated Manuscripts
Inca
Inflation
International Monetary Fund
Isabella I of Castile

Jewelry
Justinian I

Karat
Keynes, John Maynard
Klondike Gold Rush
Kublai Khan

Lace, Gold
Leprechauns

Midas, King
Mining
Mosaic

Nanotechnology
Newton, Sir Isaac
Nobel Prize Medals

Nubia
Nuestra Señora de Atocha

Olympic Gold Medal
Oscar Statuette

Pizarro, Francisco
Polo, Marco
Ponte Vecchio
Pre-Columbian Gold
Price Revolution

Ricardo, David
Rumpelstiltskin

San Esteban
Shekel
Silla, Royal Tombs of
Smith, Adam
Spanish Armada
Spanish Conquest
SS *Republic*
Sumptuary Legislation
Sutter's Mill

"The Tale of the Golden Cockerel"
Tenochtitlan
Touchstone
Tower Mint
Trans-Saharan Gold Trade
Tutankhamun

United States Bullion Depository
Ur, Royal Tombs of

Versailles, Palace of

Wat Traimit
Witwatersrand Gold Rush

Yellow Brick Road

Guide to Related Topics

ADORNMENT
Crowns, Royal
Jewelry
Lace, Gold
Sumptuary Legislation

ART, ARTIFACTS, AND ARCHITECTURE
Asante Golden Stool
Ca d'Oro
Chrysography
Chrysotype
Fabergé, House of
Gospel Books of Saint Médard de Soissons
Hagia Sophia
Illuminated Manuscripts
Jewelry
Lace, Gold
Mosaic
Ponte Vecchio
Pre-Columbian Gold
Silla, Royal Tombs of
Tenochtitlan
Ur, Royal Tombs of
Versailles, Palace of
Wat Traimit

BANKING AND FINANCE
Argentina
Banking and Credit
Baring Crisis
Bretton Woods System
Bullion Committee of 1810
Emergency Banking Relief Act of 1933
Exchange Rates
Federal Reserve System
Gold Reserves
Gold Standard
Gold Standard Act of 1925
Inflation
International Monetary Fund
Price Revolution
United States Bullion Depository

CULTURE
Bling
Field of the Cloth of Gold
Gold Album Certification
Golden Ticket
Goldfinger
Goldschläger
"The Goose and the Golden Egg"
Inca
Nobel Prize Medals
Nubia
Olympic Gold Medal
Oscar Statuette
Yellow Brick Road

GOLD RUSHES
California Gold Rush
Hargraves, Edward Hammond
Klondike Gold Rush
Sutter's Mill
Witwatersrand Gold Rush

GUIDE TO RELATED TOPICS

HEALTH AND MEDICINE

Allergies
Beauty Products
Biomedical Research
Dental Crowns
Nanotechnology

IDEAS AND MOVEMENTS

Alchemy
Gilded Age
Golden Rule
Iconoclasm
Spanish Conquest
Trans-Saharan Gold Trade

INDIVIDUALS

Aquinas, Thomas
Archimedes
Atahualpa
Balboa, Vasco Nuñez de
Briot, Nicholas
Cellini, Benvenuto
Chevalier, Michel
Columbus, Christopher
Cortés, Hernán
Cox, James
Croesus
De Gaulle, Charles
Drake, Francis
Ferdinand II of Aragon
Hamilton, Alexander
Hargraves, Edward Hammond
Hatshepsut
Henry VIII
Isabella I of Castile
Hume, David
Justinian I
Keynes, John Maynard
Kublai Khan
Midas, King
Newton, Sir Isaac
Pizarro, Francisco
Polo, Marco
Ricardo, David
Smith, Adam
Tutankhamun

MONEY AND COINAGE

Bezant
Bullion
Coin Clipping
Coin Stamping
Coinage Act of 1873
Ducat
Florin
Great Recoinage Act of 1696
Guinea
Shekel
Tower Mint

MYTHS, LEGENDS, AND FOLKLORE

Biblical Magi
El Dorado
Golden Calf
Golden Fleece
Golden Goose
Golden Rule
"The Goose and the Golden Egg"
Holy Grail
Leprechauns
Rumpelstiltskin
"The Tale of the Golden Cockerel"

RELIGION AND SPIRITUALITY

Holy Grail
Iconoclasm

Wat Traimit

SHIPWRECKS

Nuestra Señora de Atocha
San Esteban

Spanish Armada
SS *Republic*

SOCIAL AND LEGAL ISSUES

Emergency Banking Relief Act of 1933
Gold Standard Act of 1925

Sumptuary Legislation

TECHNOLOGY AND METALLURGY

Aerospace Industry
Alloying
Au
Assaying
Biomedical Research
Bonding Wire
Casting
Catalysis
Chrysotype
Cyanide
Electroplating
Filigree
Fuel Cells
Gilding
Gold Nuggets
Goldsmithing
Granulation
Karat
Mining
Nanotechnology
Touchstone

Preface

In proposing this topic to the enthusiastic editors at ABC-CLIO, I lightheartedly quipped that I like to write about shiny, valuable subjects. Having now pursued the topic to the depth of an A–Z encyclopedia across time and space, I have become even more aware of the many levels of human fascinations with gold—as an heirloom, a means of wealth, a fountain of youth, or simply a thing of beauty to be enjoyed. I recently reflected on the universal human tendency for this latter aspect in recalling a funny memory related to my eldest daughter, who, I should mention, has always been fascinated with rocks. When she was two years old, we lined our driveway with a path of small, granite stones. Among the jumbled colors and textures of white, gray, and tan, she apparently identified certain "shiny, golden" stones, and became fixated on discovering and collecting every single one of these infamous rocks to the degree that entering and exiting the car became something of an ordeal. Recalling this led me to an essential understanding of the common origins of perceptions of gold as precious in societies modern and ancient. Throughout the history of humankind, cultures from every continent have shared a certain captivation with this noble metal, an admiration that remains the essence of gold's value—the primary value of gold has been that which we have ascribed to it.

The content of this volume was organized to cover the most salient topics and themes related to the social and cultural significance of gold from prehistory to the present across the globe and to examine these themes from cultural, historical, social, religious, economic, and financial perspectives. In such a structure, important themes such as art and aesthetics; the evolution of coinage, money, and financial markets; developments in the natural sciences, chemistry, and engineering; trade, conquest, and war; and social conventions, among others, are brought to light in a multicultural, comparative context. This scope will be of interest to both general readers and students or scholars, as well as trade or industry professionals. Indeed, the big picture in the history of gold is relevant on many levels for contemporary society.

The encyclopedia is comprised of more than 130 entries along with numerous informative sidebars and quotations from throughout history. The entries are organized in alphabetical order and include cross-reference terms that appear in bold in the entry text and as subject terms at the bottom of each entry in order to point to the common threads among the topics and themes. Additionally, each entry includes a list of further readings or resources for those who wish to pursue a deeper level of research on the topic. Along with the A–Z format, the beginning of the volume contains a Guide to Related Topics that further categorizes the entries into a range of themes, including Adornment; Art, Artifacts, and Architecture; Banking and Finance; Culture; Gold Rushes; Health and Medicine; Ideas and Movements; Individuals; Money and Coinage; Myths, Legends, and Folklore; Religion and Spirituality; Shipwrecks; Social and Legal Issues; and Technology and Metallurgy. A comprehensive subject index and bibliography of print and electronic sources is provided at the end of the volume.

I conducted the research for this book with a view to including key elements from primary sources related to each topic, ranging from voices from the furthest past to present-day commentary, and data gleaned from ancient alchemical manuals or modern metallurgical reports, as these references are essential to deeper levels of understanding of the narrative explanations provided. I have also made every effort to highlight the relevant historical and cultural connections within each entry. In doing so, I sought to present the details about each topic in terms of both the historical and contemporary contexts.

Acknowledgments

In a world in which written content of all forms is being produced at lightning fast speeds to accommodate the needs of the Internet, the pursuit of writing print reference books ironically takes on a greater significance. While writing this book, I consulted many encyclopedias written from early antiquity to the present, and in doing so gained a renewed esteem for the craft. As compendia of knowledge through the ages, often organized around a specific subject or field, such as Pliny the Elder's *Natural History*, or designed as a tool for perpetuating general knowledge in wider society for the sake of education as a cornerstone of virtue, as in the case of Denis Diderot and Jean d'Alembert's *Encyclopédie*, encyclopedias have served as arbiters of the role of knowledge in human society. I am grateful to my editors at ABC-CLIO for permitting me the opportunity to follow in such an important tradition in writing this compendium of information on the many facets of gold, which in turn offers insights about every level of human experience.

Writing this book often took every bit of effort I had in me, and certainly the majority of my attention. At the same time, it made me grateful for the overwhelming prioritization of the acquisition of knowledge as a key to fulfillment and meaning that I had experienced during my own upbringing. I owe my deepest thanks to those who provided the inspiration and support that enabled me to accomplish this project: To my mother, for always guiding me to see the important connections between things; to my father, a true storyteller and encylopédiste; and to Isabella and Josephine for always believing in me.

Introduction

As early as 2.5 billion years ago, the chemical element that constitutes gold in its native form began to emerge in volcanic and sedimentary rock in sources that would later be discovered by humans on every continent on Earth except Antarctica, in addition to the gold deposits found deep under the ocean floor. Evidence for human exposure to gold dates from around 5000 BC, and it can be said that from this first encounter, humankind cherished gold as a precious symbol of wealth and majesty, thought to contain some aspect of the divine, or even to hold the key to eternal life, because of its unique physical properties and striking beauty.

From antiquity to the present day, gold has played a role in the course of human history through its influence in a wide variety of realms—spiritual, cultural, political, scientific, and economic. Access to significant supplies of gold can be credited with contributing to the rise of powerful states and empires, while devaluation of gold currency or an outflow of gold from state or royal treasuries has contributed to the downfall of mighty rulers and governments, such as those in ancient Rome, Byzantium, the Spanish Empire, and Tudor England. In the contemporary era, maintaining a gold supply and monitoring the domestic gold market remains a crucial tool utilized by policymakers in the realm of global economic foreign relations, as evidenced by the careful policies of China since 2008 in implementing financial reforms designed to maintain the strength of the yuan in the face of a weak U.S. dollar. Pursuit of gold has been the force behind significant conquests, voyages of discovery and exploration, and imperial expansion through colonization. Gold reinforced social and political boundaries by serving as a carefully regulated status symbol, restricted to only certain social groups or castes, in cultures around the world. Ancient alchemists and religious leaders held that gold offered direct access to god and constituted a symbol of nature's perfection that was also thought to have tangible healing or life-giving properties. Such beliefs contributed to the development of modern science, with alchemy providing the foundations of the field of chemistry and ancient goldsmithing techniques offering engineers a range of formulas for practical scientific applications of gold in industrial and technological advancements. Gold's purely aesthetic beauty contributed to magnificent feats of art and architecture from the shining halls of the pharaohs' palaces to the mosaics in the Hagia Sophia and the opulence of Versailles. Whether cherished as a medium of trusted monetary exchange or as a divine object, gold has captivated humankind for thousands of years.

INTRODUCTION

The Formation of Gold and its Physical Properties

Gold is a noble metal identified on the Periodical Table of the Elements as the chemical element Au based on the Romance etymology of the word "gold," which derives from the Latin word for gold—*aurum*. Gold is a unique element because it exists most commonly in its native form, not consisting of any other minerals or elements. In this form it appears with a lustrous metallic gold color. Gold is highly resistant to oxidization from exposure to oxygen or other substances, and thus is less susceptible to corrosion. It is soft and pliable, yet extremely durable and dense. Gold is biocompatible and has significant levels of conductivity, while at the same time remaining chemically inert. Each of these unique properties and attributes, along with the fact that gold is relatively rare when compared to available quantities of other metals, contributes to gold's historically high value as both a commodity and a form of money.

Geologists assert that gold formed over time in a variety of ways. Scholars continue to actively research and explore the origin and placement of gold sources in various types of rock formations and environmental settings. Some of the most ancient gold deposits occurred 2.5 billion years ago in what are known as "greenstone belts," along which geologists posit that a sizeable gold-forming event occurred in the metamorphic rock when the earth's plates collided amid volcanic and tectonic activity. The next phase in which supplies of gold formed in diverse types of rock occurred 1.5 billion years ago, and then again 400 million years ago. No significant gold formation activity was thought to occur continuously or in the human era. Scientists have identified a wide variety of ways gold potentially forms, principal among these processes are formation through a process of magmatic hydrothermal cooling systems in which water heated by magma sources is propelled upward to cooler rock levels toward the surface, during which precious metals dissolve from the surrounding rock and then precipitate into lode ore sources, or as the product of chemical reactions caused by the interaction of sulphuric vapors with diverse aqueous substances. Over time, gold was uprooted out of lode ore sources in various types of rock and deposited in alluvial sources due to weathering and water flow. An understanding of the chronology and geological origins and distribution of mineral emplacements that contributed to the formation of gold over time, along with studies of weathering patterns and typical patterns in the formation of quartz veins, assists geologists in surveying areas for potential gold mining operations and in the development of more efficient, cost-effective methods for extracting gold.

A Human Element: Gold from Antiquity to the Present

Along with other advancements such as writing and agriculture, metalworking techniques, especially those used to craft objects in gold, are a hallmark of what archeologists and anthropologists identify as "civilization." In the study of ancient societies, the quality and level of gold artifacts is often viewed as a reflection of the sophistication of a culture's social organization or religious practices, or as signifying a more advanced ideology. The 1973 discovery of a series of tombs filled with highly skilled metal work dating from the fifth millennium BC in the Varna necropolis, along the lower Danube River valley in present-day Bulgaria, represented one of the largest and most ancient assemblages of gold artifacts in human history, consisting of 3,000 pieces of gold distributed among 62 grave sites. The product of a civilization identified as Old Europe, these artifacts revolutionized the way anthropologists viewed early European societies previously thought to have consisted of simple, egalitarian social orders. Instead, the burials at the Varna necropolis signaled the existence of a stratified society and the emergence of male social dominance, social conditions evident largely from the placement and scope of the gold objects among the graves. This early culture straddling the Neolithic and the Bronze Ages appeared to have traded copper, gold, precious stones, and other items as primitive forms of commodity with other settlements in the region, evidence of more advanced and wide-reaching economic activity in which such valuables served as status symbols.

Gold artifacts excavated from a burial complex dating from the third millennium BC at the site of the Great Ziggurat of Ur, an ancient Sumerian civilization located along the Euphrates River in present-day southern Iraq, reveal the development of advanced goldsmithing techniques such as filigree, lost-wax casting, and gilding; as well as the function of gold in a hierarchical Bronze Age society as a symbol of royalty or nobility. Around the same period, these methods were also utilized among ancient pre-Colombian civilizations in the Peruvian highlands. During the period from 1500 to 1200 BC, the palaces and tombs of the Egyptian pharaohs were adorned with exquisite works of gold as goldsmiths mastered a variety of techniques and rulers took care to preserve Egypt's access to gold mines in Nubia and the trade routes that brought gold and other precious commodities considered essential for strengthening the wealth and prestige of the Egyptian dynasties. By 1000 BC, gold was officially used as a form of monetary exchange in China, and by the sixth century BC, King Croesus began to mint a form of official gold coinage in the Anatolian kingdom of Lydia that helped to foster regional trade relations and preserve the territory's legendary wealth. The Old Testament contains hundreds of references to gold and its uses in the kingdoms of Israel, Judah, and surrounding states extending as far as Africa from as early as ca. 1200 BC. In the fourth century BC, the Macedonian king Alexander the Great conquered the gold-rich lands of Persia and the Near East in a campaign that widened the power and influence of Greek culture and filled western coffers with gold, setting off centuries of expansion of Greek and, later, Roman civilizations, in which access to gold sources in African, Near Eastern, eastern European, and Iberian mines was an essential tool of foreign conquest and settlement and a critical strategy for the preservation of political power.

Throughout antiquity in cultures around the globe, gold was considered precious not only as a means of wealth but also for a range of physical and spiritual properties ascribed to the lustrous metal, often viewed as a direct product of the divine because of its beauty, durability, and apparent perfection. Beginning as early as the Bronze Age, there is evidence for the emergence of the deeply ancient protoscientific and occult practice of alchemy, a quest to transmutate base metals such as lead into gold. Ancient alchemical writings exist from the Mediterranean region, the Near East, and Asia, and by classical antiquity, were concentrated among Greek and Jewish scholars in Alexandria, where ancient Egyptian and Hermetic texts containing the keys and formulas to the mystical tradition were housed in the library. The prevalence of alchemy across cultures reflects more common perceptions of a connection between gold and the divine as manifest in nature's perfection. Gold was thus considered to possess the power to prolong or immortalize life by tapping in to this spiritual connection, an ideology that parallels the very evolution and purpose of religious practices in human cultures. From the Incan perception of gold as the "sweat of the sun" to Aristotelian theories of the elements and human life, gold's purported spiritual and mystical attributes epitomized the very aspirations of humanity to achieve life beyond the temporal world.

The economic, artistic, spiritual, and political significance and value of gold seemed to converge during late antiquity and the early Middle Ages in a wide variety of cultures—in the rise of the Byzantine Empire, among the Celtic tribes of northern Europe and Britain, the South Korean kingdom of Silla, and in the Chinese dynasties, to name only a few. Gold's range of uses supported the fabric of human life: in coronation ceremonies, it reinforced kingship and the power of the state; in religious rituals, it facilitated access to God and the ascendancy of priestly classes; it generated prestige and communicated religious and political ideologies through art and architecture; it reinforced social hierarchies as an indication of status; it contributed to the expansion of literacy and the culture of the book in the illuminated manuscripts produced by scribes in early medieval scriptoria; it perpetuated marriage and family ties as a predominant type of dowry; and it contributed to the expansion of international trade and, in turn, global navigation and exploration, as a transportable and widely accepted form of monetary exchange.

During the medieval period, gold was an essential component of the formation of the merchant republics and city-states in Europe, in which the rise of the modern state and the development of capitalist economies occurred. Advanced minting practices guaranteed standardized systems of value for gold coins, such as the Venetian ducat or the florin of the Florentine Republic, providing currencies that were accepted as money of account in international markets. Gold as a form of monetary exchange thus facilitated economic expansion and the early development of systems of banking and credit in the goldsmiths and money changer guilds. Creation of wealth during the 14th and 15th centuries, emanating from the increasingly powerful merchant and tradesman class, generated a new form of material culture, a trend that fostered much of the patronage of art and literature during the Renaissance by fueling demand with new patterns of spending disposable wealth. Gold was thus central to the circumstances behind the phenomenon of Renaissance Italy, a period when renewed appreciation for the ideologies and practices of Greek and Roman antiquity and revolutionary new attitudes about natural philosophy, civic life, and individuality coalesced to form what historian Jacob Burkhardt considered to be the nucleus of modern society.

The emergence of more sophisticated economies during the Renaissance contributed to the predominance of political centralization in Europe and the Middle East that came to be known as the "rise of the modern state" in the early modern era. Maintaining a strong treasury with adequate gold reserves; upholding authoritative monetary policy, regulating the flow of gold and silver bullion, and minting national coinage; and sustaining opulent patronage of the arts as a demonstration of power and authority in courtly life were essential components of empire building and balance-of-power politics among the great powers of Europe and the Mediterranean from the 16th to the 18th century. Demand for gold in this period created a need for greater territorial expansion and access to efficient trade routes, which set in motion centuries of discovery and exploration that resulted in European settlement and colonization of North and South America, India, Africa, and Asia. In order to support the widening bureaucratic and economic functions of the state, banking and credit institutions expanded and modern academic disciplines developed in colleges and universities to advise rulers and government administrators. Gold thus influenced the establishment of modern economic theory, chemistry, natural history, geology, and industrial technology.

Following the discovery of gold at Sutter's Mill in California's Sacramento River Valley in 1848, and subsequent 19th-century gold rushes in Australia, the Klondike, and South Africa, the global production of gold increased tenfold, a phenomenon that led to the western expansion of the United States, scientific and technological advancements in the mining industry, and a level of gold reserve assets that made the era of the gold standard possible between 1880 and the 1930s, when 59 countries around the world adopted a currency pegged to either a uniform gold standard or a gold exchange standard. During this period, statesmen and economic advisers considered gold to be integrally tied to the power and legitimacy of the state. Even after the gold standard became unsustainable in the economic environment of post–World War II global economies, economists and political leaders clung fiercely to the idea of the need to maintain currencies convertible into gold equivalents. Many leaders shared the sentiment of French economist Jacques Rueff who feared the collapse of the gold standard would undermine the preservation of free democratic society. Adherence to the gold standard had fostered credibility and cooperation among nations and was thought to have contributed to the creation of political and economic stability in developing regions.

With the final collapse of the gold standard and global pegging of foreign currencies to the U.S. dollar at exchange rates tied to fixed prices and adequate gold reserves in 1971, national currencies were no longer convertible into gold and conceptions of value in monetary policy adjusted to meet the growing needs of new dynamics in more global economic and financial systems. Yet gold reserves remained an important tool among national treasuries and supranational institutions such as the International Monetary Fund in the preservation of economic stability and international relations. Gold continued to shine; however, although no longer at the core of systems of monetary exchange, gold assumed new levels of value during the late 20th century as a traded commodity when scientific innovations in nanotechnology discovered significant applications of gold in its nanoparticulate form that could be efficiently employed in critical emergent industries such as computer microcircuit boards, aerospace technologies, information and communication technology, transportation and energy, biomedical research, and health and beauty products.

At the beginning of the 21st century, global economic crisis and the erosion of faith in banking and credit institutions triggered an unprecedented increase in the price of gold and a resurgence of investment in gold as an asset. When in doubt, it seems, humankind still retains a deep regard and faith in gold's supreme and enduring value. Indeed, it is ironic that during periods of strife or crisis, scholars, politicians, and citizens have often weighed in on gold's function in contemporary society, and in doing so engaged in an ancient debate. Perhaps gold's ultimate purpose in human history will prove to be reminiscent of the myth in which the ancient Athenian philosopher Solon reminded King Croesus of Lydia that human happiness may run a course at pace with riches but is never simply based on wealth.

AEROSPACE INDUSTRY

Coatings of gold on metallic surfaces of components of diverse technologies in the aerospace industry have been widely used and researched since the mid-1950s by NASA and its suppliers, such as Raytheon, Ball Aerospace and Technologies, and Lockheed, as a reflective coating to protect against radiant heat and light in the intense environments of space exploration. Although it is not the most reflective material under normal light conditions, in the infrared spectrum, gold is the most durable and efficient material for reflecting infrared energy as a means of temperature control, reflecting 99 percent of infrared light.

Given the space restrictions, weight requirements, and limited margin for error or failed parts in the development of space exploration technologies, the need to minimize the effects of exposure to extreme temperatures in space required efficient technologies that divert heat and light through conductivity and/or radiation. The application of electrometallurgical deposits of fine layers of gold or a variety of other methods for applying liquid gold compounds to optimize reflectivity and conductivity for the purpose of maintaining low emissions and thereby regulate thermal stability, as well as to minimize corrosion, has been essential to the success of many NASA missions.

A gold reflective film is applied to astronauts' helmets to both reduce glare and optimize light for vision and can be seen on the plastic visors covering the helmets of the astronauts in famous photos of the Apollo 11 moon landing and exploration in 1969. It was used in the subsequent lunar exploration missions in helmets and in spacecraft devices, for example, as a coating for cameras and batteries and in **fuel cells** powering moon exploration vehicles and instruments. It was also used in the Apollo aircraft to contain or maintain temperatures in radioactive **fuel cells** to ensure that the hydrogen and oxygen reactants were at the correct temperatures for the necessary electrochemical reaction powering the spacecraft.

In the present day, these technologies and others are still employed by NASA, an organization that contributes significantly to research and development projects that explore the uses of gold as in aerospace technologies. For example, sheets of gold-coated plastic film are wrapped around spacecraft parts to protect them from exposure to extreme thermal conditions. A patented formula called Laser Gold, a liquid reflective coating produced by Epner Technology, is applied to maintain temperature stability to telescope mirrors on the Mars Orbiter Laser Altimeter Mirro, which collects topographical data on Mars, and in NASA's Geostationary Operational Environmental Satellite, which collects weather data. Gold coatings were applied to the Hubble telescope to protect against corrosion and establish optimal electrical connections. The Columbia space shuttle contains gold surfaces in many elements of its engineering, from fuel cells to circuit boards and as a reflective coating. Gold has also been increasingly applied in the United States military aerospace technologies as sensors, connectors, and reflectors, with experts predicting a greater use of gold in these industries as technological advancements in terms of performance are aligned with potential cost-effectiveness and efficiency.

See also: Electroplating; Fuel Cells.

Further Reading

Blenkinsop, P. (1986, April). "Materials in *Aerospace*." Proc. RAe. Conf., London.

Langley, Robert C. "Gold Coatings for Temperature Control in Space Exploration." *Gold Bulletin*. http://www.utilisegold.com/assets/file/utilisegold/pdf/Langley_4_4.pdf.

ALCHEMY

Alchemy is an ancient scientific and philosophical practice shrouded in occultism that seeks to transmute base metals, typically lead, into precious metals, such as gold and silver, by means of an alloying agent, or solvent, often referred to as the "philosopher's stone." Because of its presumed ability to produce the perfect metal, gold, the philosopher's stone ultimately assumed a more mystical significance and was also thought to be the "elixir of life," which had the power to prolong life or even produce immortality. Alchemy is considered the precursor to the modern science of chemistry.

Philosopher's Stone

The philosopher's stone was a type of alchemical substance or elixir. At times, it held symbolic reference to transmutative or immortalizing properties or processes for tapping into the divine perfection of nature. Alchemical manuals from antiquity to the early modern era provided recipes and formulas for creating the stone, which was often also referred to as a type of matter known as "carmot." The process of making or obtaining the magical substance was referred to as the "Great Work" and was shrouded in occult secrecy. The philosopher's stone was thought to have both material and spiritual powers; it could be used to trigger the chemical reaction that would turn base metals such as lead into gold, or as a rejuvenating "elixir of life" that could heal mortal wounds or illness and prolong life or even render immortality.

Fourteenth-century Parisian bookseller Nicolas Flamel (ca. 1330–1418) was one of the most famous alchemists who reputedly succeeded in making the philosopher's stone and thereby obtaining great wealth by turning lead into gold as well as immortality for him and his wife. According to Flamel's own accounts, a mysterious stranger sold him a short book containing recipes for the philosopher's stone. Flamel spent years translating and interpreting the book, which later scholars have assumed was the ancient *Book of Abraham the Mage*. Flamel did in fact come across sudden wealth, which he used to provide food and shelter for the poor. Flamel designed his own tombstone with esoteric alchemical carvings for his burial in the Cimetiere des Innocents, which is now conserved at the Musée de Cluny. He also carved similar figures and symbols at his home at 51 Rue de Montmorency, an apartment that still stands today and remains one of the oldest houses in Paris. The legend of Flamel and the philosopher's stone inspired pilgrimages to his home and grave almost immediately after his death and informed popular culture references to his alchemy in works ranging from the novels of Victor Hugo in the 19th century to the contemporary Harry Potter youth fiction series by British author J. K. Rowling.

Ancient and medieval alchemists believed that gold constituted the perfect state of all metals and therefore was a component of all metals produced by the earth. They sought to accelerate the transformation of common metals toward their "perfect" state and, in parallel fashion, to identify the key to human perfection and, consequently, the potential for immortality.

Scholars are uncertain about the precise origins of alchemy. Some of the earliest texts that make reference to alchemical practices are from ancient Egypt and Phoenicia as early as ca. 5000 BC and from Greece and China ca. fifth century BC. Tracing the origins of the tradition is further complicated by the fact that alchemists were committed to the secrecy of their art, as well as the existence of widespread variations of alchemy across Asia, the Middle East, and Europe, each of which contain significant differences. For example, Chinese alchemy placed greater emphasis on the quest for immortality than on transmutation of base metals into gold. Practices in other regions alternately focused more on the scientific or philosophical aspects of the art.

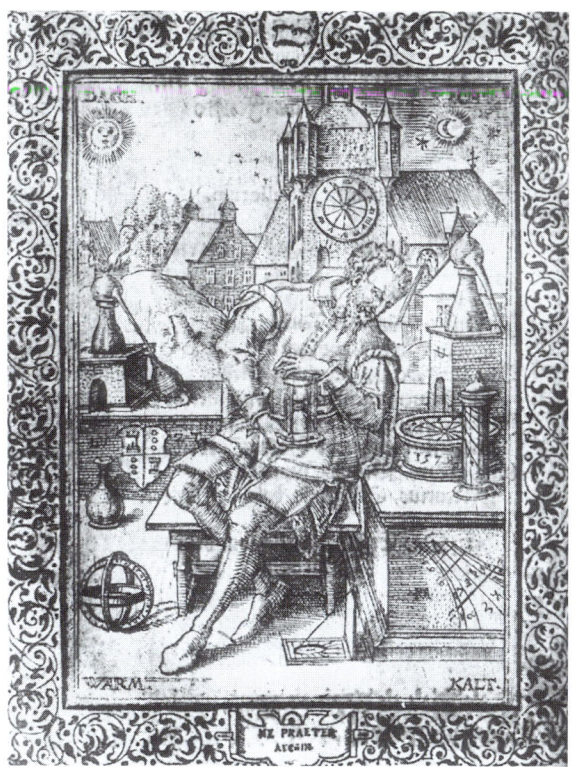

Ne Praeter Aream, an illustration from a late-16th century German book about medicine and alchemy. A man sits among various chemical apparatus, holding an hourglass. Behind him are furnaces and a city fills the background. The corners of the print are labeled "day," "night," "warm," and "cold." (National Library of Medicine)

At the core of all alchemical practices was the common worldview that all things derive from one essential substance, the universal *prima matera*. Identifying this substance therefore would enable an understanding of creation. Such a theory viewed the world in terms of matter, the

---

"Gold hath these natures: greatness of weight; closeness of parts; fixation; pliantnesse, or softnesse; immunitie from rust; colour or tincture of yellow. Therefore the sure way, though most about, to make gold, is to know the causes of the several natures before rehearsed, and the axiomes concerning the same."

—Francis Bacon, "Experiment Solitary Touching the Making of Gold" (1627)

---

core element of all things, and its form, the shape matter assumes in combination with other influences. This relationship was also perceived in conjunction with a belief in a universe comprised of interdependent elements of macrocosm and microcosm.

One of the earliest extant references to the chemical processes of alchemy is the ancient Egyptian text *The Emerald Tablet of Thoth*, known in Greek translation as *Hermes the Magician*, or, *Hermes Trismegistus*, the primary document upon which the majority of alchemy was subsequently based. Thoth was considered one of the Egyptian gods of creation, who set forth the keys to the universe in the Emerald Tablet based on the maxim "as is above, so is below," an affirmation of the duality of human existence. Other extant texts in Greek purportedly transcribed from the ancient Egyptian and preserved at the library in Alexandria include the Leydon and Stockholm papyri, formerly part of a larger collection of ancient alchemical texts.

The foundations of the philosophical aspects of Western alchemy were constructed from these ancient Egyptian sources by philosophers in ancient Greece, principle among them Aristotle, who argued that matter was comprised of four elements—earth, water, air, and fire—a postulate that would come to form the basis for the scientific practices of alchemy in the West. At the same time, craftsmen and purveyors of magic experimented with alchemical recipes to their own ends, contributing to the compilation of secret recipes for transmuting metals and accompanying magical rituals.

There is evidence that some of the earliest identifiable alchemists from the Hellenistic period were Jewish. One of the most famous works was reputedly authored by the Jewess Maria in the early third century BC. Later texts credit Maria with the invention of significant alchemical technologies and the advancement of theories of the elements.

Independent of developments in the Near East and western Europe, the practice of alchemy emerged in ancient China and India with approximately the same foundational principles, yet with a greater focus on the transformational medicinal possibilities promised by alchemy and the pursuit of the elixir of life. In ancient China, there is evidence of mystical alchemical treatises produced by Taoist priests, and the use of gold as a restorative element in Chinese medicine continues to the present day. In India, the Hindu mystical practice of Kundalini, also known as the "fire of life," sought a physical and spiritual total transformation through an elaborate system of purification processes, both physical and mental, based in part on mathematical representations of the interrelationship between the body and soul and the chemical functions of the physical body and the mind.

During the 7th and 8th centuries, Arab scholars turned to the Hellenistic sources in the development of a rich corpus of treatises in alignment with perceived clues to alchemical practices and philosophy found in Hebrew scripture. Alchemical arts grew and expanded in the Islamic world through the 12th century to form the foundation of a variety of early chemical scientific practices that would later influence alchemy in the West.

The translation of many of the Greek and Arabic alchemical texts into Latin took root in Toledo during the 11th century following the reconquest in 1085 by Alfonso VI of Castile. As word spread among European scholars about these translation efforts, alchemical texts rapidly began to circulate throughout Europe, while still retaining their occult mystique. Early medieval alchemical scholarship largely focused on the collection, translation, and interpretation of ancient or Arabic sources rather than on practice and experimentation.

Alchemy continued to develop in the direction of the occult during the High Middle Ages as the alchemical elements of the Jewish Kabbalah cast new light on the ancient practice with respect to Renaissance philosophers' fascination with the perfection of nature. The advent of the printing press made alchemical texts more widely available in the 16th century, and an increase in experimentation among both lay persons and scholars led to closer scrutiny of the practice by the political and religious authorities. For example, in 1217, Pope John XXII issued a papal bull against counterfeiting currency by means of alchemy, while in 1403, the practice was banned in England by King Henry. In *The Canterbury Tales*, Geoffrey Chaucer's Yeoman warns the Canon who dabbles in alchemy that such arts are better left to God:

> "He wills that it not discovered be,
> Save where it's pleasing to His deity . . .
> . . . Since God in Heaven
> Wills that Philosophers shall not say even
> How any man may come upon that stone,
> I say, as for the best, let it alone" (920–928)

During the early modern and modern eras, many technical elements and the foundations of scientific inquiry found in thousands of years of alchemical practice contributed to the development of the discipline of chemistry, which, ultimately, did indeed establish methods for transmuting various elements. Historians' knowledge of Enlightenment interest in the ancient art increased significantly when modern-day archival discoveries revealed that **Sir Isaac Newton** had collected and annotated a vast library of alchemical works and undertook a series of alchemical experiments. In the present day, the practice of alchemy lives on and is actively pursued in the realm of New Age spiritual mysticism, particularly in such esoteric sects as the Jewish Kabbalah and Hermeticism, and among historians and aficionados who seek to preserve the tradition through such organizations as the California-based International Alchemy Guild, which offers "certified training in alchemy" and seeks to preserve ancient and modern knowledge of alchemy and support those who pursue the craft for both technical and spiritual ends. Psychotherapeutic clinicians of Jungian transpersonal psychology also incorporate elements of ancient alchemical traditions into their approach. Alchemy lives on in non-Western traditions such as Taoism and the Kundalini Hindu practice.

The emergence of alchemy across cultures as humankind increasingly addressed philosophical viewpoints from a scientific perspective reflected a universal desire to reveal the essence of humanity through the essential tenets of nature, and thus gold was sought as the perfect element for symbolically or literally preserving life.

See also: Alloying; Newton, Sir Isaac.

Further Reading

Dobbs, B. 1991. *The Foundations of Newton's Alchemy, or, "The Hunting of the Green Lyon."* Cambridge: Cambridge Univ. Press.

Dobbs, B. 1992. *The Janus Face of Genius: The Role of Alchemy in Newton's Thought.* London: Cambridge Univ. Press.

Eisen, Arri, and Gary Laderman, eds. 2007. *Science, Religion, and Society: An Encyclopedia of History, Culture, And Controversy.* Armonk, NY: M. E. Sharpe.

Habashi, Fathi. 2009. *Gold, History, Metallurgy, Culture.* Quebec City: Metallurgie Extractive Quebec.

Healy, J. E. 1985. "The Processing of Gold Ores in the Ancient World." *Canadian Inst. Min. Metall. Bull.* 78 (874):84–88.

Láng, Benedek. 2008. *Unlocked Books: Manuscripts of Learned Magic in the Medieval Libraries of Central Europe.* Penn State University Press.

Marshall, Peter. 2001. *The Philosopher's Stone: A Quest for the Secrets of Alchemy.* London: Pan.

Newman, William R. 2004. *Promethean Ambitions: Alchemy and the Quest to Perfect Nature.* Chicago: Chicago Univ. Press.

Principe, Lawrence M. 2007. *Chymists and Chymistry: Studies in the History of Alchemy and Early Modern Chemistry.* Sagamore Beach, MA: Watson Publishing International.

Sivin, Nathan. 1968. *Chinese Alchemy: Preliminary Studies.* Cambridge. http://www.alchemyguild.memberlodge.org/.

ALLERGIES

A gold allergy, also referred to as "gold dermatitis," typically manifests as a form of contact dermatitis due to repeated exposure to gold or a gold **alloy**. Skin rashes and stomatitis (inflammation of the mucous lining of the mouth) have also been identified as allergic responses to gold in patients who were given rheumatoid arthritis treatments containing gold and medical or dental gold implants.

The symptoms associated with a gold allergy include rash, itching, swelling, blistering, and eczema. If a gold allergy is suspected based on information about exposure in the form of **jewelry** (particularly ear or body piercings), dental gold implants, gold-plated stents installed in the treatment of a coronary angiography, or gold in injectable treatments for such ailments as rheumatoid arthritis, a patch test is administered using gold sodium thiosulfate, or occasionally other gold salts, to assess whether the patient has allergies type 4, also known as cell-mediated or delayed allergies, directly tied to the activation of a T-cell response, resulting in the inflammation of the affected area of the patch test. If an allergy is identified in the patch test, patients are advised to avoid exposure to gold. The onset of gold dermatitis can occur at any age and persist for many years.

Gold allergies have historically been associated with the increasing popularity of low-purity gold and white gold containing a high **alloy** ratio of nickel, a harder metal commonly added to strengthen the gold and known to have a higher incidence of allergic response than gold, which has historically been considered as a nonsensitizing material in the metallurgical and medical research due to its immutability. Due to conceptions of gold's immutability, scientists typically considered an allergic response to pure gold as extremely rare.

During the late 1990s, however, increased attention to the incidence of allergic responses to higher purity gold among researchers around the world indicated a

10 percent occurrence rate of allergic responses to gold sodium thiosulfate in controlled studies, with an additional 4–6 percent higher rate of occurrence in women versus men. A research paper published by Dr. Thomas Fuchs of Goettingen University Hospital in Germany claimed the existence of allergies to pure gold in certain isolated cases. Additional studies also suggested that a history of eczema or previous exposure to gold in the form of dental implants or stents increased the incidence of gold allergy. Despite a high degree of controversy in the field of dermatology over the scope of the results used in these studies, the medical community generally acknowledged that despite its stability and immutability, gold should be considered a common allergen. In 2010, the American Contact Dermatitis Society cited gold as its "Allergen of the Year," and other recent studies have cited gold as among the top ten skin irritants.

See also: Alloying; Biomedical Research; Jewelry.

Further Reading

Bruze, Magnus, and Klaus E. Andersen. 1999."Gold—A Controversial Sensitizer." *Contact Dermatitis* 40, no. 6 (June):295.

Fowler, J. Jr., Taylor J., Storrs F., et al. 2001. "Gold Allergy in North America." *American Journal of Contact Dermatitis* 12:3–5.

McKenna, K. E., O. Dolan, M. Y. Walsh, and D. Burrows. 1995. "Contact Allergy to Gold Sodium Thiosulfate." *Contact Dermatitis* 32:143–146.

ALLOYING

Alloying refers to the process of combining a solution of two or more metallic or nonmetallic substances to produce a new, more desirable substance that is an amalgam, and, therefore, not a pure substance. The impurity ratio of specific alloys is commonly expressed in terms of one karat for each 24 parts, with the purest substances being 24 karat. The earliest known alloy is bronze, a metal comprised of copper and zinc.

To obtain a desired color, strength, or purity level, gold is frequently mixed with other metals, such as copper, silver, nickel, and zinc. Because of the malleability of 24-**karat** gold, jewelry makers often combine it with harder metals to produce stronger pieces. This process is also used to obtain a desired color ranging from the deep yellow color of 24-**karat** gold to paler yellows, greenish tints, rose-tinted hues, or white gold. In medical and technological applications, alloying gold with other harder metals improves its conductivity and increases resistance to heat and corrosion.

The addition of a silver alloy converts the yellow color of the gold to greenish-yellow or white tones. Nickel is often used instead of silver to produce white gold.

Debasement

Debasement is the devaluation of coinage through the process of minting coins with a lower ratio of base metals in the alloy that reduces the real value of the coins or a devaluation of other forms of monetary currency by adjusting the exchange rate or gold reserves to which the currency's value is measured in international and domestic markets.

Copper creates a reddish or rose-gold effect, and zinc can tone down such an alloy to restore a degree of the yellow color of pure gold after the addition of such alloys.

See also: Biomedical Research; Jewelry; Karat.

Further Reading

Evans, J. A. 1989. *A History of Jewellery 1100–1870*. London: British Museum Publications.

Gold jewellry alloys. http://www.utilisegold.com/jewellry_technology/colours/colour_alloys.

AQUINAS, THOMAS

Thomas Aquinas (1225–1274) was a 13th-century Italian theologian who is considered to be one of the greatest theologians and philosophers of the Catholic Church. His writings have had a profound influence on the evolution of Church doctrine and on the very fabric of Christian culture and society in the West. In particular, his contributions to theological debates on economic questions in the context of ethics, Church doctrine, and Aristotelian philosophy served as an essential foundation of modern economic theory.

Born in 1225 in Roccasecca, near present-day Naples and at that time part of the Kingdom of Sicily, Thomas Aquinas was the son of the Count of Aquino and Theodora, Countess of Theate, through whom he was related to Holy Roman Emperor Frederick II. In defiance of his family's plans to send him to the abbey at Monte Cassino to study and train for a monastic career, Thomas chose to enter the Dominican brotherhood of mendicant friars in 1244. As he set out for Paris by way of Rome to join the order, his mother arranged for relatives to kidnap him and try to coerce him into changing his mind. Thomas subsequently endured several years of imprisonment in a family-owned fortress near Monte San Giovanni Campano. Faced with Thomas's unyielding determination, Theodora relented and arranged for his release. Thomas went on to Paris to join the Dominicans and in 1245 enrolled at the University of Paris, where he studied under German Dominican friar and scholar Albertus Magnus. Perhaps on account of their scientific studies, a legend developed, claiming that Albertus Magnus had passed on his occult knowledge of alchemical practices to Aquinas.

Thomas dedicated himself to a life of teaching and study during his tenure in Paris, at the Roman Curia, in Paris once again, and, finally, in Naples. A product of the scholastic movement of the era, in which theologians sought a wider purview for their studies and commentaries that had greater relevance to secular concerns, Thomas Aquinas's writings addressed philosophical, social, and scientific questions from the perspective of Christian theology and doctrine, Aristotelianism, and what he identified as the tenets and workings of natural law. These ideas culminated in one of his greatest works, the *Summa Theologica*, written during 1267–1273, in which he explores Aristotelian philosophy relative to Church teachings in an inquiry-based style of questions, objections, and replies.

Among the vast breadth of the influence of the writings contained in the *Summa Theologica*, Thomas Aquinas's commentary on contemporary finance and economics in the context of his consideration of issues of ethics and justice provided the groundwork for the development of modern economic theory among economists such as **David Hume**, **Adam Smith**, **David Ricardo**, and many others. Thomas Aquinas viewed economic transactions in terms of their practical and necessary functions in the systems of natural law that contributed to the social order and the maintenance of human society, with an additional ethical layer governed by the Church and the state. He thus addressed questions related to the Church's position on ownership of private property, accumulation of wealth, and the charging of interest in light of the realities of an economy that was in the process of transforming from a largely feudal, agrarian system to a more urbanized, market-based economy with significantly more financial exchange and overall economic growth.

Of these economic commentaries, his theories of "just price" and the charging of interest created a structure in which Christian doctrine was reconciled with the increasingly sophisticated financial transactions that would later become the hallmark of the rise of **banking and credit** and the expansion of trade from the 14th century to the Renaissance. Concerning interest, Aquinas carefully outlined how and when the practice was permitted with an emphasis on the ethical purpose and outcome of the transaction. His theory of just price held that in the realm of natural law, goods maintained an inherent quality, or value, related to its supply, the cost of its production, and its purpose or use in society. He viewed any manipulation of the price for personal gain as unjust. Economic historians have to various degrees considered this theory as a precursor to modern economic theories such as total cost of production, market price, or Ricardian ideas of the labor theory of value.

In the Second Article of the *Summa Theologica*, in a reply to the first objection presented to the question as to "Whether a Sale Is Rendered Unlawful Through a Fault in the Thing Sold," Thomas Aquinas debates the very matter of how and why something might be considered to have a certain value, using the time-honored example of gold and silver:

> Gold and silver are costly not only on account of the usefulness of the vessels and other like things made from them, but also on account of the excellence and purity of their substance. Hence if the gold or silver produced by alchemists has not the true specific nature of gold and silver, the sale theorof is fraudulent and unjust, especially as real gold and silver can produce certain results by their natural action, which the counterfeit gold of alchemists cannot produce. Thus the true metal has the property of making people joyful, and is helpful medicinally against certain maladies. Moreover real gold can be employed more frequently, and lasts longer in its condition of purity than counterfeit gold. If however real gold were to be produced by alchemy, it would not be unlawful to sell it for the genuine article, for nothing prevents art from employing certain natural causes for the production of natural and true effects. (*Summa Theologica*, Second Article, Reply to Objection #1)

This passage stands as an artful articulation of man's relationship to gold as a precious commodity; as with other valued commodities, gold is perceived to have an intrinsic value tied to its beauty and functionality, gifts which have been received from nature, yet ultimately this value is in fact determined by human constructs and the dictates of the social order.

Aquinas died from an injury in on March 7, 1274, at the Fossanuova Abbey in southern Italy. Aquinas was canonized as a saint by Pope John XXII in 1323 and named a Doctor of the Church by Pope Pius V in 1567. Throughout modern history to the present day, the philosophy and theology of Thomas Aquinas has continued to shape Western thought in contemporary philosophy, ethics, science, economics, and spirituality.

See also: Alchemy; Banking and Credit; Hume, David; Ricardo, David; Smith, Adam.

Further Reading

Aquinas, Thomas. 1267–1273. *Summa Theologica*.
Baldwin, John W. 1959."The Medieval Theories of Just Price." *Transactions of the American Philosophical Society* 49, no. 4:15–92.
DeRoover, Raymond. 1995. "Scholastic Economics: Survival and Lasting Influence from the Sixteenth Century to Adam Smith." *Quarterly Journal of Economics* 69 (May):161–190.
Medema, Steven G., and Warren J. Samuels. 2003. *The History of Economic Thought: A Reader.* New York: Routledge.
Nichols, Aidan. 2003. *Discovering Aquinas: An Introduction to His Life, Work, and Influence.* Grand Rapids, MI: Wm. B. Eerdmans.

ARCHIMEDES

Archimedes (ca. 287–212 BC) was an ancient Greek mathematician, physicist, and engineer acknowledged through the ages among scientists and historians of mathematics as perhaps the greatest mathematician known to man due to the fact that he established a method for integration that would serve as the foundation for the achievements of future scientists and mathematicians. Archimedes's legacy also rests on his accomplishments in other fields, particularly his development of innovative war machines and essential advances in physics, such as the establishment of the principle of the lever and his experiments with hydrostatics. According to legend, when asked to verify the purity level of the gold used to make a new crown for King Hieron II of Syracuse, Archimedes discovered what later came to be known as the "Archimedes Principle," a theorem for calculating weight by immersing a solid object in a liquid.

"Water is best, but gold shines like fire blazing in the night, supreme of lordly wealth."

—Pindar (476 BC)

Archimedes was born around 287 BC in Syracuse, Sicily, at that time an independent city-state of Magna Graecia. Little primary evidence of his biography exists, and much of what we do know of his life is based on the later writings of ancient Roman authors such as Vitruvius, Plutarch, and Livy. There is substantial evidence indicating that he may have studied with Euclid's successors in Alexandria.

Woodcut by German illustrator Peter Flotner (1490–1546) depicting Archimedes's discovery of the principle of buoyancy upon stepping into his bath and observing that the water displaced was equivalent to the mass of the object placed in the water, a principle he would later use to determine the gold content of the crown commissioned by King Hiero II. (Getty Images)

One of the most famous accounts of Archimedes's genius was the story of the Golden Crown, recounted by Vitruvius, in which Archimedes was challenged by King Hieron II of Syracuse, whom Plutarch claims was Archimedes's "friend and near relation," to determine whether his new crown in the shape of a laurel wreath was indeed crafted of pure gold. The challenge, which had ensued from the king's suspicions that a dishonest goldsmith had replaced a portion of the gold with a silver alloy, was to verify the gold's purity without damaging the delicate crown. According to legend, Archimedes conceived of a solution while taking a bath: as he stepped into the water, he noticed it was displaced from the bath and overflowed, at which point he leapt up and ran out into the street unclothed, exclaiming, "Eureka!" in his excitement over this discovery. According to Vitruvius, he then devised a system in which he submerged a piece of pure gold equal in weight to the crown in a bowl filled to the top with water, after which he replaced the gold with the king's crown, only to find that the crown caused the water to overflow slightly, an indication that the crown had a greater mass than the piece of gold and, therefore, was not pure gold.

As entertaining as this story is, it has been widely repudiated by scholars and scientists, including Galileo Galilei in his 1586 treatise on the use of hydrostatic balance for weighing metals by the displacement of mass in water or air, a project built on the foundations of what is known as the "Archimedes Principle," a theorem for determining the weight of a body immersed in liquid. Archimedes's theories of hydrostatics were outlined in his *On Floating Bodies*, a 2-volume work in which he identifies the law of equilibrium of fluids and articulates a physical law of buoyancy later referred to as the Archimedes Principle; namely, that "any body, completely or partially submerged in a fluid, is acted upon by an upward force equal to, but opposite in sense to, the weight of the fluid displaced." If indeed Archimedes did perform the experiment for King Hieron II, he likely used a combination of his own law of buoyancy and law of the lever to attach the crown to one side of a lever and balance it on the other side with a piece of gold equal in weight. He could have then dipped the entire lever with both the crown and the mass of gold suspended from it into a basin of water. If the scale did not remain level, but rather adjusted in the direction of the mass of gold, this would prove that the mass of the crown was greater than that of the pure gold and therefore less dense due to the addition of an alloy of a lighter metal.

Archimedes died in Syracuse ca. 212 at the hands of a Roman soldier during the Second Punic War, despite the express instructions from Emperor Marcellus that, should Archimedes be encountered during the battle, his life should be spared. Plutarch recounts a variety of versions of Archimedes's death. In each case, the mathematician was so engrossed in study that he failed to comply with the commands of the Roman soldier, who killed him for not responding. Plutarch also writes that Marcellus was deeply distressed by Archimedes's death and that the emperor provided for his burial according to Archimedes's request that a representation of a cylinder and a sphere be placed on his tomb with an inscription of his theorem for the ratio of the relation between the two, a discovery that Archimedes considered as his own greatest accomplishment.

See also: Alloying.

Further Reading

Hirshfeld, Alan. 2010. *Eureka Man: The Life and Legacy of Archimedes*. New York: Walker.

O'Connor, J. J., and E. F. Robertson. "Archimedes of Syracuse." http://www-groups.dcs.st-and.ac.uk/~history/Mathematicians/Archimedes.html.

Plutarch. *The Lives of the Noble Grecians and Romans*.

ARGENTINA

During the 1880s, Argentina saw a decade of rapid economic expansion fueled by foreign investment from European banks, particularly in Britain, where low interest rates and rates of return led investors to seek greater profits in emerging markets such as that of Argentina. This influx of capital, primarily in the form of bonds sold in European financial markets, financed the construction of Argentinean railways, roads, and other infrastructure, which fueled the expansion of the domestic economy and, in turn, sparked the enthusiasm of foreign investors even further. By 1889, however, markets became wary of Argentinean securities as signs pointed toward their declining value, and Argentina saw a devastating decline in the availability of foreign capital. Despite a decade of financial boom, the domestic economy had not achieved an export level great enough to sustain the balance of payments on the state's foreign debt service without continued lending from international banks. These circumstances precipitated a domestic economic and political crisis that had severe repercussions in global financial markets, particularly due to their effects on one of London's most prestigious and influential merchant banks, the House of Baring Brothers, which held a major percentage of the devalued Argentinean bonds.

A large body of historical and economic scholarship has examined the circumstances behind the onset of such a dramatic financial crisis and the reasons for its far-reaching effects. At the domestic level, the Argentinean government was able to maintain their debt service on foreign loans as long as the stream of foreign investment continued to be made available by international banks as consumption of imports was paid for by a combination of foreign capital and developing domestic exports largely consisting of agricultural products in markets that saw growth due to the investment projects financed by the foreign capital. Once these sources of capital dried up, however, exports did not meet the levels required to maintain the balance of payments. In response, the government printed more paper currency and thereby devalued the peso and incited **inflation**. Yet because the bonds were required to be repaid in a currency equivalent to gold parity on international exchange markets, the securities themselves became practically worthless. The situation escalated rapidly in Argentina into a worsening economic and political crisis that culminated in a revolt in July 1890 which almost toppled the government and forced the resignation of President Miguel Juárez Celman.

At this time, Baring Brothers held a majority of the loans to Argentina that were headed toward default. When Baring executives alerted the Bank of England that their bank risked collapsing in November 1890, the British government enlisted the support of Baring competitor the House of Rothschild along with several other British banks and European financial institutions to bail out Baring Brothers in order to avert a greater crisis that would have likely damaged Great Britain's legendary power and prestige in international financial markets. This swift and decisive intervention managed to rescue Baring Brothers; however, Argentina was left to deal with a critical economic downturn, and global financial markets around the world suffered from an ensuing downturn in British foreign investment. From a monetarist perspective, the crisis also affected perceptions in foreign capital markets that emerging markets in Latin America, as well as in the United States, could not sustain currencies tied to the **gold standard**.

The example of Argentina as a case study in economic boom and bust has led to comparisons by economists of the circumstances of the crisis of 1890 and the economic crises of the 1990s and the subprime mortgage crisis of the mid-2000s. In both cases, investors ignored warning signs to pursue overinflated profits in markets where the economic conditions did not merit such high levels of investment, producing a flurry of speculative financing of unstable assets. In the contemporary global financial crisis, such speculative investment and mismanagement of capital also led to government bailouts of commercial banks, an ensuing monetary crisis with global repercussions, and calls for government reform of financial institutions with a view to stabilizing **banking and credit** to stimulate a return to economic growth.

See also: Banking and Credit; Baring Crisis; Gold Standard; Inflation.

Further Reading

Ferns, H. S. 1992. "The Baring Crisis Revisited." *Journal of Latin American Studies* 24, no. 2 (May):241–273.

Flores, Juan Huitzi. 2006. "A Microeconomic Analysis of the Baring Crisis, 1880–1890." Carlos III University, Economics History Department, Madrid. http://emlab.berkeley.edu/users/webfac/eichengreen/e211_fa05/211_baring.pdf.

Ford, A. G. 1956. Argentina and the Baring Crisis of 1890. *Oxford Economic Papers* 8, 127–160.

Marichal, Carlos. 1989. *A Century of Debt Crises in Latin America: From Independence to the Great Depression, 1820–1930*. Princeton, NJ: Princeton University Press.

Tella, Guido di, and D.C.M. Platt, eds. 1986. *The Political Economy of Argentina, 1880–1946*. Pittsburgh: Univ. of Pittsburgh Press.

Williams, John Henry. 1920. *Argentine International Trade under Inconvertible Paper Money, 1880–1900*. Cambridge, MA: Harvard University Press.

ASANTE GOLDEN STOOL

The Asante Golden Stool is the physical symbol of the monarchy of the Asante, or Ashanti, people of Ghana who inhabit an administrative region of the same name. The tradition of the Golden Stool dates back to the origins of the Asante Kingdom established in the 1670s by tribal leader Osei-tutu, who successfully formed an alliance of surrounding states into a centralized kingdom that would develop into one of the most powerful in West Africa.

According to legend, the high priest of the Asante called on the heavens to provision Osei-tutu with the Golden Stool and thereby legitimate his position as king and symbolically inaugurate the Asante Kingdom. The heavens complied, as the stool appeared in the sky and then floated down to the new king. Similar to other non-Western cultures such as that of the **Inca**, the Asante revered gold both as a spiritual reflection of the essence of nature, specifically the sun, and as an indication of social status.

The stool is a curved seat cast in pure gold and decorated with gold ornaments such as bells. It is revered by the Asante people as the soul of their nation and has historically remained hidden amid a high level of secrecy for security, with only the king and his close personal advisors informed of its location. The stool is not permitted to touch the ground and can only be carried by an official royal standard-bearer for ceremonial purposes.

The Asante Kingdom held one of the richest supplies of gold in West Africa, a fact that lured early European explorers in search of this wealth. Asante chiefs played a significant role in the regulation of the **Trans-Saharan Gold Trade** that passed through their kingdom. The Golden Stool was later implicated in the fall of the kingdom to the British following a series of wars in the late 19th century. In 1896, fearing a defeat that might risk the removal of the Golden Stool, the Asante cooperated with the deportation of their king, Prempeh I, to avoid the confiscation

The Asante king Otumfuo Opoku Ware II sits next to the Golden Stool (on right), symbol of the Asante Nation, with his court at Kumasi, 200 miles north of Accra, Ghana, August 13, 1995. (AP/Wide World Photos)

of the Golden Stool by the British. As a result, the kingdom was annexed to the Gold Coast colony. A few years later in 1900, the colony's governor, Sir Frederick Hodgson, visited the Asante capital of Kumasi and sought to claim the Golden Stool for the British. In response, the Asante people mounted a sustained uprising that came to be known as the War of the Golden Stool. Despite losing the sustained conflict, the Asante considered that they had won by virtue of the fact that they still possessed the Golden Stool, the physical symbol of their claim to rule the region, of their intact nationhood.

The stool was at risk once again in 1920 when several Africans working on a road-building project came across its hiding place and stole many of the gold ornaments adorning the stool. The potential political unrest from the extreme reactions among the Asante people led the British authorities to grant the local chiefs authority to try the thieves in their own court. Yet when the defendants were sentenced to death for treason, the British reduced their punishment to lifetime exile.

The confederation of the states of the original Asante Kingdom was restored as an administrative district of the Gold Coast colony in 1935, with Asante chiefs representing the region in executive and legislative councils. Following the decolonization of the region in the mid-1950s, the Asante retained their claims to the land as a political region in the modern state of Ghana, and the Asante king still wields significant influence in south Ghana. The stool remains a symbol of royal succession and the power of the king.

See also: Inca; Trans-Saharan Gold Trade.

Further Reading

Bortolot, Alexander Ives. "Gold in Asante Courtly Arts." http://www.metmuseum.org/toah/hd/asan_2/hd_asan_2.htm.

Cole, Herbert M., and Doran H. Ross. 1977. *The Arts of Ghana*. Los Angeles: Museum of Cultural History, University of California.

Lystad, Robert A. 1969. *Ashanti: A Proud People*. Westport, CT: Greenwood.

McLeod, Malcolm D. 1981. *The Asante*. London: Trustees of the British Museum.

Wilks, Ivor, ed. 1993. *Forests of Gold: Essays on the Akan and the Kingdom of Asante*. Athens: Ohio Univ. Press.

ASSAYING

The **mining** industry has employed the use of fire assaying for centuries as a method for identifying whether the amount of precious metals such as gold or platinum contained within a rock sample assessed as suitable for **mining** as ore are adequate for determining the purity level, or grade, of the gold extracted from the sample. Archeological evidence from Asia and the Middle East indicates that as early as 2500 BC these methods already existed much as they are employed today. In the early modern period, fire assaying was described by German scientist Georgius Agricola in his *De Re Metallica* (1556). This process continues to be an essential stage of any **mining** project as it is employed to evaluate whether the value of the mine's precious metals provides an adequate profit relative to the cost of **mining** the ore.

Fire assaying is used to extract precious metals from impurities and other base metals by adding a flux material, commonly borax, soda, or silica, depending on the metal, to the pulverized sampling of ore, along with lead. This alloyed sample is then put into a crucible and fired in an assaying furnace at temperatures around 2000° Fahrenheit in order to melt the substances into a liquid that is then poured into a mold and allowed to cool. The lead portion of the alloy hardens with the gold to form the heaviest portion of the new substance. The remaining lighter materials from the molded piece are removed, leaving the lead sample, which is placed into a ceramic cupel that absorbs the lead when the cupel is again placed in the furnace and brought to the melting point, leaving a sample of extracted gold that can then be analyzed for its purity levels.

See also: Mining; Touchstone.

Further Reading

Bugbee, E. E. 1940. *A Textbook of Fire Assaying*. New York: John Wiley and Sons.

Hafferty, J., L. B. Riley, and W. D. Goss. 1977. *A Manual on Fire Assaying and Determination of the Noble Metals in Geological Materials*. U.S. Geological Survey Bulletin 1445, Washington, D.C.

Marsden, John, and Iain House. 2006. *The Chemistry of Gold Extraction*. Littleton, CO: Society for Mining Metallurgy and Exploration.

Philips, J. P. 1965. "16th-Century Texts on Assaying." *Chem. Educ.* 12, no. 7:393–394.

Pliny the Elder. 1893. *The Natural History of Pliny*, vol. 6, eds., trans. John Bostock and Henry Thomas Riley. London: George Bell & Sons.

Stockdale, D. 1924. "Historical Notes on the Assay of Gold." *Sci. Prog.* 18:476–479.

ATAHUALPA

Atahualpa (1497–1533) was the last surviving emperor of the **Inca** Empire who perished during the legendary Battle of Cajamarca, a watershed event in the Spanish Conquest. His father Emperor Huayna Capac ruled until his death in the late 1520s, at which time Atahualpa's brother Huáscar succeeded to the throne in Cusco, the capital of the **Inca** Empire, while Atahualpa ruled a regional territory based in Quito to the north. A civil war ensued among the brothers, most likely over a disputed succession. Atahualpa defeated his brother in 1532 in the Battle of Quipaipan and mobilized his large army for the journey to Cusco to claim the throne. It was on this fateful journey that he would encounter the army of Spanish conquistador **Francisco Pizarro** in Cajamarca, a meeting that fueled the interest of the Spanish in the rumored plentiful gold sources of the **Inca** Empire.

In 1531 Pizarro landed in what is now northern Ecuador to embark on his third expedition to conquer Peru, equipped with only 180 men and 37 horses. After establishing a fort at San Miguel de Piura and receiving additional reinforcements in the fall of 1532, Pizarro learned of the civil war between the royal brothers and advanced inland in pursuit of Atahualpa. Informed of the Spaniards' approach, the Incan ruler dispatched a retinue of emissaries to greet them. According to 19th-century historian William H. Prescott, the emissaries offered the Spaniards a drink of the traditional Peruvian beverage *chicha*, served in an opulent golden goblet of particular interest to the conquistadors. Based on the emissaries' report that the Spaniards' numbers were few, Atahualpa invited them to Cajamarca, with the likely intention of capturing them there.

Incan emperor Atahualpa is taken prisoner by Spanish soldiers under Francisco Pizarro and sentenced to death. (Library of Congress)

After witnessing the strength and numbers of the Incan army, Pizarro devised a plan to ambush and capture Atahualpa as it would have been impossible to engage him in battle. Pizarro sent his brother, Hernando Pizarro, and Hernando de Soto along with a native interpreter to invite the emperor to meet with him at the Spanish camp the next day. That morning, Pizarro instructed his men to lay in wait concealed in the barracks around the plaza in anticipation of the Incas' approach. According to contemporary Spanish chronicler Pedro Pizarro, the approaching emperor and his attendants made such a sumptuous spectacle that they "blazed like the sun." In addition to the gold adorning their garments and head ornaments, the Incas carried their emperor on a litter of gold, upon which he sat in a throne of gold.

When the Incas entered the plaza unarmed, they were alarmed to find it empty. At this point Dominican Friar Vicente de Valverde, the Spaniards' chaplain, accompanied by the interpreter, approached Atahualpa carrying a cross and a breviary and attempted to explain the tenets of the Christian faith to the Incan leader and obtain his consent to convert, which would be an implicit acknowledgment that the **Spanish Conquest** was God's will. Atahualpa replied in anger, and, when the friar presented the breviary itself as evidence of the authority of his words, Atahualpa grabbed the sacred book and threw it to the ground. In response, Pizarro signaled his men to begin their attack.

Unfamiliar with gunfire and cavalry, the Incas were stunned by the siege. Trapped in the square and unarmed, Atahualpa's attendants and thousands of **Inca** soldiers were massacred, and the emperor was taken prisoner. During his imprisonment, Atahualpa observed the Spaniards' great pleasure in pillaging the treasures of the **Inca** encampment, particularly their lust for gold. In an effort to secure his freedom, he attempted to strike a bargain with Pizarro, offering to fill a large room once with gold and a smaller adjoining room twice with silver over a period of two months in exchange for his release or perhaps simply to be spared from execution. Pizarro agreed, and the terms of the bargain were recorded by his notary. The emperor promptly dispatched his available men to the corners of the empire to collect gold from religious temples, royal palaces, and other buildings. The room was filled with all manner of gold in the form of various vessels and plates, decorative panels, musical instruments, and elaborate figurines. A small portion of the treasures was shipped intact back to King Charles V of Spain, while the remainder was melted into gold bricks reported to total more than 6,080 kilograms of gold of approximately 22 karats. Pizarro claimed for himself the golden throne from the emperor's litter along with a large percentage of the treasure.

Once the pact was carried out, however, Pizarro still feared the emperor's existence posed too great a risk to the conquest. Atahualpa was thus tried for a series of dubious charges, including sedition for allegedly attempting to incite a rebellion against the Spanish, idolatry, and murder and was sentenced to burn at the stake. According to the account of Pedro Pizarro, Atahualpa responded to his conviction with an emotional outburst directed at **Francisco Pizarro**, exclaiming, "What have I done, or my children, that I should meet such a fate? And from your hand, too, you who have met with kindness and friendship from my people, with whom I have shared my treasures, who have received nothing but benefits from my hands!" In desperation, he pleaded for his life, offering an even greater ransom to no avail.

Friar Vicente de Valverde persisted in his efforts to convert the emperor, now enticing him with an offer to commute his sentence to execution by strangling rather than by fire. Already tied to the stake, Atahualpa agreed in desperation, denouncing the **Inca** faith and accepting baptism from the friar with the Christian name Juan de Atahualpa, in honor of St. John the Baptist.

A building from this era known as the "Ransom Room," which was infamously filled with gold intended for Atahualpa's ransom, still stands today in Cajamarca, Peru. It is widely believed by historians that this building is in fact where Atahualpa was held prisoner and eventually executed, while the room filled with gold was elsewhere in the compound.

See also: Inca; Pizarro, Francisco; Spanish Conquest.

Further Reading

Hemming, John. 1970. *The Conquest of the Incas*. London: Macmillan.
Kubler, George. 1945."The Behavior of Atahualpa, 1531–1533." *The Hispanic American Historical Review* 25, 4 (November):413–427.
Prescott, William H. 1847. *History of the Conquest of Peru*. Boston: Harper & Bros.
Stern, Steve J. 1993. *Peru's Indian Peoples and the Challenge of Spanish Conquest: Huamanga to 1640*, 2nd ed. Madison: Univ. of Wisconsin Press.
Xeres, Francisco de. 1530–1534. *Narrative of the Conquest of Peru*.

AU

Au is the symbol for gold on the periodic table of the elements, a chart developed in the late 19th century as a means of categorizing those elements known to science according to atomic number. Grouping the elements in this manner helped scientists to detect patterns among them and predict the existence of as yet unidentified metals, metalloids, and other elements. The identifying symbol Au originated from the Latin word for gold, *aurum*.

The creation of the first periodic table of the elements is attributed to Russian scientist Dmitri Mendeleev (1834–1907). In his *The Principles of Chemistry* (1869), Mendeleev published a table classifying the 63 elements according to their atomic mass and grouped according to similar properties. Although other

> "We can never know what is the precise number of properties depending on the real essence of gold, any one of which failing, the real essence of gold, and consequently gold, would not be there, unless we know the real essence of gold itself, and by that determined that species."
> —JOHN LOCKE, "AN ESSAY CONCERNING HUMAN UNDERSTANDING" (1690)

scholars such as German scientist Johann Dobereiner and English scientist John Newlands had also created versions of a periodic table, Mendeleev's was the predecessor to the present-day modern table, which was developed by British chemist Henry Mosely in 1913. Mosely modified Mendeleev's table by recategorizing the elements according to atomic number, rather than atomic mass.

The elements in the modern chart are organized by rows and groups in ascending order based on their atomic number. The square for each element indicate its symbol, atomic number, and atomic weight. Gold is categorized among the transition elements in group 11, with the atomic number 79 and an atomic weight of 196.9665.

See also: Mining.

Further Reading

Bachman, H. G., and Robert B. Cook. 2003. *Gold: The Noble Metal*. Denver, CO: Lapis International.

BALBOA, VASCO NUÑEZ DE

Vasco Nuñez de Balboa (1475–1519) was a Spanish conquistador and explorer credited with discovering and claiming the Pacific Ocean for the Spanish Crown, on September 13, 1513. Balboa named the newly discovered sea the Mar del Sur, or South Sea, because he had traveled south along the Isthmus of Panama to reach the reputed sea on the other side while exploring the region in search of gold and other valuable commodities.

Balboa was born in 1475 in the city of Jerez de los Caballeros in the Spanish province of Badajoz, in the southwest border region known as Extremadura, to a family of the Hidalgo class, a gentleman class of lesser nobility. Balboa was sent out as a youth to be trained under Lord of Moguer Don Pedro Puertocarerro in the arts of war, a common vocation for young men of his social position, who lacked wealth and needed to make their fortune as soldiers on campaigns or expeditions.

In 1500, he embarked on his first expedition to the New World led by Rodrigo de Bastidas, which arrived on the eastern side of the Isthmus of Panama in 1501 after sailing along the coast of present-day Colombia in search of gold and other treasures. On the voyage back to Spain, the badly damaged ships ran aground in Hispaniola, where Bastidas was arrested by the infamous governor Francisco de Bobadilla, who was also responsible for the arrest of Christopher Columbus. Bastidas was sent back to Spain in chains, but Balboa stayed on in Hispaniola to try his hand at farming a plot of land in the town of Salvatierra on the west coast. His efforts were largely a failure, and he quickly assumed large debts, which he sought to evade by joining another expedition to South America led by Alonso de Ojeda in 1509, but his creditors had the local authorities prohibit him from leaving the island without first settling his debts. Balboa evaded their efforts by stowing away in a barrel on a later voyage led by cartographer and surveyor Martín Fernández de Enciso in September 1510 to replenish troops and supplies to the Spanish colony in San Sebastian, on the coast of Urabá, in present-day Colombia, which had been established by Ojeda and would come to be acknowledged as the first permanent European settlement in South America. In the words of British historian Clements Markham, Balboa was "a penniless fugitive, with no authority, no official appointment of any kind, one who was in Enciso's ship, headed up in a cask to escape from his creditors" (p. 518). Upon initially discovering his infamous stowaway, Enciso had Balboa arrested and chained, but fellow soldiers and sailors convinced their captain that Balboa's experience with the local terrain and natives would be an asset to their expedition, and he set him free.

Balboa assisted the struggling settlers, at that point heading back to Hispaniola under the command of **Francisco Pizarro**, by relocating the settlement to the other side of the Gulf of Urabá to the province of Darien, on the Isthmus of Panama, and establishing the new settlement of Santa Maria de la Antigua with Enciso as *alcalde* (magistrate). From his new base, Balboa built his reputation and his wealth by accumulating gold through both bartering with and plundering the local Indian communities. The colonists quickly became disaffected with the poor leadership of Enciso and turned against him with accusations that he misappropriated the authority entrusted to him by the crown. Enciso was subsequently sent back to Hispaniola, and Balboa and Martin Zamudio were elected jointly to take his place as *alcalde*. In 1511 Balboa was appointed by King Ferdinand II as captain general and governor of the entire province of Darien.

Balboa continued to organize expeditions among the various native chiefdoms of his territory to search for gold and consolidate his authority through a policy of making treaties and forming alliances with some tribes while attacking hostile native communities. According to contemporary chroniclers, he was known to use particularly cruel methods of torture against the Indians, such as ordering their execution by means of being mauled by his pack of vicious war dogs. During these campaigns, he learned from the local natives of a mighty empire with cities of gold to the south, likely in reference to the **Incas**, and the existence of another sea on the other side of the continent. Balboa sent news of these rumors back to Spain and requested troops to launch an expedition in search of gold. Yet Balboa's enemies had filled the king's ears with stories of Balboa's treachery and alleged misrule, turning the sovereign against him. Ferdinand sent an expedition of 2,000 men in April 1514 led by Pedro Arias Dávila, known as Pedrarias, who was to replace Balboa as governor of Darien and commander of the armada on the expedition to the south. Prior to that, however, Balboa had already set out on his own expedition in September 1513, sailing to the narrowest point of the isthmus. He then marched overland, where he ascended a mountain and glimpsed the Pacific Ocean for the first time. Balboa reached the sea a few days later, and claimed it and all the lands touched by the sea, naming it Mar del Sur, or South Sea.

By the time Balboa returned to Santa Maria de la Antigua in January 1514, his letters informing the crown of the newly subjugated territory along with substantial gold and other treasures had arrived in Spain, at which point the king had also learned the truth about the intrigues among Enciso and the other authorities who had accused Balboa of misdeeds. His royal favor now restored, Balboa was appointed as governor of the provinces of Panama and Coiba and the entire territory of the Mar del Sur. With the arrival of Pedrarias in Darien, now known as the province of Castilla del Oro, Balboa was met with animosity and mistrust by the new governor, who proceeded to plot his demise while Balboa continued his expeditions of discovery throughout the new territories during 1517–1518. In late 1518, Pedrarias lured Balboa into a trap by requesting his return to Castilla del Oro to report on his findings, ensuring that Balboa returned without the support of his loyal men. Upon his return, Balboa was arrested and tried on false charges of treason, along with charges of rebellion and mistreatment of Indians. In a staged trial, Balboa was convicted of all charges and condemned to death along with four of his men. In January 1519, Balboa was executed by beheading on a scaffolding in the square of Acla.

See also: Ferdinand II of Aragon; Inca; Pizarro, Francisco.

Further Reading

Anderson, Charles Loftus Grant. 1914. *Old Panama and Castilla del Oro: A Narrative History of the Discovery.* Berkeley: Univ. of California Press.
Anderson, Charles Loftus Grant. 1941. *Life and Letters of Vasco Nuñez de Balboa.* New York: Fleming H. Revell.
Bohlander, Richard E., ed. 1992. *World Explorers and Discoverers.* New York: Macmillan.
Markham, Clements R. 1913. "Vasco Nuñez de Balboa, 1513–1913." *The Geographical Journal* 41, no. 6 (June):517–527.
Sterne, Emma G. 1961. *Vasco Nuñez de Balboa.* New York: Knopf.

BANKING AND CREDIT

Banking is a system of financial institutions that accept deposits, make loans, and facilitate currency exchange among individuals, businesses, and governments. Banks earn revenue through service fees and the interest generated by lending and issuing securities. The origins of modern banks and credit systems can be traced to the commercial revolution of the 13th century, when the transition to market-based economies centered on regional trade fairs created a need for more sophisticated payment methods and forms of money that could be exchanged easily on an international level. Contemporary banking and credit institutions form the cornerstone of national economies and have fueled the expansion of financial globalization.

There are several types of banking institutions that serve various functions in the modern economy. Commercial banks are profit-generating private sector businesses that primarily accept deposits and make loans to individuals and businesses. Central banks are public sector financial institutions that operate in coordination with the national government to help regulate commercial banks, issue paper currency, guide and implement monetary policy, and maintain deposits from member banks.

The roots of banking and credit derive from early practices related to money storage, lending, and exchange activities in early antiquity, from Mesopotamian societies to ancient Greece and Rome, when financial transactions often occurred in religious temples or public buildings considered as sacred or secure. Economic contraction during the Middle Ages and Christian and Islamic prohibitions against usury, or the charging of interest, led to a general disappearance of such activities and a reversion to a feudal, agrarian economy based largely on barter exchange or strictly commodity forms of money.

> "Gold was an objective value, an equivalent of wealth produced. Paper is a mortgage on wealth that does not exist, backed by a gun aimed at those who are expected to produce it."
> —Ayn Rand, *Atlas Shrugged* (1957)

During the 13th century, the expansion of the merchant class and commercial activity tied to the development of independent Italian city-states such as Venice, Florence, and Genoa, in which a ruling class of merchants and tradesmen governed as opposed to the traditional feudal nobility, and the emergence of trade fairs in urban commercial centers throughout Europe signaled a transition to market-based economies that required internationally recognized currencies of exchange, a need met by the minting of gold coins such as the **ducat** in Venice, the **florin** in Florence, and Genoa's genovino d'oro, and a system for extending credit as an easier means of accepting payments for goods without having to physically transport gold **bullion**. Notaries and money changers, who were often goldsmiths, originally served this function by completing contracts that served as bills of exchange, a form of promissory note for a future payment. These contracts formed the basis for the creation of double-entry bookkeeping methods of accounting for credits and debits and maintaining balance sheets. By the 15th century, money changers and lenders were organized into formal professional guilds in cities such as Florence, the birthplace of the Medici family's banking enterprise. In fact, the etymological origins of the word "bank," derive from the term *banco*, a portable form of wooden shelf or bench used by money changers in their shop fronts. Examples of these still exist today on the **Ponte Vecchio**, in Florence, a medieval bridge lined with goldsmiths' shops where these transactions occurred. Medieval and Renaissance money changers used elaborate loopholes to circumvent the Christian prohibition against usury, often by concealing the interest charged by categorizing it as a profit earned from fluctuating **exchange rates**.

The first modern banks evolved out of these early guild corporations in the 17th century to serve the needs of an expanding merchant economy and more centralized governments that required more sophisticated financial services, particularly loans and large flows of capital to fund prolonged wars. The Bank of Amsterdam, Bank of Venice, and Bank of England were some of the earliest formal modern banking institutions established to meet the needs of commerce and government finance, originally serving as a type of clearinghouse where deposits were received and bank notes, a type of IOU, were issued as a credit redeemable against the hard currency deposit. A similar system was documented to have existed in China as well, where the issuance of paper money tied to gold was monitored by the emperor. According to economist **Adam Smith**, the flow of capital made possible by progress in early banking and increasing capacities to extend credit was a central factor in the formation of industrialized, capitalist society.

During the 18th century, many currencies were issued according to a bimetallic system of both gold and silver coinage. Following the financial pressures of the Napoleonic Wars, such a system threatened certain nations, notably the Spanish Empire and Great Britain, with a devaluation of the national currency and potential **inflation** due to an outflow of gold **bullion** and increase in price of silver that undermined the central banks' ability to convert bank-issued currencies into gold. Concerns voiced by economists such as **David Ricardo** led to the investigation of the situation in Great Britain by a parliamentary committee in 1810 that would become known as the **Bullion Committee**. The committee's findings and recommendations relative to the significance of the correlation between the quantity and value of money in circulation, foreign **exchange rates**, and **gold reserves** would be foundational for monetary policy and the progression of central banking during the subsequent era in which the **gold standard** model emerged as predominant.

From the 1870s to the 1930s, banking systems throughout the industrialized world became aligned by various degrees to the **gold standard**, a system in which governments assigned a fixed value to the paper currency issued by the national central bank or treasury that was tied to a determined level of **gold reserves** with the gold fixed at a set price relative to international rates of foreign exchange. By the early 20th century, however, periods of financial crisis and economic strain from war undermined this system. Following diminished consumer confidence in the commercial banking system in the United States after a banking crisis in 1907, the Federal Reserve Bank was established in 1913 to attend to monetary policy more vigilantly and regulate commercial banks. After suspending the **gold standard** out of necessity during World War I, Great Britain attempted to restore it with the **Gold Standard Act of 1925**, which restored the Bank of England's obligation to convert bank notes to gold at prewar prices, a short-lived attempt that was reversed by 1931. Other nations including the United States were soon constrained to abandon the **gold standard** as well amid the monetary crisis of the Great Depression. With the Banking Act of 1933, U.S. President Franklin D. Roosevelt abandoned the classical **gold standard**; nationalized private gold stores, which were held in reserve at the newly created federal depository in Fort Knox, Kentucky; and implemented a system of international gold exchange that was formalized as the **Bretton Woods System** in 1945.

Under the **Bretton Woods System**, a plan for global monetary policy that was significantly influenced by British economist **John Maynard Keynes, the International Monetary Fund** (IMF) and World Bank were established to facilitate the regulation of the banking industry and the flow of money and credit on a global level among a large membership of sovereign states who contributed to the reserve levels of the IMF on a sliding scale and agreed to peg their currencies to a floating exchange rate based on the U.S. dollar, which in turn was tied to gold. By the 1960s, however, this system was perceived as unsustainable by U.S. policymakers concerned about the devaluation of the dollar based on levels of national **gold reserves** and escalating **inflation**. In 1971, President Richard Nixon suspended the **gold standard** exchange rate for the dollar and converted to a floating exchange rate system regulated at the level of international markets, the system that prevails today.

With the globalization of national economies, reduced government regulation, and the expansion of traditional banking operations into additional finance operations such as foreign investment and securities trading, the banking industry faced a series of crises during the late 20th and early 21st centuries, such as the Savings and Loan Crisis in the United States during the 1980s and 1990s, when hundreds of savings and loan associations, a type of banking institution known as a "thrift," failed, triggering a government bailout, economic recession, and escalating budget deficits. In 2006, an abrupt slowdown in the real estate market triggered the subprime mortgage crisis in the United States, which in turn led to sustained severe economic recession on a global level due to the interconnectivity of global banking and financing operations, in part due to the rise of multinational banking institutions and new methods of data and information processing and money transfer facilitated by advances in technology and the rise of the Internet. Governments around the world were constrained to intervene with measures that involved bailing out national and regional banks at risk of failing and worsening the crisis. The prolonged and wide-reaching effects of the subprime mortgage crisis have led to efforts to regulate banking and credit more vigilantly and align contemporary institutions with the needs of the global market in the technological age.

See also: Baring Crisis; Bretton Woods System; Bullion; Bullion Committee of 1810; Ducat; Emergency Banking Relief Act of 1933; Exchange Rates; Federal Reserve System; Florin; Gold Reserves; Gold Standard; Gold Standard Act of 1925; Inflation; International Monetary Fund; Keynes, John Maynard; Ponte Vecchio; Ricardo, David; Smith, Adam; United States Bullion Depository.

Further Reading
Ball, Laurence M. 2008. *Money, Banking, and Financial Markets.* New York: Worth Publishers.
Bernstein, Peter L. 1968. *A Primer on Money, Banking, and Gold*, 2nd ed. New York: Random House.
Chown, John F. 1994. *A History of Money: From AD 800.* London: Psychology Press.
Davies, Glyn. 1995. *A History of Money from Ancient Times to the Present Day.* Cardiff: Univ. of Wales Press.
Eichengreen, Barry. 1996. *Globalizing Capital: A History of the International Monetary System.* Princeton, NJ: Princeton Univ. Press.
Grossman, Richards. 2010. *Unsettled Account: The Evolution of Banking in the Industrialized World since 1800.* Princeton, NJ: Princeton Univ. Press.
Heffernan, Shelagh. 2005. *Modern Banking.* Oxford: Wiley and Sons.
Kindleberger, Charles. 1993. *A Financial History of Western Europe.* New York: Oxford Univ. Press.
Lane, Frederic, and Mueller, Reinhold. 1985. *Money and Banking in Medieval and Renaissance Venice.* Baltimore: Johns Hopkins Univ. Press.
Miskimin, Harry A. 1989. *Cash, Credit, and Crisis in Europe, 1300–1600.* London: Variorum Reprints.
Rothbard, Murray N. 2002. *A History of Money and Banking in the United States: The Colonial Era to World War II.* Auburn, AL: Ludwig von Mises Institute.
Teichova, Alice, Ginette Kurgan van Hentenryk, and Dieter Ziegler. 1997. *Banking, Trade, and Industry: Europe, America, and Asia from the Thirteenth to the Twentieth Century.* Cambridge: Cambridge Univ. Press.
Vilar, Pierre. 1976. *A History of Gold and Money, 1450–1920.* London: New Left Books.

BARING CRISIS (1890)

The Baring Crisis of 1890 was a severe financial crisis resulting from the involvement of the House of Baring Brothers, one of London's largest and most influential merchant banks since the mid-18th century, in an excessively risky foreign investment in **Argentina** during the late 1880s along with a variety of other global economic and political factors that brought the prestigious firm to the brink of bankruptcy in 1890. It precipitated one of history's most legendary financial rescue efforts by the Bank of England with cooperation from the principal financial institutions of Great Britain and Europe, a financial bailout that many economists consider to have been motivated largely by Western political leaders' perceptions of the need to preserve the authority and legitimacy of the **gold standard**.

As one of the largest financial institutions providing foreign investment to Latin America in the aftermath of the region's independence from Spain, Baring Brothers had a relatively long history of business dealings with **Argentina** during the mid-19th century, a period when many of the firm's initial investments were fairly traditional and conservative and focused on development and infrastructure. In the 1880s, however, exuberance about the prospects for growth in **Argentina** increased, and in 1885, Edward Charles Baring, Lord Revelstoke (1828–1897) became the head of the firm and underwrote excessive sovereign debt to **Argentina**, over £28 million in 1888–1889 alone backed by securities in the form of bonds, which Barings sold largely in European markets.

In 1889, however, concern that overseas investment was becoming too speculative in the face of devalued currencies led the Bank of England and the Reichsbank of Germany to raise their key interest rates, which had an immediate impact on the availability and cost of credit on the global market. A series of unforeseen events, such as the failure of the wheat crop and political riots in the summer of 1890, weakened Argentina's fragile situation even further. Over the course of the previous decade, **Argentina** had overborrowed to fuel its development in the face of increasingly competitive and cheaper loans. As long as the national treasury continued to receive foreign investment funds in the form of such liabilities, **Argentina** was able to maintain a balance of payments on its import consumption and overseas debt service through a combination of capital from current exports and additional loans. With the sudden halt in available credit, however, proceeds from domestic exports were not sufficient to stay current on debt obligations. Additionally, adverse political and economic conditions devalued the national currency, which worsened the situation from a monetarist perspective, as the loans were required to be repaid in a currency equivalent to gold parity. **Argentina** plunged quickly into a position of having to default on its loans, the majority of which were held by Barings.

With the understanding that the failure of one of Britain's largest and most reputable merchant banks with holdings all over the world might result in a national economic crisis, Bank of England governor William Lidderdale intervened in November 1890, enlisting assistance from Barings' primary competitor in the London banking industry, the Rothschilds, along with other national banks and European financial institutions with a vested interest in preserving economic stability, particularly with respect to the integrity of the **gold standard**. Lidderdale formed an international consortium into which funds from these institutions were deposited to guarantee the bad Barings debts and prevent the firm's bankruptcy, a rescue effort that is infamous in financial history for the swiftness and secrecy of the international response, a success story that is often interpreted as a reflection of the power and prestige of England's international financial prowess during that era.

Baring Brothers was thus saved from ruin and the bank was subsequently reorganized as a limited financial institution so that the firm could begin the process of repaying its debts. Despite the successful outcome of the Bank of England's intervention in the crisis, it had an irreparable effect on the flow of foreign investment to Latin America in the immediate aftermath and contributed to increasing doubts in foreign capital markets about the viability of adherence to the **gold standard** among North and South American currencies.

See also: Argentina; Banking and Credit; Gold Standard.

Further Reading

Ferns, H. S. 1992. "The Baring Crisis Revisited." *Journal of Latin American Studies* 24, no. 2 (May):241–273.
Flores, Juan Huitzi. 2006. "A Microeconomic Analysis of the Baring Crisis, 1880–90." Carlos III University, Economics History Department, Madrid. http://emlab.berkeley.edu/users/webfac/eichengreen/e211_fa05/211_baring.pdf.
Ford, A. G. 1958. "Argentina and the Baring Crisis of 1890." *Oxford Economic Papers* 8: 127–149.
Marichal, Carlos. 1989. *A Century of Debt Crises in Latin America: From Independence to the Great Depression, 1820–1930.* Princeton, NJ: Princeton University Press.
Pressnell, L. S. 1968. "Gold Reserves, Banking Reserves, and the Barings Crisis of 1890." In *Essays in Honor of R. S. Sayers*, ed. C. R. Whittlesey and J. S. G. Wilson, 167–288. Oxford: Clarendon Press.
Williams, John Henry. 1920. *Argentine International Trade under Inconvertible Paper Money, 1880–1900.* Cambridge, MA: Harvard Univ. Press.
Wirth, Max. 1893. "The Crisis of 1890." *Journal of Political Economy* 1, no. 2:214–235.
Ziegler, Philip. 1988. *The Sixth Great Power: A History of One of the Greatest of All Banking Families—The House of Barings, 1762–1929.* New York: Alfred A. Knopf.

BEAUTY PRODUCTS

As a result of innovations in **nanotechnology** and nanoscience research at the turn of the 21st century, the high-end cosmetics industry has increasingly developed a wide range of beauty products that incorporate nanoparticularate, or colloidal, gold into formulas for the face and body. These gold-laced cosmetics claim to have powerful therapeutic benefits due to certain attributes of gold in this form such as its biocompatibility, which makes gold a good carrier of other active ingredients that can thus be absorbed better on a cellular level; its oxidating and conduction capacities, which are said to promote circulation and thus have an antiaging effect as increased circulation promotes new cell growth; as well as its anti-inflammatory and antibacterial properties. Gold flake is also used in certain cosmetics and beauty products as a color additive and for supposed therapeutic benefits.

The unique chemical and electrical attributes of gold are used in upscale products by cosmetics companies who claim to harness its regenerative capacities, such as LaPrairie, who sells their Cellular Radiance Concentrate Pure Gold serum for $580 an ounce; in Chantecaille's Nano Gold Energising Cream, priced at $420 for 1.7 ounces; and in Guerlain's L'Or Radiance Concentrate with Pure Gold, a

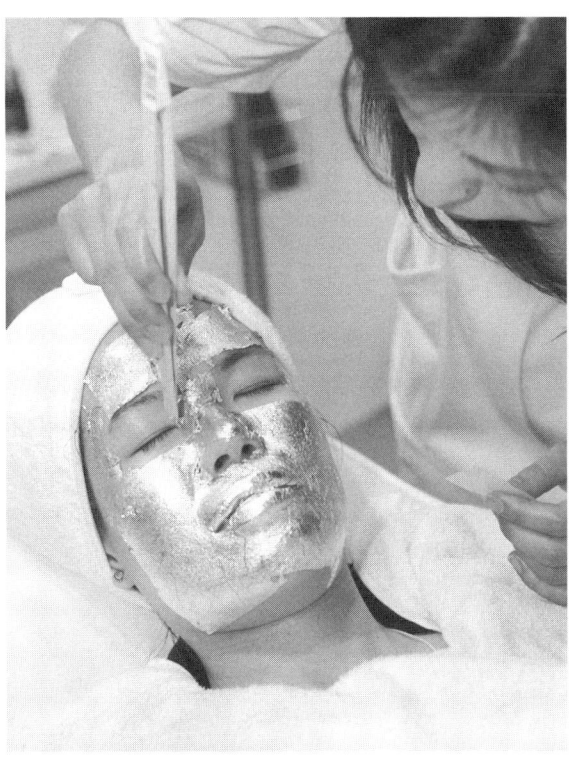

A guest is covered with pure gold foil as she receives a gold facial treatment by Japanese beauty company Umo in Tokyo, on May 19, 2008. (AFP/Getty Images)

makeup primer that is, according to the label, infused with 24-karat gold "complexion perfecting flakes." Luxury salons and spas offer specialty facials that include masks of gold in these two forms; for example, the 24-Carat Gold facial system is produced by Japanese cosmetics company UMO, which claims it "utilizes a proprietary new technology of Gamma PGA, Ions, and Ultrasonic Nano Mist combined with pure 24 Karat Gold to deliver glowing, radiant skin" (http://www.umouniverse.com/spa-treatments.html), with the benefits of lifting and firming the skin, minimizing appearance of fine lines and wrinkles, treating sun damage and age spots, and combating free radicals that age the skin. UMO's Ultrasonic Nano Mist Spray was designed to deliver the 24-karat gold solution along with the Gamma PGA formula in an ionically charged form that stimulates the surface of the skin through the electrical current of the negative ions that gives the product's active ingredients a boost and helps them to penetrate more deeply. The SongSing Nanotechnology firm in Taiwan produces a variety of nano form materials, including gold, for use in a range of commercial and industrial applications and contributes to research and development of such products for diverse enhancements and greater effectiveness.

Despite widespread claims of the therapeutic benefits of colloidal gold for the skin, dermatologists typically dismiss such claims and warn of the risk of developing allergies or contact dermatitis as a result of their use. In the United States, the Federal Drug Administration does not acknowledge gold as an active ingredient in cosmetics, and the Personal Products Council, a national trade organization, only recognizes gold as a potential additive for the purpose of achieving color or sheen. Although many manufacturers of beauty products, industry suppliers, distributors, and marketers refer to centuries-old traditions of using gold in beauty regimens dating back to antiquity, there is scarce historical evidence of such practices, which must be distinguished from biomedical uses of gold in ancient times.

See also: Allergies; Nanotechnology.

> ### Gold Face Lift
>
> In 1999, a Russian physician at the Vishnevsky Institute in Moscow developed a cosmetic surgery procedure to rejuvenate the appearance of aging skin by implanting a web of gold threads under the upper surface of the skin on the face, neck, chest, and other areas of the body. In addition to providing a physical network of support that prevents the skin from sagging further, the gold implants are thought to stimulate collagen production, which further creates a firmer appearance of aging skin. The treatment's effects can last for up to a 10-year period.
>
> ---
>
> Bellamy, Caroline. 2008. "Medicine Is Going for Gold As Experts Rediscover the Ancient Properties of the Precious Metal." *Mail on Sunday* (London), June 7. *Source*: http://www.mailonsunday.co.uk/health/article-1024865/Medicine-going-gold-experts-rediscover-ancient-properties-precious-metal.html#ixzz0yPm3Jb3d.

Further Reading

Keel, Trevor, Richard Holliday, and Tim Harper. 2010. "Gold for Good: Gold and Nanotechnology in the Age of Innovation." *World Gold Council Bulletin* (January).

Kingson, Jennifer A. 2010. "Gold Face Cream: A Costly Leap of Faith." *New York Times*, May. http://www.nytimes.com/2010/05/27/fashion/27skinWEB.html?src=twt&twt=nytimesfashion.

Ogino, Kazuo, and Masakatsu Ohta. 1994. "Application of Colloidal Gold to Cosmetics." *Japan 21st* 39, no. 2 (Dec.):51–53.

BEZANT

Bezant is the term widely used since the Middle Ages for the gold coin minted by the Byzantine Empire from the reign of Constantine I during 324–337 to the fall of the empire's capital of Constantinople to the Turks in 1453. After Constantine relocated the capital of the Roman Empire from Rome to Byzantium, he began to mint a new version of the traditional Roman coinage known as the gold solidus. Under the reign of Emperor **Justinian I** (527–565), a series of successful financial reforms undertaken to consolidate the wealth and influence of the empire included the establishment of an official version of the gold solidus as the primary currency. Justinian utilized the empire's abundant gold reserves to produce a coin that would come to be the primary currency of Byzantium and western Europe for over 700 years. Later in the early Middle Ages, the gold solidus coin was also known in Europe and the Mediterranean as the bezant from the Latin derivative *byzantius nummus*, or, "Byzantine coin," evidence in and of itself of the coin's supremacy.

The first coins minted by Justinian depicted the bust of the emperor on the obverse and victory holding a large cross on the reverse. The coins were also one of the very first to include a date stamp, beginning with Anno XII in 538 in reference to the 12th year of Justinian's reign. Later versions of the coin were minted with religious images on the reverse, even a bust of Jesus Christ in 695 during the reign of Justinian II, an act that is credited with provoking a reaction that launched the **iconoclasm** controversy in protest against sumptuous religious iconographic displays.

According to historian Judith Herrin, Justinian and later Byzantine emperors "grasped the simple but fundamental importance of a stable currency for radiating influence and exercising hegemony over [the empire's] opponents" (http://www.opendemocracy.net/judith-herrin/back-to-eleventh-century). Herrin notes that the military influence and economic viability of the Byzantine Empire can largely be attributed to the reputation and value of its currency. By the 12th century, however, new mintings of the bezant were limited by a shortage of gold supplies. As a result, the coin's value had become increasingly debased, down to as low as 6 karats. Other coins such as the Italian **florin** and the Venetian **ducat** thus grew to greater dominance on the international market. Not surprisingly, Byzantium's power also began to wane. The last Byzantine emperor to mint the bezant was John VIII (1392–1448). The subsequent emperor, Constantine XI, did not have the opportunity to mint the coin as he died in the Turks' siege of Constantinople in May 1453.

See also: Ducat; Florin; Iconoclasm; Justinian I.

Further Reading

Doty, Richard G. 1982. *Macmillan Encyclopedic Dictionary of Numismatics.* New York: Macmillan.
Grierson, Philip. 1982. *Byzantine Coins.* Berkeley: Univ. of California Press.
Hendy, Michael F. 2008. *Studies in the Byzantine Monetary Economy c. 300–1450.* Cambridge: Cambridge Univ. Press.
Herrin, Judith. 2009. "Back to the Eleventh Century?" November 17. http://www.opendemocracy.net/judith-herrin/back-to-eleventh-century.
Herrin, Judith. 2009. *Byzantium: The Surprising Life of a Medieval Empire.* Princeton, NJ: Princeton Univ. Press.
Hobson, Burton, and Robert Obojski. 1970. *Illustrated Encyclopedia of World Coins.* Garden City, NY: Doubleday & Co.
MacDonald, George. 1916. *The Evolution of Coinage.* Cambridge: Cambridge Univ. Press.
Oddy, Andrew, and Susan La Niece. 1986. "Byzantine Gold Coins and Jewellery: A Study of Gold Contents." *Gold Bulletin* 19, no. 1:19–27.
Room, Adraian. 1987. *Dictionary of Coin Names.* London & New York: Routledge & Kegan Paul.

BIBLICAL MAGI

In the Christian biblical tradition, the magi were three wise men who came from the East bearing gifts of gold, frankincense, and myrrh to present to the infant Jesus Christ, whose birth had been signaled to them through the appearance of a brilliant new star over Bethlehem, a sign in local prophecy of the arrival of a great man or prophet. The only biblical source to reference the story of the magi is the Gospel of Matthew, which recounts the tale as follows:

> In the time of King Herod, after Jesus was born in Bethlehem of Judea, wise men from the East came to Jerusalem, asking, "Where is the child who has been born king of the Jews? For we observed his star at its rising, and have come to pay him homage." When King Herod heard this, he was frightened, and all Jerusalem with him; and calling together all the chief priests and scribes of the people, he inquired of them where the Messiah was to be born. They told him, "In Bethlehem of Judea; for so it has been written by the prophet: And you, Bethlehem, in the land of Judah, are by no means least among the rulers of Judah; for from you shall come a ruler who is to shepherd my people Israel."
>
> Then Herod secretly called for the wise men and learned from them the exact time when the star had appeared. Then he sent them to Bethlehem, saying, "Go

and search diligently for the child; and when you have found him, bring me word so that I may also go and pay him homage." When they had heard the king, they set out; and there, ahead of them, went the star that they had seen at its rising, until it stopped over the place where the child was. When they saw that the star had stopped, they were overwhelmed with joy. On entering the house, they saw the child with Mary his mother; and they knelt down and paid him homage. Then, opening their treasure-chests, they offered him gifts of gold, frankincense, and myrrh. And having been warned in a dream not to return to Herod, they left for their own country by another road. (Gospel of Matthew 2:1–12)

Vast biblical exegesis and scholarly speculation exists on the phenomenon of the star of Bethlehem; the geographical origin of the magi, thought to be Zoroastrians of a priestly caste from Saba, in Persia, or perhaps from Babylon; and the significance of their presentation of the gifts to Jesus and interaction with Herod. In fact, the Gospel of Matthew does not explicitly refer to the number of wise men or their names, but rather, over time, it has been assumed that there were three, each bearing an individual gift, with their names cited most commonly as various versions of Gaspar, Melchior, and Balthasar, among other names typical of the Near East. Interestingly, the gifts themselves began to assume greater significance as the

Adoration of the Magi from *Meditationes seu Contemplationes devotissimae* by Juan de Torquemada, 1479. The illustration shows the opulence imagined by Europeans of African and Asian rulers, a perception that derived in part from the historical gold sources in those regions. (Library of Congress)

story evolved, with the wise man bearing gold being the first to present, and the others following. What may have simply been practical, customary gifts in the culture from whence the magi came developed in Christian stories into gifts of great significance and symbolism, with the most predominant interpretation being that if the infant accepted the gift of gold, this proved he was a king; the gift of frankincense would indicate he was a priest; and the gift of myrrh would indicate he was a healer or prophet. The fact that Jesus accepted all three was interpreted as a sign that he was all three at once. The mystical meaning of the gifts developed in particular in the early Christian era and the Middle Ages as the story of the magi became an essential part of the Nativity celebrations, as evidenced by an early Christian hymn pronouncing: "Incense doth their God disclose, Gold the King of Kings proclaimeth, Myrrh his sepulchre foreshows."

Images and stories of the magi were remarkably predominant throughout the Christian realms, yet the legend also existed in the cultures of the Middle East and Asia Minor. In an account by ninth-century Arab chronicler al-Tabari (838–923), who cites an earlier, seventh-century text as his source, the significance of the gifts is likewise contemplated:

> What is the meaning of the gold, the myrrh and the frankincense, which you are offering in preference to all other gifts? And they said: These are symbolic of Him, for gold is the lord of the material world, and this prophet is the lord of the people of his time; and myrrh is used to heal wounds and sores and thus God through this prophet will heal the crippled and the sick; and the smoke of incense reaches heaven as does no other smoke, and thus this prophet will be raised to God in heaven as no other prophet of his time shall be.

Upon reaching Persia during the late 13th century, **Marco Polo** encountered what was considered to be the tombs of the three kings, preserved and venerated as part of local traditions and prophecies about the kings' journey. Yet another legend claims that the magi are buried in the cathedral at Cologne, Germany, a gift from Saint Helena who brought the relics home with her in the 12th century on her journey back from Constantinople.

There is evidence that through a convergence of sources ranging from the Hebrew bible, Arab sources, the New Testament, biblical exegesis, oral legend, and travel narratives such as that of **Marco Polo**, that story of the three wise men of the East and the gifts they bore for the infant Jesus also fueled the legends of the riches of the East that inspired the era of discovery and exploration in the late 15th and 16th centuries. Writings by explorers such as **Christopher Columbus**, whose marginal notes make reference to the magi, and later writings on the legend of **El Dorado** suggest that medieval depictions of the sumptuous dress, customs, and gifts of the wise men, clad in gold-embellished dress and adornments, contributed to the greater cosmology of the quest for a land of gold during the early modern era. Historian Richard Trexler describes the historical significance of the magi's gold as reflective of a sacred geography evolving around mankind's relationship to a precious metal perceived of as eternal, while at the same time very much of the earth.

See also: Columbus, Christopher; El Dorado; Polo, Marco.

Further Reading

"Earth Has Many a Noble City" *Hymns Ancient and Modern*. London: William Clowes and Sons, Ltd., 1889, #76, p. 75.

Gospel of Matthew 2:1–12. *Bible, New Revised Standard Version*. http://bible.oremus.org/.

Jackson, A. V. Williams. 1905. "The Magi in Marco Polo and the Cities in Persia from Which They Came To Worship The Infant Christ" *Journal of the American Oriental Society* 26:79–83.

Molnar, Michael R. 1999. *The Star of Bethlehem: The Legacy of the Magi*. Piscataway, NJ: Rutgers Univ. Press.

al-Tabari, Muhammad ibn Jarir. *History of the Prophets and Kings*. New York: State University of New York Press, 2007.

Trexler, Richard. 1997. *The Journey of the Magi*. Princeton, NJ: Princeton Univ. Press.

BIOMEDICAL RESEARCH

Since the earliest evidence for ancient practices of medicine, gold has been employed by medical practitioners of Eastern and Western cultures in the treatment of a variety of ailments related to both mind and body. In a wide range of traditions the medical properties of gold were often viewed from an alchemical point of view, as potential elixirs with restorative, revitalizing properties based on perceptions of gold as a perfect element of nature.

The oldest recorded evidence for the medicinal use of gold exists from ca. 3000 BC in Alexandria, Egypt, where archeological discoveries indicate that gold was ingested as a purifying tonic. Ancient written sources in alchemy transmitted through the ages suggest that the Egyptians believed that gold itself held mystical properties that could restore youth and vitality when administered through the correct amalgam. Evidence also exists from practices in ancient Chinese medicine around 2500 BC that gold dust and flakes were used in medicines and tinctures for healing.

In Europe and the Near East during the Middle Ages, medical treatises and pharmacological manuals provide recipes for gold-coated medicines and tonics used for numerous ailments, especially swelling or discomfort in the joints or limbs. In his medical treatises on pharmacological remedies, Renaissance physician Paracelsus (1493–1541) describes his use of precious metals such as gold for several types of medicine, stating "Many have said of Alchemy, that it is for the making of gold and silver. For me such is not the aim, but to consider only what virtue and power may lie in medicines" (*Fragmenta Medica*, 149). During the 19th and 20th centuries, gold was used by physicians for pain management and to reduce swelling.

All of these traditions combined with modern advances in the industrial uses of gold contribute to the escalation of the use of gold in the technology and pharmacology of modern medicine, particularly with the employment of gold in biomedical research technologies. Whereas gold has relatively consistently been used on a variety of levels as a drug to treat medical conditions such as rheumatoid arthritis and digestive ailments, and to increase circulation due to its anti-inflammatory properties, contemporary researchers are making great strides in employing it in noninvasive diagnostic and treatment technologies to cure cancer and to identify and track the body's cellular reactions on an advanced level.

Gold's nonreactive qualities make it a substance that does not interfere with other elements of biological experimentation and can therefore be used to form compounds or map trajectories of cells or tumors within the body without disrupting the targeted cell or tissue. For example, researchers are developing procedures to employ gold as a means of identifying malignant tumors and targeting radiation therapies more effectively. Gold nanoparticles are also presently used in anticancer drugs such as paclitaxel, manufactured by Bristol-Meyers Squibb, which prevents the spread of malignant cells, as a delivery and binding agent that optimizes the delivery of the drug to the identified cells within the body by passing through blood vessels. The U.S. company Cytimmune is developing similar pharmaceuticals. Nanospectra, also in the United States, has made advancements in biomedical research employing gold nanoparticles as injectables administered prior to laser surgery to optimize removal of cancerous tumors by improving the laser's penetration of tissue due to the attraction of the nanoparticles as well as improving success rates in the removal of the malignant cells.

Gold compounds continue to be used in the clinical treatment of rheumatoid arthritis in a variety of ways among mainstream medical practitioners despite a certain unpredictability in the specific mechanism of this type of treatment with respect to the ailment and a high degree of side effects. Further research into the delivery, absorption processes, and clinical mechanism of such drugs as applied to arthritic diseases is needed to develop the effectiveness of these drugs.

See also: Alchemy; Nanotechnology;

Further Reading

Bellamy, Caroline. 2008. "Medicine Is Going for Gold as Experts Rediscover the Ancient Properties of this Precious Metal." *Daily Mail*, June 7.
Georgia Institute of Technology. 2010. "Using Gold Nanoparticles to Hit Cancer Where It Hurts." *Science Daily*, February 18.
Gibson, Jacob D., Bishnu P. Khanal, and Eugene R. Zubarev. 2007. "Paclitaxel-Functionalized Gold Nanoparticles." *Journal of the American Chemistry Society* 129, no. 37:11653–11661. http://pubs.acs.org/doi/abs/10.1021/ja075181k.
Hartfall, S. J., H. G. Garland, and W. Goldie. 1937. "Gold Treatment of Arthritis: A Review of 900 cases." *Lancet* 233:838.

Gold as a Flu Vaccine Breakthrough

In May 2010, scientists at the University at Buffalo and the U.S. Center for Disease Control published the results of their research on drug delivery dilemmas in vaccines targeting potential viral pandemics in the *Proceedings of the National Academy of Sciences*. In this study, gold nanorods were successfully used as the delivery agent to inject a single strand RNA molecule proven to trigger an immune response in the body capable of warding off severe strains of influenza and even H1N1 by triggering the body to produce interferons, a type of protein that limits the spread of the virus in human cells. In previous scientific trials, these RNA molecules became unstable and ineffective in the process of transmitting them to the cells using other delivery agents. The scientists credit gold's biocompatibility as the primary factor in its effectiveness. These findings have charted new ground for researchers developing antiviral therapies and for disease prevention on the public health level.

Source: University at Buffalo. 2010. "To Attack H1N1, Other Flu Viruses, Gold Nanorods Deliver Potent Payloads." Press release, University at Buffalo, May 24. http://www.buffalo.edu/news/11387.

Holmyard, Eric. 1957. *Alchemy.* New York: Courier Dover Publ., 1990.
Keel, Trevor, Richard Holliday, and Tim Harper. 2010. "Gold for Good: Gold and Nanotechnology in the Age of Innovation." *World Gold Council Bulletin* (January).
Messori, Luigi, and Giordan Marcon. 2004. "Gold Complexes in the Treatment of Rheumatoid Arthritis." *Met Ions Biol Syst* 41:279–304.

BLING

The term "bling," also commonly referred to as "bling-bling," is a slang reference to large, ostentatious accessories and **jewelry**, typically made of gold and diamonds, worn in a fashion popularized among hip-hop artists and other celebrities during the mid-1990s. The origin of the term is commonly identified in the lyrics of songs by such hip-hop and rap music artists as Lil Wayne and Outkast.

> Bling bling, I know
> And did you know I'm the creator of the term
>
> —"Hollywood Divorce," by Outkast

The use of the term bling spread more widely in popular culture during the early 21st century, particularly among professional athletes with ties to hip-hop culture and as a phrase of cultural reference within the mainstream media. "Bling" was included in the 2002 edition of the *Oxford English Dictionary* and the 2006 edition of *Merriam Webster's English Dictionary*.

Rapper Soulja Boy shows off his bling during an interview in New York, December 15, 2008. (AP/Wide World Photos)

Bling fashions spread widely among the popular black urban youth culture to include a terminology that referred to "blinged out," "iced," or "pimp" items of adornment and conspicuous display ranging from hats and headwear; ladies handbags; cuff links; bracelets and rings; clothing adornment such as chains; **dental crowns** and grillwork of gold and diamonds; household items such as cups, trophies, and lamps; sunglasses; dog tags; and watches.

By 2008, however, a dramatic decline in consumer spending due to a prolonged global economic crisis effectively led to the demise of the culture of bling and produced a backlash in popular opinion against the conspicuous consumption of the late 1990s that had given rise to the term.

Scholars of African American pop culture have interpreted the "aesthetic of bling" as an assertion of visibility, power, and identity for the urban underclass of the early hip-hop generation.

See also: Dental Crowns; Jewelry.

Further Reading

Delevinge, Lawrence. 2009. "10 Ways Sports Stars Go from Riches to Rags." *BusinessInsider.com* Sept. 18.
Duffy, Jonathan. 2003. "How Bling Became King." *BBC News.* Oct 15.
Lindhe, Jane. 2009. "Bling Is Dead." *BRW* 31, no. 12:17.
Thompson, Krista. 2007. "Bling!: Reflections on the Surface of the Image in Black Youth Culture." Lecture, Department of Art and Archaeology, Princeton University, Princeton, New Jersey, April 24.

BONDING WIRE

Bonding wire is used in the manufacture of electronics to achieve electrical connection between contact pads on microchips and external circuitry to create semiconductor devices such as the integrated circuits that are common in most electronics today, including television sets, computers, cell phones, GPS devices, and many other small electronics. The manufacture of these products represents the largest industrial application of gold in technology, and 90 percent of the bonding wire produced in the world today is made with gold.

Gold bonding wire was developed as early as 1957 in the Bell Labs, the engineering research and development arm of AT&T in New Jersey. Since 1970, the demand for bonding wire in the developing electronics industry has increased an average of 20 percent per year.

Gold is the most efficient material for producing bonding wire in these small electronic applications due to its versatility, its high electrical conductivity, the ease with which it can be bonded onto the desired surface at the required temperatures, and its resistance to corrosion. Gold bonding wire is manufactured at very high levels of purity in order to perform optimally in extreme conditions. Bonding wire can be used to create two types of bonds, a wedge bond and a ball bond, which offers great versatility with respect to the development of new microchip designs as the bonding wire can be welded to different size base-metal pads, often those created with gold **electroplating** to create the desired electronic charge,

Urban Gold Mining in Japan

In November 2009, the Japanese Ministry of Economy, Trade, and Industry announced a program in which consumers were invited to turn in their used cell phones for recycling of the precious metals contained in the phones at thousands of participating electronics retailers and supermarkets throughout Japan in exchange for lottery tickets to win gift certificates, to be given away at the end of the campaign in February 2010. The program brought in 567,000 cell phones, which yielded 22 kilograms of gold at an approximate value of $784,313.00; 79 kilograms of silver, 5,670 kilograms of copper, and 2 kilograms of palladium. Lottery tickets were issued to 158,800 people, and the total face value of the prizes given away was over $2.5 million. This experimental project confirmed that cell phones contain a level of precious metals significant enough to pursue organized recycling efforts. Government estimates in Japan indicate that approximately 200 million used cell phones still exist that could potentially be recycled, which would yield precious metals of an estimated value at over $3 million. Similar campaigns have been launched since 2009 in cooperation with Japanese telecom companies, electronics retailers, and recycling operations.

Recycling of precious metals in Japan from electronics became a growing industry in the mid-2000s with the rise in gold prices and consumer cell phone turnover. Traditional gold mining operations produce an average of 0.18 of an ounce from every metric ton of ore extracted—urban gold mining produces 5.3 ounces of gold from every metric ton of cellular phones recycled. The Eco-Systems recycling plant near Tokyo generates an average of 440–660 pounds of 24-karat gold bars a month with an estimated value of $6–9 million. According to Eco-Systems plant manager Nozumu Yamanaka, "To some it's just a mountain of garbage, but for others it's a gold mine."

Source: Belson, Ken. 2002. "Mining Cell Phones, Japan Finds El Dorado." *New York Times*, February 28. http://www.nytimes.com/2002/02/28/technology/mining-cellphones-japan-finds-el-dorado.html.
Yoshikawa, Miho. 2008. "Urban Miners Look for Precious Metals in Cell Phones." *Reuters*, April 27. http://uk.reuters.com/article/idUKT13528020080427.

using different methods that conform to the needs of the specific device.

The present-day development of smaller, more efficient electronic devices with greater functionality and the need to manufacture such devices in more cost-effective ways to meet the demands of the market is contributing to an increase in research and development and registered patents for the use of gold in circuitry and microchips, particularly with a view to creating more durable and efficient forms of bonding wire through such innovations as the application of an additional layer of insulation to the wire to minimize short circuits. Along with this increased use, however, is the concern that electronics such as cell phones, which contain an average of 50 cents worth of gold per phone, are not being recycled, and the gold is thus going to waste after perhaps centuries of being refashioned or reused.

See also: Electroplating.

Further Reading

Humpston, Giles, and David M. Jacobson. 1992. "A New High Strength Gold Bond Wire." *Gold Bulletin* 25, no. 4. http://www.goldbulletin.org/assets/file/goldbulletin/downloads/Humpston_4_25.pdf.

Ramsey, H. 1973. "Metallurgical Behaviour of Gold Wire in Thermal Compression Bonding." *Solid State Technology* 16, no. 10:43–37.

BRETTON WOODS SYSTEM

The "Bretton Woods System" is a term used to refer to the international monetary guidelines and protocols established by the delegates at the July 1–22, 1944, economic summit in Bretton Woods, New Hampshire. Economic advisors, treasury secretaries, and statesmen from the 44 Allied countries as well as neutral Argentina met to formulate a strategy for overcoming the challenges to monetary policy that characterized the interwar period and craft a plan for the international regulation of foreign **exchange rates** in the absence of a gold exchange standard by a multilateral, cooperative organization comprised of countries of every degree of economic development. As a result of the conference, the **International Monetary Fund** and the World Bank (International Bank for Reconstruction and Development) were created to facilitate the regulation of the standards agreed to by member sovereign states with voting rights, with the U.S. dollar emerging as the currency maintained at a fixed exchange rate to the price of gold, valued at $35.00 per ounce, and member nations establishing an adjustable pegged rate tied to the dollar based on various factors related to the size and strength of their economies. This system of international currency exchange prevailed until 1971 in global markets after the U.S. dollar became the primary benchmark currency of foreign exchange.

As an end to World War II approached, the United States and Great Britain took the lead in organizing the Allied nations to attend the summit in Bretton Woods to

Participants of the United Nations Monetary and Financial Conference meet in Bretton Woods, New Hampshire, in 1944. The result was an agreement between 45 countries to ensure worldwide financial stability after World War II. (Library of Congress)

prevent a recurrence of the pitfalls of the prewar **gold standard** and protectionist wartime monetary and trade policies that contributed to the economic crises of the 1930s. During the conference, 730 delegates from the 44 Allied nations deliberated proposed solutions, with the most significant proposals being presented by **John Maynard Keynes**, representing the British Treasury, and Harry Dexter White of the U.S. Treasury Department. Despite certain substantial differences in how they hoped to implement a plan for maintaining economic stability through monetary policy, Keynes, White, and the majority of the delegates generally agreed on the fundamental issues that needed to be addressed, a consensus that is reflected in the document drafted at the summit, the Articles of Agreement for the creation of the **International Monetary Fund** (IMF), which state the organization's essential mandate as providing a vehicle for promoting international monetary cooperation, facilitating the global expansion of trade and stable trade balances, maintaining stable currency **exchange rates**, minimizing protectionist restrictions prohibiting the expansion of international trade, and supporting the creation of multilateral payment systems and short-term balance of payments assistance to member nations in need.

Wary of the potential problems of both a fixed exchange rate and a floating exchange rate in the postwar political and economic environment, policymakers agreed upon the creation of a hybrid exchange rate, the par value system, in which participating member nations pegged their currencies at adjustable yet regulated par values based on the gold value of the U.S. dollar at that time, which was agreed to be set at $35.00 per ounce. The delegates also agreed on the importance of maintaining international liquidity in sovereign states experiencing deficits, especially among weaker economies, to support political and economic stability as well as cooperative international relations. Keynes and White disagreed on the mechanism for providing the liquidity, with the former advocating for the creation of an independent world bank with its own reserves and the latter proposing the creation of borrowing protocols for IMF member nations. White's plan ultimately succeeded on this issue, and the Articles of Agreement were signed over a year later on December 27, 1945, by the original 29 member nations, who subsequently convened a committee the following year to appoint a board of directors, elect executive directors, and create bylaws. The operational aspects of the agreements outlined at Bretton Woods took force beginning in March 1947 from the organization's new headquarters in Washington, D.C., with the IMF monitoring the par value of floating exchange rates among member nations' currencies relative to the dollar and developing a system of quotas and subscription levels assigned according to member nations' economic conditions. Each nation was obligated to pay a combination of 75 percent national currency and 25 percent gold or assets convertible into gold (such as the U.S. dollar) into the fund's reserves, which entitled them to borrow from the fund accordingly as needed. Voting rights were also assigned according to these quota levels, with the United States maintaining the highest percentage of voting rights as the largest contributor to the fund.

During the 1950s–1970s, membership in the **International Monetary Fund** expanded significantly to include many nations from Africa and Latin America and other less-developed regions, and the bylaws evolved to accommodate the changing global economic conditions. During this period, the U.S. dollar effectively became the primary currency of international exchange. Yet as central banks and treasuries around the world increasingly held their currency reserves in dollars during the 1950s, the dollar became devalued. A confluence of several factors, including doubt about the ability of the United States to meet its convertibility obligations, a sizeable increase in U.S. balance of trade deficits since 1960, and troubling levels of **inflation**, caused an outflow of **gold reserves** from the United States as concerns about the dollar led international financial institutions to convert their reserves to gold equivalents. In the face of such pressures, on August 15, 1971, President Richard Nixon announced that the United States was suspending its **gold standard** exchange rate and converting to a floating rate currency exchange, effectively ending the par value system and gold exchange standard that comprised the core of the Bretton Woods System. In the aftermath of Nixon's announcement, however, the institutional structures of the **International Monetary Fund** and World Bank proved to be adaptable enough to preserve the multilateral international cooperation that Bretton Woods policymakers had envisioned and continued to expand the fund with a view toward maintaining global monetary stability.

See also: Banking and Credit; Exchange Rates; Gold Reserves; Gold Standard; Inflation; International Monetary Fund; Keynes, John Maynard.

Further Reading

Bordo, M. D., and B. Eichengreen, eds. 1993. *A Retrospective on the Bretton Woods System: Lessons for International Monetary Reform.* Chicago: University of Chicago Press.
Buchan, James. 2008. "When Keynes Went to America." *New Statesman*, November 10.
Eichengreen, B. 1996. *Globalizing Capital: A History of the International Monetary System*, Princeton, NJ: Princeton University Press.
Solomon, R. 1982. *The International Monetary System, 1945–1981: An Insider's View.* New York: Harper and Row.
Strange, S. 1976. *International Economic Relations of the Western World 1959–1971, Vol. 2: International Monetary Relations,* London and New York: Oxford University Press.
Triffin, Robert. 1960. *Gold and the Dollar Crisis.* New Haven, CT: Yale Univ. Press.

BRIOT, NICHOLAS

Nicholas Briot (1579–1646) was a renowned French coin engraver and medalist credited by some historians with inventing the coining press and known for advancing the mechanization of coin production at the Paris Mint and the English royal **Tower Mint** during the early 17th century.

Briot was born in 1579 to a Huguenot family known for their work as engravers in the village of Damblain, in the present-day Vosges department of the Lorraine region of France, an area historically associated with metallurgical production. Son of local merchant and medalist Didier Briot, Briot was appointed at the age of 27 as master engraver under King Louis XIII at the Paris Mint, where he served until 1625. During this period, Briot increasingly dedicated himself to the development of mechanization processes for coin engraving, publishing the pamphlet "Raisons, Moyens, et Propositions pour Faire Toutes les Monnaies du Royaume" in 1617, in which he argued for converting from the traditional hammering method of minting to a coining press and experimenting with a variety of mills and rolling machines intended to speed up production at the mint and minimize counterfeiting and **coin clipping**.

In 1625 Briot fled France for England due to financial trouble with his creditors and disfavor at court and secured the support of King Charles I, who engaged him in 1626 to design a series of puncheons and dies for a set of coronation medals honoring the king. In 1632 Briot was sent to Scotland to oversee production of coinage in the capacity of Master of the Mint from 1635 to 1637, during which time he was the first to introduce mechanized production of coins in Edinburgh. During his tenure in Scotland, he also produced a series of commemorative coins dated June 18, 1633, which marked the coronation of Charles I to the Scottish throne and is the only gold coin known to have been produced in Scotland.

In January 1634 the king appointed Briot as chief engraver of the **Tower Mint**, granting him a regular income of £50 a year, which he continued to receive until his death. When the English Civil War erupted in 1641, the king ordered Briot to collect his minting tools from the **Tower Mint**, including his puncher, roller instruments, and coining presses, which he wrapped in his saddle and took with him as he fled to join the king in exile in York, where he was able to continue minting coins for the English king. During 1641–1645, Briot also delivered mechanized presses to his brother, Isaac, who was at that time affiliated with the Paris Mint.

Briot died on December 4, 1646, in London, leaving his second wife Esther Petau as his heir. Among his most famous works was the Dominion of the Seas medal produced in 1630 and conserved today at the British Museum in London as a testament to both his artistry as an engraver as well as his technical acumen in developing and employing superior coining machinery.

See also: Coin Clipping; Tower Mint.

Further Reading

C. E. Challis. 2008. "Briot, Nicholas (1579–1646)." *Oxford Dictionary of National Biography*. Oxford University Press. http://www.oxforddnb.com/view/article/3444.

Jones, M. 1988. "Nicolas Briot." *The Medal*, 12:4–9.

BULLION

Gold bullion is by definition a form of gold produced as bars, ingots, or coins according to internationally recognized standards of weight and purity to function as a means of investment rather than of currency exchange. The value of gold bullion is therefore tied exclusively to its mass and fineness, the price of which is tied to fluctuations in international commodity markets. The standard weight of a bar of gold bullion is 400 troy ounces and bullion purity levels must be 99.9 percent gold, the equivalent of 22 to 24 karats.

The majority of the gold held as reserves by state governments and central banks is comprised of bullion in brick form, although contemporary mints continue to produce gold bullion coins that often have a value that is a degree higher than the bricks due to the convenience of storing and trading coins, as well as some commemorative numismatic value, such as the American Eagle, the Austrian Philharmonic, the South African Krugerrand, and the Canadian Gold Maple Leaf, all weighing one ounce or less. Among the largest gold coins minted is the one-kilogram Australian Gold Nugget coin. Such coins are generally purchased by investors as a simple, secure means of acquiring gold as a commodity. Mints around the world also commonly produce ingots, smaller bricks of gold of varying weights.

The most widely produced form of gold bars adhere to the standards of the London Good Delivery Bars, which must be produced by mints certified by the London Bullion Market Association with a minimum gold content of 350 troy ounces and a maximum of 430 troy ounces and a minimum purity of 995.0 parts per 1,000 parts fine gold. The bars are required to be stamped with a serial number, manufacture year, assay mark of the refinery, and percentage of fineness. The London Good Delivery Bars form the basis of the London Gold Bullion market, the largest wholesale trade clearinghouse for the exchange of gold as a commodity in the world.

> ### Troy Ounce
>
> The weight of gold is measured in present-day international commodities markets in troy ounces, a unit of measurement that equals 31.1034768 grams, also quoted as equal to 480 grains. The troy ounce is approximately 9.7 percent heavier than the traditional avoirdupois ounce and originated as the regional unit of weight used at the popular medieval trade fair in Troyes, France, hence the name. During the Middle Ages, systems of weight and measurement were regional, and not universally standardized. With the growth of international commerce in the 12th and 13th centuries, however, growing channels of market exchange created the need for more standardized units of measurement. The troy ounce was thus adopted as the common means of identifying the weight of precious metals and gems. In this capacity, it also had a significant influence on medieval money systems as coinage was minted according to weights articulated in troy ounces.

See also: Bullion Committee of 1810; Karat.

Further Reading

Green, Timothy. 1993. *The World of Gold: The Inside Story of Who Mines, Who Markets, Who Buys Gold*. London: Rosendale Press.

Grendon International Research. "London Good Delivery Bars." http://www.goldbarsworldwide.com/PDF/GB_6_LondnGooDelivBars.pdf.

Temple, R. C. 1899. "Beginnings of Currency." *Journal of the Anthropological Institute of Great Britain and Ireland* 29(1–2).

BULLION COMMITTEE OF 1810

The Bullion Committee of 1810 was a parliamentary committee comprised of financial experts appointed by the House of Commons in Great Britain in February 1810 to "enquire into the cause of the high price of gold **bullion** and to take into consideration the state of the circulating medium and of the exchanges between great Britain and foreign parts." Concerns about the depreciation of Britain's currency and potential **inflation** due to an increase in the currency supply and an outflow of gold **bullion**, along with the Bank of England's temporary suspension of the convertibility of the currency directly to gold due to the financial pressures of prolonged military engagement with France during the Revolution and the subsequent Napoleonic Wars, prompted what has become known as the "Bullionist Controversy," in which economic theorists, bankers, and statesman engaged in a debate over the degree to which the quantity of bank-issued money is the result or the cause of price changes relative to factors such as **exchange rates**, local market conditions, and **bullion** reserves. Commentaries on the problem by economists such as **David Ricardo**, Thomas Malthus, and Henry Thornton, the member of Parliament who chaired the committee, gave rise to the monetary theory that would provide the foundations for the era of the **gold standard** in central banking.

A 50-gram, pure, gold brick stamped with a picture of Hunyadi János, a Hungarian hero. (Fatusi/Dreamstime.com)

A series of letters written by **David Ricardo** to the *Morning Chronicle* during 1809–1810 urged for a restoration of the convertibility of bank notes into their equivalent value in gold. These letters, along with his 1810 pamphlet *The High Price of Bullion, a Proof of the Depreciation of Bank Notes*, contributed to the political environment that led financial expert and member of Parliament Henry Thornton to call for the creation of a parliamentary committee to hear testimony from economists, politicians, financiers, and Bank of England directors on the state of the national currency with respect to the supply of gold **bullion**; bank policies in issuing currency and extending credit; prices; and the economic crisis. In response to the questions posed by the committee, the Bank of England directors staunchly maintained their position that their increase in extensions of credit were not related in any way to the depreciation of the currency or price levels, in effect denying that there is any correlation between the quantity and value of the money in circulation. To the contrary, the committee's report and recommendations to Parliament called for an acknowledgment on the part of the bank of the connections between foreign **exchange rates**, **gold reserves**, and the money supply and urged for a return to the convertibility of the currency to gold within six months of the war's end or within two years if the war should continue, as well as coordination between the state and the bank over the monitor of the national money supply. When put to the house for a vote, however, the committee's recommendations were rejected. Despite this, the *Report of the Bullion Committee of 1810* would continue to influence the development of public policy with respect to central banking and political economy and the gradual adoption of a form of **gold standard** for the British sovereign during 1816–1819.

See also: Banking and Credit; Bullion; Exchange Rates; Gold Reserves; Gold Standard; Inflation; Ricardo, David.

Further Reading

Bain, Francis William. 1896. *The Bullion Report and the Foundation of the Gold Standard.* Oxford: James Parker.

Fetter, Frank Whitson. 1942. "The Bullion Report Reexamined" *Quarterly Journal of Economics* 56, no. 4 (Aug.):655–665.

Report of the Bullion Committee of 1810. Great Britain. Parliament. House of Commons. Select Committee on the High Price of Bullion. Adam Smith Institute, 1984.

Ricardo, David. 1810. *The High Price of Bullion, a Proof of the Depreciation of Bank Notes.* Pamphlet printed in London by John Murray.

Wood, John Harold. 2005. *A History of Central Banking in Great Britain and the United States.* Cambridge: Cambridge Univ. Press.

C

CA D'ORO

The Ca d'Oro is one of the most famous of the Renaissance palazzi along Venice's Grand Canal. Known as the "House of Gold" by contemporaries because many features of the ornate façade were gilded in gold leaf, this archetypal example of the Venetian Gothic style popular during the 15th century was built during 1421–1437 by Marino Contarini, a member of one of the city's most illustrious patrician families.

Contarini received the property in 1412 from the family of his first wife, Soradamor Zen, who died only five years later in 1417. There are indications that he constructed the palazzo both as a reflection of his family's social status and as a monument to his deceased wife. Contarini himself was responsible for the design of the entire project, both building and decoration, which perhaps accounts for some of the more unconventional features of the palazzo, such as its asymmetry. According to art historian Deborah Howard, this may have also reflected a shift in aesthetic tastes in Venice toward a more ornate, elaborate architecture that

Ca d'Oro on the Grand Canal in Venice. (iStockPhoto)

privileged ostentatious decoration over classical elements such as harmony and symmetry.

The most famous feature of the palazzo, the lavish painted decoration of the façade, was completed during 1431–1437 by French painter Zuan di Franza. In his contract with Zuan, Contarini stipulated to the very last detail how the façade was to be painted, with directions about which elements to gild with gold leaf; where to apply the ultramarine pigment, a very rare and expensive material made of crushed lapis lazuli; and how to oil the inlaid sections of rare precious marbles to make them appear more brilliant. This undertaking required a total of over 22,000 sheets of gold leaf at a cost of approximately 8.3 **ducats** per thousands sheets, which came to be roughly 8 percent of the entire cost for the project. Such an expense is particularly remarkable given the fact that Contarini must have realized that the gold leaf and pigment would not last long in the harsh weather conditions of Venice's Grand Canal.

Pandoro

Pandoro (gold bread) is a traditional cake prepared in the Italian city of Verona during the Christmas holiday season. The modern cake is a rich, buttery sponge cake covered in a dusting of powdered sugar; however, the recipe is thought to have originated during the Renaissance in Venice when a luxurious gold-leaf-covered dessert known as the *pan de oro* was served at noble banquets. Recipes from Venice during this period also record the fact that beyond serving as a symbol of decadence and sumptuous display, edible gold was considered to have digestive and fortifying properties. Candies made of gold-leaf-encrusted dried fruits and sugared nuts were thus often served at the end of lavish banquets.

Indeed, the present-day Ca d'Oro stands as stately as ever on the canal, yet the façade reveals little of its original polychromatic sheen. During the modern period, the palazzo underwent a series of unfortunate renovations that stripped it of many of its important original features, such as the inner courtyard's Gothic stairway, ornate balconies, and the well head sculpted by Bartolomeo Bon. In 1894, Baron Giorgio Franchetti purchased the palazzo and worked to restore the building to its original splendor. In 1922 the state acquired the Ca d'Oro in a bequest from Franchetti's estate along with his art collection. The palazzo is now preserved as a museum and houses the Galleria Giorgio Franchetti.

See also: Ducat; Gilding.

Further Reading

Goy, Richard J. 1992. *The House of Gold*. Cambridge: Cambridge Univ. Press.
Howard, Deborah. 2002. *The Architectural History of Venice*. New Haven & London: Yale Univ. Press.

CALIFORNIA GOLD RUSH

The discovery of gold in the American River in northern California in January 1848 sparked the California Gold Rush, a watershed event for the history of the state of California and the course of American history during the era of westward expansion and development. In a period of only a few years, an estimated 500,000 people moved to California from North America and around the world, an internal migration that remains the largest and most significant in American history. The California Gold Rush proved to be a foundational event in the birth of the state in 1850 and in the nature of the state's uniquely diverse population, as well as the impetus for significant developments in the national infrastructure such as the transcontinental railroad and the expansion of mail and telegraph services.

On January 24, 1848, carpenter James Wilson Marshall discovered several small **gold nuggets** in the American River in the central Californian settlement of Coloma while building a sawmill for developer John Sutter, founder of the fort town of New Helvetia, present-day Sacramento. Concerned that news of this discovery would delay the completion of the mill and derail his efforts to develop New Helvetia further, Sutter attempted to keep the news a secret. By March, however, local newspapers such as San Francisco's the *Californian* began to report that gold had been found at **Sutter's Mill**. Over the next month, the streets and shops of San Francisco were emptied as gold seekers left their jobs and families in search of fortune, especially motivated due to the severe economic recession that had crippled businesses and industry since the late 1830s. By May 29, an editorial in the *Californian* described the changes in the city over the short period since the discovery of gold:

A prospector sluices for gold in California, late 19th century. (Getty Images)

> The whole country from San Francisco to Los Angeles, and from the sea shore to the base of the Sierra Nevadas, resounds with the sordid cry of "Gold, gold, gold!" while the field is left half-planted, the house half built, and everything neglected but the manufacture of shovels and pickaxes. (PBS "Timeline")

Nine days after Marshall's discovery, California was ceded to the United States by Mexico with the signing of the Treaty of Guadalupe Hidalgo on February 2, 1848, which marked the end of the Mexican-American War and the culmination of westward expansion and fulfillment of American Manifest Destiny to complete the conquest of the American West from coast to coast. This course of events had a profound effect on the unraveling of the gold rush and the development of California, as during the onset of the gold rush in 1848–1849, the territory was in many ways lawless, with citizenship status, local law enforcement, and property claims in an intermediary period. It was essentially a rugged frontier land where fending for oneself and a strong dose of luck were necessary to achieving success.

By August 1848, news of gold in California was announced in East Coast newspapers and spread among port communities in Central and South America, prompting a significant migration from areas of Mexico and Chile. Whereas the news initially met with skepticism on the East Coast, in his December 5, 1848, State of the Union Address, President James K. Polk confirmed that government officials had verified the scope of the reported findings in California gold fields. With the president's announcement, what had been a relative trickle of prospectors thus far now burgeoned into the largest gold fever the world had yet seen, as gold seekers known as Forty-Niners flooded into California by the thousands over the following year. Throughout 1849, prospectors identified gold fields throughout north and central California, and a few in Southern California. **Mining** engineers also began to identify quartz veins with significant gold deposits.

In the initial months of 1849, transportation networks expanded to meet the needs of growing numbers of passengers heading to California. Yet passage to the West was not easy by any route, was dependent on weather and the season, and took at least five months if not longer. Migrants set out overland by horseback or wagon train, a journey of over 2,200 miles from the overland trail heads in the east, or departed from East Coast steamships that navigated around Cape Horn and up the Pacific to San Francisco. Others sailed to the Isthmus of Panama, then cut across at the shortest point to catch another ship on the Pacific side. It was also possible to arrive at various Mexican ports and head north on land through difficult territory. Each of these choices presented significant dangers ranging from disease, harsh conditions and difficult terrain, banditry, and hostile Native American tribes. Despite these hardships, the non-Native population of California grew from 20,000 in 1848 to 100,000 by the end of 1849. While boom towns seemed to spring up overnight along transportation routes to the gold fields, the city of San Francisco grew from an estimated 1,000 inhabitants to 25,000 and Sacramento grew from a population of ca. 150 to 6,000, a level of growth that strained the cities' infrastructures and resources to their limits. Among these migrants was a significant percentage, nearly 40 percent, of non-American gold seekers from all over the world, particularly from Mexico and South America and, notably, from China, as well as from various regions of Europe. This cosmopolitan mix would place an indelible mark on the multicultural development of the population of the state of California.

Once the gold seekers arrived in California, they found **mining** for gold to be back-breaking hard work amid dangerous, uncomfortable living conditions with scarce, overpriced provisions. Diaries, letters, and the legacies of the Forty-Niners reveal that entrepreneurs who positioned themselves to take advantage of the mass migration by offering essential goods and services profited much more than the miners themselves, among them bankers Henry Wells and William Fargo, founders of Wells Fargo Bank; San Francisco millionaire Samuel Brannan; and denim manufacturer Levi Strauss. Many such entrepreneurs were women who made their fortunes serving the domestic needs of the miners and businessmen. In spite of these hardships, however, diaries and correspondence among both men and women expressed an appreciation of the freedom and possibility offered in these western gold rush encampments in comparison to the more rigid constraints of American society on the East Coast.

On September 9, 1850, California was admitted as the 31st state in the Union as a free state due to provisions of the Compromise of 1850, a significant event in the path to the Civil War. In the course of this year, placer **mining** yielded increasingly scarce output in gold, and an atmosphere of ethnic tension erupted against non-American gold seekers. Statehood and the establishment of legislation and state law enforcement exacerbated such tensions with the enactment and enforcement of racist legislation that imposed taxes and fines on foreign miners, limited their claim and property rights, and provided for the expulsion of Native American populations from their lands. By 1851, large **mining** operations had begun to replace individual miners, many of whom now began to work as wage laborers for corporate mines. These larger-scale mines increased the gold output significantly, which by 1852 peaked at a record-breaking $81 million per year and led to important advancements in **mining** technology, such as the use of hydraulic **mining**, which tragically had severe environmental consequences in a period of only a few years. Statehood was also accompanied by important transportation and communication projects such as the Transcontinental Railroad and the Western Union telegraph service linking commerce and trade from east to west.

The legacy of the California Gold Rush endures in the subsequent history and the course of American history itself as the overwhelming output of gold created a gold supply that launched the United States into a new era of financial history and contributed to the establishment of the **gold standard** for the U.S. dollar. The event has influenced all aspects of California society to the present day in terms of patterns of settlement, transportation networks, demography, agriculture, government, industry, and social organization. Perhaps above all, the California Gold Rush reinforced the American dream of the promise of good fortune in an era of expansion.

See also: Gold Nuggets; Gold Standard; Klondike Gold Rush; Mining; Sutter's Mill; Witwatersrand Gold Rush.

Further Reading

Brands, H. W. 2003. *The Age of Gold: The California Gold Rush and the New American Dream*. New York: Anchor Books.

Clay, Karen, and Randall Jones. 2008. "Migrating to Riches? Evidence from the *California Gold Rush*." *Journal of Economic History* 68, no. 4 (December):997–1027.

Endicott, William, ed. 1998. "Gold Rush: The Series" *The Sacramento Bee*, January 18. http://www.calgoldrush.com/.

Goodman, David. 1994. *Gold Seeking: Victoria and California in the 1850's.* Stanford, CA: Stanford Univ. Press.

Holliday, J. S. 1999. *Rush for Riches: Gold Fever and the Making of California.* Oakland, Berkeley, and Los Angeles: Oakland Museum of California and University of California Press.

PBS. "The Gold Rush: American Experience." http://www.pbs.org/wgbh/amex/goldrush/index.html.

Rohrbough, Malcolm J. 1997. *Days of Gold: The California Gold Rush and the American Nation.* Berkeley: Univ. of California Press.

Rohrbough, Malcolm J. 1998. "The California Gold Rush as a National Experience." *California History* 77, no. 1 (Spring):16–29.

Walker, Richard A. 2001. "California's Golden Road to Riches: Natural Resources and Regional Capitalism, 1848–1940." *Annals of the Association of American Geographers* 91, no. 1 (March):167–199.

White, Richard. 1998. "The Gold Rush: Consequences and Contingencies." *California History* 77, no. 1 (Spring):42–55.

CARAT. See Karat.

CASTING

From early antiquity among a variety of civilizations around the world, a method known as lost wax casting, also often referred to using the French term *cire perdue*, emerged as a common technology among metalworkers for the production of decorative and functional items made from precious metals such as bronze, silver, and gold. In the modern era this process is also known as "investment casting."

The basic process of lost wax casting involves the creation of a model out of wax that is then covered in a mold of clay or plaster with small holes in it, typically at the base of the object. The item is then heated to the point at which the wax melts and pours out of the holes, leaving an empty form behind, into which the desired molten precious metal is poured. When the metal cools and hardens, the outer mold is broken, leaving the metal version of the original wax model. In effect, the mold "loses" the wax to leave a cavity for the item to be cast. When cool, the object would then be filed or polished to create a sheen and remove any imperfections. To craft larger objects, an additional inner mold was often used as well. Archeological evidence indicates that early castings were produced using wax from wild beehives. Despite the obvious flaw that this process regrettably destroys the original wax model (and therefore each creation is unique), due to the efficiency of this primitive technology, this technique has persisted in roughly the same steps to the present day. Variations on this basic method include hollow casting versus solid casting, indirect casting, and casting on to affix a decorative cast item onto another metal item such as a cup or a sword.

Although there is evidence that the lost wax casting method was employed in the manufacture of copper objects as early as 4000 BC, the earliest surviving gold artifact created using the lost wax technique was discovered in the tombs of Queen Puabi, in the Royal Tomb complex at the ancient Mesopotamian metropolis of Ur in modern-day Iraq. Dating to ca. 2600 BC, the object was created using electrum, a natural alloy of gold and silver and consists of chariot rein-rings adorned with a detailed cast of a wild ass, one of Queen Puabi's symbolic animals. The rein-rings are now conserved at the British Museum.

Despite the lack of local sources of precious metals in Mesopotamia, the lost wax casting method flourished in the region during the third millennium BC as advanced methods evolved for the creation of dress pins and other ornaments, vessels, and practical and decorative items made from gold imported from other areas such as Anatolia and Africa. The technology spread quickly along trade routes, emanating out in all directions to the Near East, the Aegean area, and central Europe. There is widespread evidence for advanced goldsmithing works in lost wax casting dating to the second millennium BC from excavations in settlements throughout the Aegean and in central Europe along the Danube River in modern-day Romania, Hungary, Czechoslovakia, and Germany. Egypt was relatively late to assume the technology, with the earliest evidence of lost wax casting dating to around the 14th century BC. Among the most famous examples of Egyptian gold casting from this early period are the solid gold cast animal goddess figures of Nekhbet the vulture and Buto the cobra affixed to pharaoh Tutankhamun's funerary mask.

During the first millennium BC the sophisticated production of the art flourished throughout the Middle East Mediterranean region among the ancient civilizations of the Phoenicians, the Greeks, and the Assyrians; in Asia Minor among the Scythians; in Italy among the Etruscans; and in the Celtic civilizations of northern Europe and the British Isles. The earliest extant written evidence for lost wax casting is contained in a clay tablet from the ancient city of Sippar, in Babylonia, during the reign of King Hammurabi I (r. 1792–1750 BC). The tablet recorded a transaction dated 1789 BC in which a "mina of wax for a bronze key for the temple of Shamash" is furnished to a metalworker by a merchant.

Methods of lost waxing casting were also practiced in ancient India following the conquest of Alexander the Great in the 4th century BC, and were observed among natives of West Africa by Western traders in the 15th century AD, where the technology may have developed independent of Western influence or it may have been transmitted by ancient seafarers along the trade routes. Spanish conquistadors discovered sophisticated, skilled examples of lost wax decorative arts among the gold treasures of the pre-Columbian civilizations of South and Central America, where the technique originated among the **Inca** in Peru around the 7th–8th centuries and spread north to the area of present-day Colombia and Mexico by the 9th century.

During the Middle Ages, the art continued among the Celtic traditions of northern European goldsmiths and metalworkers, who created intricate designs for **jewelry**, decorative ornaments, vessels, and religious reliquaries and altar items. An early 13th-century manuscript conserved at the British Library contains the treatise "De Diversis Artibus," written by a Benedictine monk from a monastery in Germany under the pen name Theophilus, considered by scholars to have been the work of metalworker Roger of Helmarshausen, ca. 1110–1130. The craft later flourished among the master goldsmiths of Renaissance Italy and was employed in the creation of masterpieces by such artists as **Benvenuto Cellini**, who described the technique in his "Treatise on Goldsmithing and Sculpture," published in 1568 in Florence.

The process continued to be employed for the creation of unique works in the modern era among such famous goldsmiths as **Fabergé**; however, the development of processes of mechanization during the 18th and 19th centuries gradually increased in popularity to produce many pieces from one mold inexpensively. Industrial technological advancements ultimately occurred most profoundly in the field of dentistry following Chicago dentist W. H. Taggart's 1907 invention of a mechanized casting process for the production of gold dental inlays. In 1936, a patent was registered by Danish engineer Thoger Gronber Jungersen on behalf of a Canadian **jewelry** manufacturer for a reusable rubber mold for making multiple small pieces of jewelry. This technology was later integrated into industrial casting machinery capable of producing inexpensive gold and silver jewelry and other small objects on a larger scale with significantly less waste of materials in the process.

See also: Cellini, Benvenuto; Dental Crowns; Fabergé, House of; Inca; Jewelry; Pre-Columbian Gold; Ur, Royal Tombs of.

Further Reading

Bruhns, Karen Olsen. 1972. "Two Pre-Hispanic *Cire Perdue* Casting Moulds from Colombia." *Man* 7, no. 2 (June):308–311.
Evans, Joan. 1982. *A History of Jewelry, 1100–1870*. London: British Museum Publications.
Hunt, L. B. 1980. "The Long History of Lost Wax Casting: Over Five Thousand Years of Art and Craftsmanship." *Gold Bulletin* 13:63–79. http://www.goldbulletin.org/assets/file/goldbulletin/downloads/Hunt_2_13.pdf.
Maryon, Herbert. 1954. *Metalwork and Enamelling: A Practical Treatise on Gold and Silversmiths' Work and Their Allied Craft*. London: Chapman & Hall.
Noble, Joseph Veach. 1975. "The Wax of the Lost Wax Process." *American Journal of Archaeology* 79, no. 4 (October):368–369.
Philips, Clare. 1996. *Jewelry: From Antiquity to the Present*. London: Thames and Hudson.
Tait, H. 1986. *Seven Thousand Years of Jewelry*. London: British Museum Publications.

CATALYSIS

Catalysis is a process that induces a chemical reaction by means of a catalyzing substance or material that is not altered or affected by the reaction. Whereas scientists and researchers have traditionally not considered gold as a potential catalyst, recent advancements in **nanotechnology** have opened the door for a wide range of application of nanoparticulate gold as a catalyst in a variety of industrial processes and pollution reduction technologies due in particular to gold's effectiveness in oxidation processes and its resistance and durability upon contact with toxins or poisons. During the 1990s and 2000s, research and development activities and pending and registered patents for processes employing gold as a catalyst increased rapidly.

According to the World Gold Council, gold has been proven as an effective catalyst in a wide range of commercial industrial processes, with great potential for improvements in terms of cost, efficiency, purity levels, and environmental sustainability. Among the reactions supported by gold are carbon monoxide oxidation, catalytic combustion of hydrocarbons, the hydrochlorination of ethyne, production of hydrogen peroxide as a result of a hydrogen and oxygen reaction, purification of hydrogen sulphide and sulfur dioxide, the oxidation of glucose to gluconic acid, decomposition of dioxins by oxidation, mercury removal by oxidation, minimization of ozone decomposition, water gas shift, nitrogen oxide reduction, hydrogenation of certain hydrocarbons to mono-olefins, and vinyl acetate synthesis. These chemical reactions and processes can be utilized in a range of industries, such as industrial chemical processing, reducing fuel emissions in the automotive industry through the use of **fuel cells** and other technologies, and manufacturing.

At the 2008 meeting of the World Economic Forum, the California **nanotechnology** research firm Nanostellar, Inc., was awarded the Technology Innovator of the Year award for developing NS Gold, a cost-effective gold catalyst product that reduces diesel engine exhaust emissions in a process that improves the efficiency of gold as a catalyst by 25–30 percent. Other research firms are currently working to develop an effective catalysis that can be sustained in the higher operating temperatures of regular fuel combustible engines, as gold functions as a catalyst largely in ambient or low temperature environments. Ionic gold shows significant potential for this latter application as well and has been applied as a catalyst to purify exhaust emissions in Toyota vehicles. In the chemical processing industry, gold has been utilized as a catalyst since the 1970s in the production of vinyl acetate monomer, a common substance in emulsion-based paints, wood glue, and wallpaper paste. In the early 21st century, Japanese research firm Nippon Shokubai developed a process for using gold as a catalyst in the fusion of ethylene glycol and methanol to produce methyl glycolate, a solvent used as a base for **beauty products**, industrial cleaning processes, and in semiconductor manufacturing. Recent patents have also been registered by the Federal Agricultural Research Center in Braunschweig, Germany, for the use of gold as a catalyst to produce gluconic acid, a common additive in food and beverage products as well as household cleaners, by oxidating glucose to produce the desired chemical reaction. Gold's oxidative properties also have great potential for the development of efficient air and water filters for both industrial (for example, in gas masks) and household purposes to reduce common pollutants in the air and water such as carbon monoxide and VOCs (volatile organic compounds), chlorofluorocarbons, and dioxins. These types of filters have additionally proven to be highly effective in the removal of undesirable odors and wastewater treatment.

Current research on the application of gold in catalysis demonstrates vast potential for applications in a variety of industries that stand to benefit from the economic and environmental efficiency of such technologies and promises to yield higher more sustained levels of performance in the industrial manufacture of chemical compounds while also reducing toxic waste and emissions.

See also: Beauty Products; Fuel Cells; Nanotechnology.

Further Reading

Bond, Geoffrey C., Catherine Louis, and David T. Thompson. 2006. *Catalysis by Gold.* London: World Scientific.

Carabineiro, S. A. C., and D. T. Thompson. 2007. "Catalytic Applications for Gold Nanotechnology." *Nanoscience and Technology* (2007):377–489.

Haruta, M. 1992. "Preparation and Environmental Applications of Supported Gold Catalysts." *Now and Future*:7, 13.

Hutchings, Graham J. 1996. "Catalysis: A Golden Future." *Gold Bulletin*:29, 123.

Hutchings, Graham J. 2002. "Gold Catalysis in Chemical Processing." *Catalysis Today* 72:11.

Hutchings, Graham J. 2008. "Nanocrystalline Gold and Gold Palladium Alloy Catalysts for Chemical Synthesis." *Chem. Commun*:1148–1164.

Pattrick, G., E. van der Lingen, C. W. Corti, R. J. Holliday, and D. T. Thompson. 2004. "The Potential for Use of Gold in Automotive Pollution Control Technologies: A Short Review." *Topics Catal* 30/31:273

Thompson, David T. 2007. "Using Gold Nanoparticles for Catalysis." *Nanotoday*:2, 40.

CELLINI, BENVENUTO

Benvenuto Cellini (1500–1571) was an Italian artist and author of the Renaissance period known for his talents as a goldsmith and for his detailed autobiographical account of his life and experiences designing gold and silver works and sculpture for powerful patrons among the Catholic Church hierarchy and nobility of Italy and France. Cellini's greatest artistic legacy is his contribution to the establishment of the mannerist style in the decorative arts. Contemporary chronicler of Renaissance artists Giorgio Vasari hailed Cellini's exceptional abilities as a goldsmith in his *Lives*, writing, "When he applied himself to goldsmiths' work in his youth he was unequalled. . . he set jewels and decorated them with marvelous collets composed of figures of such fine workmanship and sometimes of such original and capricious design that one could not imagine anything better."

Cellini was born in Florence in 1500, son of Maria Lisabetta Granacci and Giovanni Cellini, a musician and instrument maker. Despite his father's hope that Cellini would also pursue a career as a musician, Cellini's interest and talent in the metallurgical arts was apparent very early on and he was sent to study **goldsmithing** under a series of Florentine masters from ages 13 to 18, primarily as an apprentice to the goldsmith Antonio di Sandro, known as Marcone. He did learn to play the flute, however, and at one point was employed as a musician at the Papal court in Rome.

Despite his eventual reputation as a master craftsman, beginning in his teenage years Cellini also gained a certain infamy as a notorious rogue, a temperament he himself chronicled proudly at times in his *Autobiography*, shamelessly recounting his violent behavior and the personal motivations behind it. At the age of 16, he had his first encounter with the Florentine magistrates for his involvement in a skirmish among other young hooligans and subsequently avoided punishment by fleeing to Siena, where he continued his studies at the workshop of local goldsmith Francesco Castoro. After periods of study in Pisa and Bologna, Cellini was permitted to return to Florence.

In 1519 Cellini sought to advance in his career by moving to Rome to work for goldsmiths Giovanni de Georgis of Firenzuola and Lucagnolo di Ciccolini of Jesi, known primarily for making larger gold and silver cast pieces such as basins and ewers (water vessels such as jugs or vases). Cellini remained in Rome for two years and then returned to Florence for a brief period, forced to flee once again for his arrest for involvement in a violent attack on members of the Guasconti family. With the assistance of powerful friends, Cellini returned to Rome in 1523.

During his second stay in Rome from 1523 to 1527, Cellini received some of his most significant early commissions, notably a pair of gold candlesticks for the altar commissioned by Don Francesco di Cabrera, Bishop of Salamanca. In 1524 he joined the workshop of Milanese goldsmith Giovan Piero della Tacca, and soon thereafter opened his own studio. He also benefited greatly from his family's personal ties to Giulio de'Medici after de'Medici's election as Pope Clement VII in 1523. It was during this period that Cellini joined the orchestra at the Papal Court as a paid musician for a brief time. Amid fierce competitions and rivalries with other artists in the Eternal City, Cellini honed his art and produced a prolific number of notable works. During this period he was implicated in a number of skirmishes and intrigues that are amusingly recounted in his *Autobiography*. According to his own account, during the siege of Rome by Charles III, Duke of Bourbon, in 1527, when Pope Clement VII and the cardinals took refuge in the fortifications of Castel Sant'Angelo, Cellini fought in the battle as a bombardier and claims to have killed the Prince of Orange, Philibert of Châlon, and may have killed Charles III. During the weeks of the siege, the pope hired Cellini to remove various jewels from the papal tiara and other pieces in the Treasury of St. Peter and stitch them into the lining of the pope's vestments for safekeeping, a task that would later lead to his arrest in 1537 on charges that he stole some of the jewels and subsequent imprisonment in Castel Sant'Angelo.

Following the siege, Cellini returned once again to Florence with a pardon from the Florentine magistrates on account of his heroics at the Medici papal court. To avoid an outbreak of plague in the city, he left to work at the court of the Duke of Mantua for a short period, ultimately returning to Rome in 1529. Over the next several years, Cellini was implicated in a series of murders and violent attacks on rivals in the name of blood revenge. In his *Autobiography* he brazenly admits to involvement in five homicides. Yet Cellini managed to avoid imprisonment on account of his good graces with the new pope, Paul III, until he was sentenced to prison in 1537 for allegedly pocketing some of the papal jewels, a charge he may likely have been set up for. Upon receiving a pardon on behalf of Cardinal d'Este of Ferrara in 1540, Cellini left Rome to work for French King Francis I at his courts in Fontainebleau and Paris.

Cellini's *Autobiography* provides particularly detailed accounts of his artistic production during his years in France from 1540 to 1545. Interestingly, he appears to have made pieces largely for members of the Italian nobility or Church hierarchy, rather than the French monarchy. From 1540 to 1543 Cellini worked on his most legendary creation, the *Saliera*, or saltcellar, for Francis I, cited by art historians as one of the most significant pieces for the history of Renaissance **goldsmithing**. The 26 centimeter–high sculpture is cast in gold on a base of ivory with decorative enamel, with representations of the god Neptune as the sea and the goddess Ceres as the earth sitting with their legs intertwined. The *Saliera* avoided the melting pot in 1566, when it was included in an inventory by French King Charles IX among a list of items to be melted down for gold **bullion**. Instead, the salt cellar was given to Archduke Ferdinand of Tirol in 1570 to honor his marriage to Elizabeth of Austria, and has remained in Austria to the present day. On May 11, 2003, it was stolen from the Kunsthistorisches Museum and was later famously recovered by Austrian police in 2006. At the time of the theft the piece was valued at approximately $50 million.

In 1545 Cellini returned to Florence, where he would remain until his death. During the latter half of his career, Cellini's own personal emphasis on sculpture as an artist was developed, as he affiliated himself with Cosimo de'Medici's Accademia delle Arti del Disegno and produced his most famous sculpture, the bronze cast *Perseus Holding the Head of Medusa*, on display to the present day in the Loggia dei Lanzi in Florence's Piazza della Signoria. Cellini died in Florence in 1571, leaving six daughters as his heirs.

On account of his colorful *Autobiography* and robust personality, Cellini continued on through the modern period to assume a place in contemporary pop culture in such works as the Gershwin Broadway musical *The Firebrand of Florence* (1945) and as a subject of inspiration or intrigue for many other artists and authors.

See also: Bullion; Casting; Goldsmithing.

Salt cellar or *Saliera*, belonging to King Francis I of France depicting the earth and sea united represented by a female earth goddess and a male sea god, 1540–1543 (gold and enamel), by Benvenuto Cellini (1500–1571). (AFP/Getty Images)

Further Reading

Cellini, Benvenuto. *The Autobiography of Benvenuto Cellini.* Translated by John Addington Symonds. Project Guttenberg. http://www.gutenberg.org/catalog/world/readfile?fk_files=4622.
Hayward, J. F. 1976. *Virtuoso Goldsmiths and the Triumph of Mannerism.* New York: Rizzoli International.
Hobart Cust, Robert Henry. 1912. *Benvenuto Cellini.* London: Metheun.
Pope-Hennessy, John. 1985. *Cellini.* New York: Abbeville Press.
Vasari, Giorgio. *Lives of the Artists.*

CHEVALIER, MICHEL

Michel Chevalier (1806–1879) was a French politician, engineer, and economic theorist whose research and diplomacy supported the cause of free market liberalism and contributed to the adoption of the international **gold standard**.

Chevalier was born on January 13, 1806, in Limoges, of the Haute-Vienne department of France, where he studied at the École Polytechnique. He later earned a degree in engineering from the École des Mines in Paris in 1829 and subsequently pursued a career as a civic engineer. He emerged from the July Revolution of 1830 as a supporter of Saint-Simonianism, serving as editor of the sect's paper the *Globe.* In 1832, Chevalier was arrested for "moral outrage" and sentenced to serve one year in prison for advocating the Saint-Simonian case for the sexual liberation of women. He received a pardon after serving only seven-and-a-half months of his sentence through the intervention of Minister of the Interior Adolphe Thiers, who wished to send Chevalier on a research mission to the United States and South America to evaluate the development of land and water transportation networks in the Americas.

In 1841 Chevalier received an appointment as professor of economics at the Collège de France, the only scholar holding that title during the period. Chevalier began his political career in 1845 when he was elected deputy for the Aveyron department, and later senator in 1860. During his first year as senator, he successfully worked with Richard Cobden and John Bright to negotiate a free trade agreement with Great Britain, the Anglo-French Treaty of Commerce of 1860, commonly referred to as the Cobden-Chevalier Treaty. He also served as an economic advisor to Napoleon III during this period and testified in a government investigation on the status of money and banking in France.

Chevalier published a prolific number of works of economic theory related to fiscal policy and public works, a vital combination in Saint-Simonian thought, advocating such projects as a trans-Andean railroad, a trans-Siberian railroad, and a tunnel beneath the English Channel. Based on many of his observations from his 1834–1835 visit to the Americas, in 1859, Chevalier published *On the Probable Fall of the Value of Gold: The Commercial and Social Consequences which May Ensue, and the Measures which it Invites*, in which he argues that the dramatic increase in the international gold supply produced by the gold rushes of North America in the mid-19th century would lead to a decline in the price of gold relative to silver, resulting ultimately in **inflation**. Chevalier presents his arguments in the context of industrial development and advancements in transportation, arguing that the expansion of global economies and faster more prolific manufacturing processes, along with more sophisticated banking instruments not tied to the direct exchange of gold currency, would also reduce the demand for gold, driving the price down even further. Although many of Chevalier's predictions did not come to be, as prices remained relatively stable during the late 19th century, such arguments greatly contributed to the adoption of the international **gold standard** in the 1870s.

Chevalier died at his chateau in Montpelier on November 28, 1879, after receiving many public honors and widespread recognition in both England and France for his contributions to the field of political economy.

See also: Banking and Credit; Gold Standard; Inflation.

Further Reading

Drolet, Michael. 2008. "Industry, Class, and Society: A Historiographic Reinterpretation of Michel Chevalier." *The English Historical Review* 123, no. 504:1229–1271.

Kindleberger, Charles P. 1985. "Michel Chevalier (1806–79), the Economic de Tocqueville." In *Keynesianism versus Monetarism and Other Essays in Financial History*, 24–40. London: George Allen and Unwin.

CHRYSOGRAPHY

Chrysography is a term derived from the Greek roots *chrysos* (gold) and *grapho* (writing) to refer to the art of writing in gold ink, a practice with origins dating from early Hebrew religious texts that is associated largely with scribal arts from the early Christian era to the High Middle Ages, reaching a peak in the illustrious works of Carolingian Renaissance calligraphers in northern Europe during the eighth and ninth centuries, among the most significant examples of which are the so-called Golden Gospels, notably the Godescalc Gospels from ca. 781, and the **Gospel Books of Saint Médard de Soissons** of the early-ninth-century Ada School, both conserved at the Bibliothéque Nationale in Paris. Commissioned by Charlemagne himself to be produced at the legendary Frankish scriptorium at Aachen, the Godescalc Gospels demonstrate the incorporation of early Christian, Insular (scripts of early medieval Britain and Ireland), and Byzantine styles into what would come to be known as the Carolingian Illumination style.

The early Christian practice of chrysography was largely used to produce copies of the Gospels and may have had the dual purpose of honoring the sacred texts with a likewise precious material and highlighting the words on the page so that they would glow more brightly in the candlelight. Gold inks were made of crushed gold mixed with a binding agent such as glair to produce a form of egg white tempera or gum from a variety of botanical adhesives. The writing was then burnished with a smooth stone or animal tooth to create sheen. During the High Middle Ages, another technique known as mordant **gilding** developed around the 12th century in which the script was written initially using a tinted adhesive, and then gold leaf was subsequently applied to the letters before the adhesive dried. Medieval manuscripts often consisted of purple-dyed parchments rather than white or off-white pages because the gold letters appeared more prominently on a background of red, purple, and black hues.

The prevalence and quality of chrysographical manuscripts from late antiquity and the Middle Ages is remarkable considering the great expense necessary to produce these masterpieces, both in terms of the precious materials and the labor. By the 15th century, chrysography in the classical sense saw a decline due to the advent of new materials and technologies, although it continues to the present day as a decorative art form among bookmakers and artists.

See also: Gilding; Gospel Books of Saint Médard de Soissons.

Further Reading

Alexander, Jonathan J.G. 1992. *Medieval Illuminators and Their Methods of Work*. New Haven, CT: Yale Univ. Press.
Calkins, Robert G. 1983. *Illuminated Books of the Middle Ages*. Ithaca, NY: Cornell Univ. Press.
De Hamel, Christopher. 1992. *Scribes and Illuminators*. London: British Museum Press.
Dodwell, C. R. 1993. *The Pictorial Arts of the West, 800–1200*. Pelican History of Art. New Haven, CT: Yale Univ. Press.
Pächt, Otto. 1986. *Book Illumination in the Middle Ages*. Oxford: Oxford Univ. Press.
Thompson, Daniel Varney. 1956. *The Materials and Techniques of Medieval Painting*. New York: Courier Dover.
Whitley, Kathleen P. 2000. *The History and Technique of Manuscript Gilding*. New Castle, DE: Oak Knoll Press.

CHRYSOTYPE

Chrysotype is a modern-day method of alternative photography in which a sheet of archival quality paper is first treated by hand with a solution of light-sensitive chemicals, such as ferric ammonium citrate, and then exposed through a negative to an ultraviolet light source, such as a sun lamp or the sun itself. The photographic image is then developed by the application of a neutral solution of gold chloride (colloidal gold) and potassium iodide. The application of gold in the form of nanoparticles in this chemical solution offers unique benefits to the practice of the art of photography, particularly in the longevity of the image and depth of tonality.

Colloidal gold had been used since antiquity to color stained glass and other glass and ceramic creations, and the development and experimentation with gold chloride during the 18th and 19th centuries in the practice of **alchemy** gave rise to understandings of the functions and processes of gold in this form in the field of modern chemistry during the mid-19th century, when chemist John Herschel developed a photographic process in 1842 that he referred to as "chrysotype," in which gold was used to develop an image from a negative onto paper. Yet Herschel's method faced many obstacles and ultimately was not as financially feasible or reliable as the method developed by his colleague William Henry Fox Talbot around the same time (in 1841). Talbot's method, called the "silver nitrate method," uses silver iodide as a solution for developing an image from a negative onto paper, which remains the most common practice to this day.

While the contact printing process invented by Talbot would prove to be the most practical and widely employed method in the history of photography and the challenges of Herschel's method left it to languish in relative obscurity until the later 20th century, gold compounds and solutions such as sodium bisthiosulphateaurate have been applied over images created with silver solutions both to gild an image in order to deepen its hues and as a method of stabilizing the silver images to prevent them from fading or tarnishing. Such practices reinforce the advantageous elements of the application of gold compounds in photography. Because the properties of gold are less catalytic than silver, gold images are less vulnerable to breaking down on contact with acid or oxidization and are therefore longer lasting.

Examples of Herschel's experimentations with gold prints are conserved today at the Museum of the History of Science in Oxford and the National Museum of Photography in Bradford, United Kingdom. Over the course of the 19th century, other scientists continued to experiment with this medium, yet it never achieved much prominence.

In the 1980s, advances in modern chemistry allowed scientists to control the oxidation levels of the gold salts employed in the chrysotype process and thereby develop a method that was both more economical and easier to control. While researching and exploring the further development of platinotype and palladiotype photographic imaging processes in the 1990s, Scottish chemist and professor of photography Mike Ware developed a method for printing photographic images in gold, which he referred to as the "New Chrysotype," an art form that he has since pursued and promoted by virtue of both its aesthetic and archival merits.

See also: Alchemy.

Further Reading

Schaaf, Larry J. 1992. *Out of the Shadows: Herschel, Talbot & the Invention of Photography*. New Haven: Yale Univ. Press.
Ware, Mike. 1991. "Prints of Gold: The Chrysotype Process Reinvented." *Scottish Photography Bulletin* no. 1: 6–8.
Ware, Mike. 2006. "Chrysotype: Photography in Nanoparticle Gold." *Gold Bulletin* 39, no. 3:124–131.
Ware, Mike. 2006. *Gold in Photography. The History and Art of Chrysotype*. Brighton, UK: Cromwell Press.
Wiveleslie Abney, Sir William de. 1878. *A Treatise on Photography*. D. Appleton & Co.

COIN CLIPPING

Coin clipping was a practice in which counterfeiters and criminals would clip or file shavings off the edges of gold or silver coins in small enough amounts so as not to alter the immediate appearance of the coin. These shavings would then be accumulated and melted down into gold **bullion** for resale to goldsmiths or even to mints to be recast as new coins. The practice occurred from the earliest appearance of regular minted coinage in antiquity, and became rampant during the Middle Ages with the expansion of **minting** of certain universally traded coins such as the **bezant** and the **ducat**, to the degree that increased legislation during the 15th and 16th centuries treated the practice as a capital offense, the equivalent of high treason, punishable by death sentence. The risk that widespread coin clipping posed to national treasuries by causing **inflation** through the accrued devaluation of the currency led to innovations in **minting** technology and the advent of mechanized, milled coins by the 17th century, which effectively curtailed the practice.

Evidence of coin clipping exists in surviving gold and silver from throughout antiquity, yet the practice became particularly widespread with the expanding manufacture of certain globally traded hammered gold coinage, such as the **bezant** and the **ducat**, which assumed such financial prowess as a currency of international exchange that the coins began to be accepted at their very face value, rather than on the value of the weight of the precious metals at the original minted size. Nevertheless, government devaluation of the currency by manipulating the minted weight of the coinage coupled with the gradual devaluation of the coinage from clipping, estimated at a rate of 20–40 percent over periods of only a few years, ultimately led to potential **inflation** and, at times, diminished confidence in the value of gold coinage.

During periods of economic crisis amid the European wars of the later Middle Ages and early modern period, the financial effects of coin clipping and counterfeiting had such potentially disastrous political consequences that legislation toughened the pursuit and punishment of offenders. For example, the Treason Act of 1415 in England deemed coin clipping an act of high treason punishable by death. During the early 17th century, Paris Mint engraver **Nicholas Briot** developed a new technology that would become known as "milling," in which coins were produced with an engraved pattern or inscription along their edges to make clipping practically impossible as even the slightest filing or clipping would be apparent. Yet Briot's attempts to implement this new technology met with resistance in both Paris and also at the **Tower Mint** in London, where traditionalists preferred to maintain the time-honored hammering method.

Economic and political upheaval in England at the end of the 17th century would ultimately lead to the demise of coin clipping in the wake of the **Great Recoinage Act** of 1696. As the English treasury became strained to the point of bankruptcy due to large capital outflows of both gold and silver, the situation was worsened by an overall increase in coin clipping that further deteriorated the value of English gold and silver coinage. Beginning in 1695, in an urgent attempt to curtail the practice, King William III issued a series of alarming proclamations forbidding transactions in currency that had been clipped, which incited panic in British commerce. To address the problem, a parliamentary committee of the House of Commons convened in 1695 to hear testimony from noted political economists and intellectuals, among them John Locke, Charles Montagu, and **Sir Isaac Newton**, on how to remedy the crisis. Based on the committee's proposals, on January 17, 1696, parliament passed "An Act for Remedying the Ill State of the Coin of the Kingdom," known as the **Great Recoinage Act**, which ordered that all coinage be remitted and collected and recoined using the new mechanized technologies, such as milling, to prevent future clipping and counterfeiting. **Sir Isaac Newton** was subsequently appointed as Warden and later Master of the **Tower Mint** to oversee the persecution of clippers and counterfeiters and the implementation of the new **minting** technologies. Despite a period of economic backlash and financial chaos in the aftermath of the legislation, within a period of only a few years, these reforms had for all intents and purposes largely eliminated the problem of coin clipping in England and influenced the spread of the new technologies to the Continent.

See also: Bezant; Briot, Nicholas; Bullion; Ducat; Great Recoinage Act; Inflation; Minting; Newton, Sir Isaac; Tower Mint.

Further Reading

Becker, Thomas W. 1970. *The Coin Makers.* New York: Doubleday.
Craig, John. 1963. "Isaac Newton and the Counterfeiters." *Notes and Records of the Royal Society of London* 18, no. 2 (December):136–145.
Lewis, Bernard. 1958. "Some Reflections on the Decline of the Ottoman Empire." *Studia Islamica* no. 9:111–127.
Li, Ming-Hsun. 1963. *The Great Recoinage of 1696 to 1699.* London: Weidenfeld and Nicolson.
MacDonald, George. 1916. *The Evolution of Coinage.* New York: Univ. Press.

COIN STAMPING

Coin stamping, also referred to as "coin striking," is the process used since earliest antiquity to stamp the image to be struck on both sides of a metal coin using two coin dies, one for each side, that are then simultaneously hammered by hand or struck by a machine onto the blank piece of metal in the desired shape.

A coin press fitted with two matrices. One matrix prints the head of Louis XV on one side of the coin, and the other the "escutcheon," corresponding to our "tails." Blanks from basket (P) are stamped one by one and tossed into the second basket (Q). (Gillispie, Charles C., ed. *A Diderot Pictorial Encyclopedia of Trades and Industry.* New York: Dover Publications, Inc., 1959)

From the earliest evidence of coinage in the Anatolian kingdom of Lydia during the 7th century BC to the mid-16th century AD, coins in Europe and the Near East were struck by hand using individual, hand-crafted dies created by skilled artisans. The earliest method for marking the cast blank disks, also known as "planchets," was a form of stamping known as a "punch," in which the metal blank of a gold or silver alloy was "punched" with a bronze or iron implement that made a unique hole or impression on the blank. This process evolved into the creation of individually engraved upper and lower dies, which were made also by carving the image or design with steel tools into bronze or iron dies that were used to strike coinage. The major disadvantage of these individually engraved dies was that they could not be mass produced. An anvil was used to secure the lower die, onto which was placed the blank, and then, finally, the upper die. The impression was then hammered into the blank by hand. The technique varied among regions according to how many hammer blows were needed to achieve the impression, typically one to three, and whether the blank was hammered cold or slightly heated for an easier impression. The durability of the lower die was longer, whereas the upper die tended to wear out about 50 percent faster due to the fact that it received the blow from the hammer directly. It is estimated that a single upper die could be used to stamp approximately 10,000 coins, while the lower die would last for 20,000 coins. To speed up the process, a team of up to four men would work together, each assigned to a separate task, to produce more coins faster. Medieval coin stampers were able to strike up to 12 coins per minute using these methods.

During the Middle Ages, the engravers who created the dies formed into guilds and developed the art of coin stamping from the combination of punching and graving characteristic of the early and High Middle Ages into a skilled craft using largely iron dies, some of which could be reproduced by using a mold, an innovation that helped advance the technology and meet the ever-increasing need for more coinage. A variety of other innovations related to hammering methods and the thickness and size of the blanks permitted almost twice as many coins to be struck from one die before it became worn out.

The expansion of commerce and larger national mints during the 14th century and the accompanying need for more coins led to the development of machinery that would, by the mid-16th century, largely substitute the traditional hammering methods with mechanized processes using simple machines such as the screw press, which was operated by man power or a water mill; or the rotary press, which allowed mints to strike several coins at a time using a water mill, horse power, or, later, steam power. This new mechanized process was thus often referred to as "milling." Variations on these machines were developed by inventors throughout Europe, among them Leonardo da Vinci, who created a type of screw press in the 15th century. With the advantages of mechanizations, mints grew into larger scale operations that employed many more laborers to produce coins at a higher capacity, while at the same time maintaining dies and stamping processes that minimized counterfeiting.

In 1790, Matthew Bouton, Scottish owner of the Soho mint, near Birmingham, partnered with James Watt, inventor of a more efficient steam engine, to patent the two technologies into the manufacture of a new and improved steam-powered coin press. The machine was a great success, and was exported widely to mints in Europe and North America. By the 19th century, mints were consolidated and industrialized, implementing these mechanized processes on a larger scale.

Modern coin stamping is done through highly mechanized processes largely fueled by electric power in which a design created by an artist is formed into a mold that is then used to create a steel master punch. In addition to the working die that stamps the impression on either side of the coin, a steel collar also stamps an impression on the edges, further insurance against fraud such as **coin clipping** and counterfeiting. Rigorous inspection processes provide quality control, and some errors are corrected by hand. Throughout history, mint errors or irregularities from the stamping process have resulted in coins of significant value among collectors.

See also: Alloying; Coin Clipping.

Further Reading

Becker, Thomas W. 1970. *The Coin Makers*. New York: Doubleday.
Hill, George F. 1977. *Ancient Methods of Coining*. New York, Attic Books.
MacDonald, George. 1916. *The Evolution of Coinage*. London: Cambridge Univ. Press.
Vermeule, Cornelius C. 1954. *Some Notes on Ancient Dies and Coining Methods*. London: Spink.

COINAGE ACT OF 1873

The Coinage Act of 1873 was passed in the United States Congress on February 12, 1873, as the continuation of a series of antebellum monetary legislation intended to restore currency values through reestablishing specie payments on the wartime issue demand notes, after a period in which their redemption was suspended during the Civil War. The law effectively converted the United States to a de facto **gold standard** of currency exchange by officially demonetizing silver coinage.

Although the United States had for the most part been on a de facto **gold standard** since 1834, when the convertibility of gold was fixed at a price of $20.76 per ounce, with countries throughout Europe abandoning bimetallism for the **gold standard**, U.S. economic analysts and politicians feared that an increased demand for gold and an increased supply of silver would lead to a disadvantageous bimetallic ratio for the U.S. currency that might lead to **inflation**. The Coinage Act of 1873 was presented to Congress in May 1872 and debated actively on the floor of both the House of Representatives and the Senate, both of which passed it with majority votes, with the former vote coming in at 110 to 13 and the latter at 36 to 14. The law specified which coins were permitted to be minted to include official gold coins and ancillary silver coins, but the law failed to include the traditional standard silver dollar. Thus, the legislation did not explicitly convert the domestic currency to the **gold standard**. Yet, by ending the legal basis for bimetallism, currency in the United States switched to a **gold standard** in practice.

In his report to Treasury of the Secretary in November 1872 in defense of the proposed legislation, Director of the Mint Henry R. Lindeman wrote that "the fluctuations in the relative value of gold and silver during the last hundred years have not been very great, but several causes are now at work, all tending to an excess of supply over demand for silver and its consequent depreciation." He further explained the importance of the fact that the weight of opinion in the United States and Europe rested "in favor of a single **gold standard**." Such views apparently contributed to the swift passage of the act; however, opponents of the measure vehemently attacked the exclusion of the silver dollar and launched an active campaign for the resumption of bimetallism that mobilized those with a vested interest in maintaining a silver-to-gold ratio, such as silver miners, farmers, and other working-class voters, who cast the legislation as the "Crime of 1873," perpetrated by the monied class of financiers and bankers against the working and middle classes. The expression of discontentment using the term "crime" was premised on the perceived lack of diligence on the part of the congressman who passed the act so quickly and on the suggestion of an actual crime, as noted in an 1877 editorial in the *Nation*, that suggested British policymaker Ernest Seyd, a "bullionist and agent of foreign bond holders," visited the United States in 1873 to bribe members of Congress with $500,000 for their votes for gold. There is no evidence to support this rumor, however.

Criticism of the legislation gave rise to the Free Silver Movement and produced an issue that would define the Democratic Populist Party platform in the following decades, which argued for the resumption of free coinage of both gold and silver at the specified ratio. In response to political attacks, U.S. Senator John Sherman, Chairman of the Senate Finance Committee, declared in a speech to the Ohio Republican Convention on August 15, 1895, that: "There have been a great many battles fought against gold, but gold has won every time. Gold never has compromised. Gold has made the world respect it all the time. The English people once thought they could get along without gold for a while, but they had to come back to it" (Bryan 1906). Although persistent campaigning by opponents of the issue, principal among them Democratic Party presidential candidate William Jennings Bryan during the 1895 campaign, succeeded in obtaining certain concessions for the silver interests, such as an agreement to mint $2–4 million worth of silver dollar coins, such policies had little effect. In fact, the consequences of the Coinage Act indicate that perhaps policymakers were overly concerned about the potential increases in silver, and estimates for what the fluctuating silver-to-gold ratio would have been during this period indicate that a silver-to-gold ratio may have been the more prudent choice. By 1900, the United States officially converted to the **gold standard** de jure.

The controversy over the Coinage Act of 1873 and the ensuing Free Silver Movement in the following decades are viewed by some scholars as the symbolic focal point of Frank L. Baum's children's novel *The Wizard of Oz* (1900), in which the **Yellow Brick Road** and the Land of Oz are considered to be representative of the false promises of the government's adherence to the **gold standard**, with the burden of the consequences of the measure falling on the common man, or, the little people.

See also: Exchange Rates; Gold Standard; Yellow Brick Road.

Further Reading

Bryan, William Jennings, ed. 1906. *The World's Famous Orations. Vol. X: America, III (1861–1905)*. New York: Funk and Wagnalls. http://www.bartleby.com/268/10/23.html.

Friedman, Milton. 1990. "The Crime of 1873." *Journal of Political Economy* 98, no. 6:1159–1194.

Friedman, Milton, and Anna Schwartz. 1963. *A Monetary History of the United States, 1867–1960*. Princeton: National Bureau of Economic Research and Princeton Univ. Press.

Rockoff, Hugh. 1990. "The Wizard of Oz as a Monetary Allegory." *Journal of Political Economy* 98 (August):739–760.

COLUMBUS, CHRISTOPHER

Christopher Columbus (1451–1506) was an Italian navigator and explorer who is credited with the discovery of the New World of the Americas in 1492 during a voyage supported by Spanish monarchs King Ferdinand II of Aragon and Queen Isabella I of Castile, a discovery that sparked the initial colonization of South and Central America by the Spanish Crown, undertaken in large part with the intention of exploiting the new territory's reputed gold sources. Columbus completed four transatlantic voyages during 1492–1504 and established a series of settlements in Hispaniola.

> "Many of these people, all men, came from the shore. . . and I was anxious to learn whether they had gold."
> —CHRISTOPHER COLUMBUS, *TRAVEL DIARY*, OCTOBER 13, 1492

Columbus's exact date of birth is unknown, but he is presumed to have been born ca. 1451 in Genoa. He was the eldest son of a Genoese wool worker, Domenico Colombo, of modest means, and Susanna Fontanarossa. The family also had close ties to Savona and may have lived there for a period. Columbus received a limited but adequate education and was sent to school in Pavia, where he demonstrated a passion for navigation at an early age. In his own memoirs, Columbus claims to have been engaged as a sailor by the young age of 14. As a young man, he worked as an agent for several principal Genoese merchant families. In 1476, while aboard a Portuguese merchant marine fleet as a passenger, he survived a shipwreck off the coast of Cape St. Vincent in Portugal, after which he and his brother Bartolomeo settled in Lisbon to work as cartographers. There is also evidence that during this period, he sailed to northern European ports in England, Ireland, and Iceland.

In 1479 he married Felipa Perestrello e Moniz, the daughter of Portuguese nobleman Don Bartolomeo Moniz de Perestrello, a cavalier with Genoese heritage who had served Portuguese King Henry. The marriage earned Columbus a title of nobility but with little wealth. He had a son with Felipa, Diego, born in 1480. During the 1480s, Columbus continued to accompany Portuguese and Genoese fleets as a merchant agent on routes along the coast of West Africa, further developing his understanding of navigation with respect to Atlantic currents and wind. During this period, he developed a plan to explore a possible westward passage to Asia from Europe in search of a swifter, more secure route to the riches of the East. Columbus's calculations and motivations were based on a variety of materials and myths, among them the travelogue of **Marco Polo**, whose tales of the wealth in gold and spices in the lands of Cathay (China) and Cipangu (Japan), where he had seen rooftops of gold, along with medieval legends of the existence of seven cities of gold to the east on the island of "Antilia," noted on medieval maps as being located far west of Spain and Portugal. Columbus's plan was further motivated by geopolitical events in Europe and the Near East, as war with the Ottomans disrupted typical overland and marine trade routes to Asia and the Spanish Reconquest of Granada from the Moors had both drained Spanish royal coffers and incited a missionary zeal among Spanish Catholics. Columbus thus viewed his plan as one that he was destined by god to fulfill, yet also with significant financial and political benefit.

In 1484, Columbus presented his plan and a request for support to navigate a westward route to Asia by way of the Atlantic to King John II of Portugal, who turned him down based on doubts about its feasibility relative to Columbus's calculations of the distance of the crossing. Over the following year, Columbus inquired with Genoa, the Venetian Republic, and perhaps even England in search of a benefactor, to no avail. In 1486, he left Portugal for Spain to present his plan and appeal for support to King **Ferdinand II of Aragon** and Queen **Isabella I of Castile** at the Spanish court in Seville. The king and queen postponed their decision until a committee of experts could be convened to examine its viability. In the meantime, they offered Columbus room and board and settled him in Cordoba. It is assumed his wife Felipa died, as Columbus arrived in Spain with his son Diego as an apparent widower, although historians also suggest he may have left his first wife. A few years later in 1488 he met Beatriz Enriquez de Harana in Cordoba, a woman of low social status who became his mistress and with whom he had his second son, Fernando, around that same year.

The Spanish monarchs convened in Salamanca a group of experts, largely clergyman, to evaluate Columbus's proposal. Many expressed doubts about Columbus's calculations of the distance across the Atlantic, and the Genoese navigator answered them carefully, taking care not to divulge his plan in its entirety and avoiding any accusations of heresy. Yet Ferdinand and Isabella remained preoccupied with the Reconquest campaign in Granada, so Columbus stayed on as a guest of the Spanish court during 1487–1492, even accompanying a military campaign. During this period, Columbus established important relationships with several powerful Franciscan friars, among them the queen's confessor Juan Pérez of La Rábida, who is credited with arranging Columbus's successful royal audience at the Alcazar Palace in Cordoba on the tails of the Spanish victory in Granada in January 1492.

In April 1492, Ferdinand and Isabella granted Columbus a contract naming him the right to the titles of viceroy and governor general over all lands he discovered, as well as the title of "Admiral of the Ocean Sea," titles to which his heirs could succeed, along with a fully outfitted fleet of three ships. His contract also provided that he had a right to 10 percent of any proceeds from the newly discovered lands. Columbus's small fleet, comprised of the *Niña*, the *Pinta*, and the *Santa Maria*, departed on August 3, 1492, from the port of Palos de la Frontera, financed by additional funds from Italian financiers. After a stop in the Canary

Islands to restock supplies, the fleet headed out into supposed uncharted waters on September 6, sailing for over five weeks until they spied land on October 12 as they approached the island of San Salvador in the Caribbean. Here Columbus had his first encounter with the native peoples of the area, observing in his journal that they greeted him kindly and wore gold adornments in their noses. While exploring the island further, Columbus inquired with local chiefs about the gold and was informed of a local abundance of gold. Eager to explore further with the goal of reaching Cipangu, Columbus staked his claim with the Spanish flag and departed on October 28, sailing to present-day Cuba and then on to Hispaniola in December, where the *Santa Maria* was shipwrecked on Christmas Day in the Bay of Caracol, where Columbus established the first Spanish fort in the New World as La Navidad. According to his memoirs, the local inhabitants sought to trade gold objects with the Spanish sailors. Observing this, Columbus inquired with the local chief, who told him of an abundance of gold in the region and presented him with a gold-inlayed mask as a gift.

Having discovered at least a hint of wealth in gold, although certainly not the treasures he had hoped for, Columbus set sail for Spain in January 1493, sending a letter along to the king and queen with details about the treasures he had discovered. Following a journey plagued by bad weather, Columbus was forced to come to port in Portuguese territory, where they were seized and escorted to Lisbon to be interrogated by King John. Ironically, this encounter, along with the increasing disloyalty and disaffection of Columbus's crew, aroused a high degree of suspicion of Columbus's loyalties to the Spanish court. Nevertheless, his fleet was greeted with much pomp and circumstance on March 15 as it returned to the Spanish port at Palos, and tales of the curiosities and treasures he brought back with him, including ornately dressed natives, parrots, spices, and gold, spread rapidly across Europe.

The king and queen felt these treasures merited another voyage, and on September 25, 1493, Columbus set sail once again from the port at Cadiz for the New World in command of a fleet of 17 ships that included provisions and men for establishing settlements. The ships also carried a group of Franciscan friars with plans to establish missions. Proceeding along a more southerly route, the fleet reached the island of Dominica on November 3, then continued on to Hispaniola, where they found the La Navidad settlement completely destroyed as a result of conflicts with the native Tainos. Columbus continued his explorations of the island in pursuit of gold, establishing additional forts near potential goldfields. In February 1493, he sent 12 ships back to Spain carrying cargo of spices, gold, and slaves. Yet this bounty remained disappointing to the king and queen, who had hoped for more, and the returning sailors also brought reports of Columbus's treachery toward the native peoples and ineffective rule of the forts.

Columbus's journals indeed reflect his apparent obsession with the pursuit of gold and belief that his mission to discover the "West Indies" was divinely ordained. His daily diary entries contain significant references to gold, and his letters to the Spanish court are marked by this quest. In a 1494 letter to Ferdinand

and Isabella outlining statutes for the colonization of the islands, seven provisions are related to prospecting and collecting gold, while three are related to regulating its transport. Columbus even assumes the possibility that the pursuit of gold may prove a distraction to colonization, informing the king and queen that:

> As, in the eagerness to get gold, every one will wish, naturally, to engage in its search in preference to any other employment, it seems to me that the privilege of going to look for gold ought to be withheld during some portion of each year, that there may be opportunity to have the other business necessary for the island performed. ("Letter to Ferdinand and Isabella," 1494)

After exploring the area as far as Jamaica and failing to find further passage to Asia, Columbus announced that Cuba was in fact Cathay and departed for Spain in March 1496 to organize materials and provisions for a third voyage intended to establish the primacy of Spain's settlements in the New World, explore the region further in search of gold and other commodities, and continue his pursuit of a route to India. He set sail from Spain on May 30, 1498, reaching Trinidad on July 31, and continuing on to reach the South American continent at the Gulf of Paria, claiming the territory of modern-day Venezuela for the Spanish Crown in early August. Failing to find a passage through to India, he returned to Hispaniola, where he discovered the fort of La Isabela in shambles due to dissent among both colonists and the indigenous peoples over their mistreatment by the colonial authorities, particularly with respect to their disappointing efforts to exploit the region's gold sources. In response to complaints about the colonists' treatment and the treatment of the natives, the king and queen had dispatched Francisco de Bobadilla, a royal envoy, to investigate. Upon arrival and after reviewing the case, Bobadilla arrested Columbus and his two brothers in October 1500 on accusations of mistreatment of natives and colonists during their rule of the settlements on Hispaniola.

After Columbus and his brother had served several weeks in prison, King Ferdinand responded to the Genoese explorer's letters by freeing him and his brothers. Based on the promise of further treasures to be discovered, the king and queen then supported Columbus's fourth and final voyage to the New World, although Columbus was stripped of his governorship over the new territories. Columbus departed from Cadiz on May 11, 1502, with a fleet of four ships, headed for the New World with explicit orders from the Spanish monarchs to continue his pursuit of a westward passage to India and greater riches but to avoid Hispaniola. In the face of bad weather, however, Columbus sought to enter the port at Santo Domingo and was denied entrance by the governor. Columbus continued on exploring the Caribbean, Nicaragua, and Costa Rica, reaching Panama by October. Amid bad weather, Columbus continued his exploration, encouraged by natives' reports of cities of gold and another sea beyond the peninsula. Forced by bad weather to turn back to Hispaniola for provisions and repairs, however, Columbus's fleet shipwrecked on the coast of Jamaica in the winter of 1503, where Columbus and his crew awaited rescue for over a year.

They were finally rescued on June 29, 1504, and arrived in Spain in late fall of that year. Despite his tarnished reputation and resentment at the Spanish Crown's failure to honor his contract in its entirety, Columbus had access to the wealth he had accumulated in gold and other commodities from the New World through accounts in Italian banks. Yet he persisted in his final years to appeal to King Ferdinand for full acknowledgment of his discovery and the attendant rights of the original contract so as to be able to pass these honors and income on to his heirs. Columbus died on May 20, 1506, in Valladolid, Spain. His body was originally laid to rest at the local Franciscan monastery, then transferred to Seville. By virtue of his son Diego's will, Columbus's remains were then transferred to Santo Domingo in Hispaniola. When the Spanish conquered the island in 1795, his remains were moved to Havana, Cuba, and then reportedly shipped back to the Cathedral of Seville in 1898. The archeological evidence for his final resting place remains in question, however, as an excavation uncovered a set of remains marked as those of Christopher Columbus in Santo Domingo in the late 19th century.

Despite the disappointing end to his life, Columbus's legacy assumed the status of myth over the course of modern history as his discovery of the New World was quickly acknowledged as a watershed event in human history. In 1992, commemorations of the 500th anniversary of Columbus's voyage of discovery examined the significance of these events from a wide variety of angles and disciplines: geographic, economic, anthropological, historical, and biological. Columbus's life and achievements continue to be a hotly debated topic among scholars, who scrutinize the sometimes scarce evidence for his early and later years and analyze the continued significance of his encounters with the native cultures and ecosystems as an example of how exploitation and competing value systems in the cultural contact between the explorers and native peoples shaped colonial history and continues to shape contemporary politics and society.

See also: Ferdinand II of Aragon; Isabella I of Castile; Polo, Marco; Spanish Conquest

Further Reading

Charalambous, Demetrio. 1994. "The Enigma of the Isle of Gold." *Revista de Historia de América* no. 118 (July–December):33–49.

Columbus, Christopher. "Letter to the King and Queen of Spain, 1494." Medieval Sourcebook. http://www.fordham.edu/halsall/source/columbus2.html.

Davidson, Neiles H. 1997. *Columbus Then and Now: A Life Reexamined.* Norman: Univ. of Oklahoma Press.

Fernandez-Armesto, Felipe. 1991. *Columbus.* Oxford: Oxford Univ. Press.

Jones, Julie. 2002. "Gold of the Indies." *The Metropolitan Museum of Art Bulletin* 59, no. 4 (Spring):3–4.

Morison, Samuel E. 1942. *Admiral of the Ocean Sea: A Life of Christopher Columbus.* 2 vol. Boston: Little & Brown.

Morison, Samuel E., trans. & ed. 1963. *Journals and Other Documents on the Life and Voyages of Christopher Columbus.* New York: Heritage Press.

Rouse, Irving. 1992. *The Tainos: Rise and Decline of the People Who Greeted Columbus.* New Haven, CT: Yale Univ. Press.

Sale, Kirkpatrick. 1990. *The Conquest of Paradise: Christopher Columbus and the Columbian Legacy.* New York: Plume-Penguin.

Sauer, Carl Ortwin. 1966. *The Early Spanish Main.* Berkeley: Univ. of California Press.
Starrs, Paul F. 1992. "Looking for Columbus." *Geographical Review* 82, no. 4 (October): 367–374.
Wright, Louis B. 1970. *Gold, Glory and Gospel: The Adventurous Lives and Times of the Renaissance Explorers.* New York: Atheneum.

CORTÉS, HERNÁN

Hernán Cortés (1485–1587) was a Spanish conquistador whose quest to discover the reputed wealth of the sophisticated inland city of **Tenochtitlan** in Central America led him to conquer the lands of the Aztec Empire in present-day Mexico and claim them for the Spanish Crown in 1521 under the rule of Emperor Charles V.

Cortés was born in Medellin, in the Extremadura area of Castilian Spain, in 1485, son of *hidalgo* (lesser nobility) parents Martin Cortés de Monroy and Catalina Pizarro Altamarino, through whom Cortés was related as a cousin to conquistador **Francisco Pizarro**. Following a period of two years of study in law and Latin in Salamanca at the age of 14, Cortés returned home and soon became intrigued with stories of the wealth and adventure of the newly discovered Indies. In 1504, he departed for Hispaniola, where he remained and worked as a farmer and public administrator until 1511, when he joined Diego Velazquez on a successful expedition to conquer Cuba. Cortés received a significant land grant and Indian slaves to work the land in Santiago as well as a prestigious appointment as clerk to the treasurer in the new colonial administration. As he consolidated his leadership and authority on the island, he would go on to serve two terms as mayor of Santiago and marry Catalina Juárez, a relative of Velazquez, who bore him no children. Following Catalina's death, he married noblewoman Juana de Zuniga with whom he had four children.

Spaniard Hernán Cortés (1485–1547) conquered the Aztec Empire in Mexico and later led expeditions into Guatemala and Honduras. He was thus a prime mover in the establishment of Spain's vast American empire. (Library of Congress)

Based on news of a wealthy, sophisticated native empire on the mainland in the region of present-day Mexico, in October 1518, Velazquez appointed Cortés as captain general of a new expedition to explore the coast of the Yucatán peninsula. Yet political tensions had arisen between Cortés and Velazquez, who is thought to have feared Cortés's increasing power and authority, and Velazquez rescinded the orders a few months later. Cortés ignored the revocation and set out anyway in February 1819 with a fleet of 11 ships, around 500 soldiers, 100 sailors, and a dozen or so horses, effectively constituting a mutiny by going against Velazquez' orders. Upon landing in the Gulf of Mexico region, Cortés acquainted himself with the native peoples, who assisted him as interpreters and told him news of a great, wealthy, inland city ruled by a powerful emperor. After claiming the coastal territories for the Spanish Crown and founding the city of Veracruz as his base, and, by dealing directly with the crown, establishing independence from the jurisdictional authority of Velazquez, Cortés embarked on an expedition inland in search of the city, educating himself along the way about the structure of the Aztec Empire and taking advantage of the tenuous system of vassal states established by the Aztec emperors by appealing to the resentments among the leaders of the conquered native territories and establishing them as allies in his campaign to conquer the Aztec capital of **Tenochtitlan** in the central highlands.

As Cortés and his growing force continued inland on the route toward **Tenochtitlan**, Aztec Emperor Montezuma II initially sought to prevent his arrival through attempts to bribe him with gold and other riches as a form of tribute for staying away. When such efforts failed, he sent out vassal lords to organize ambushes and avert Cortés militarily, yet to no avail. Cortés and his expedition arrived in the Aztec capital on November 8, 1519, and were greeted by Montezuma with customary diplomacy, including the presentation of lavish gifts of gold and silver objects of many types. According to accounts by Cortés himself as well as other observers, the emperor also declared that he viewed Cortés's arrival in light of the Aztec prophecy that the fair-skinned god Quetzalcóatl would return one day from a faraway kingdom to rule the Aztecs once again.

Cortés and his men were housed graciously in the city that they declared in letters and memoirs to be among the great cities of the world, with elaborate architecture, infrastructure, religious temples, palaces, and a population estimated at around 200,000. Yet Cortés remained doubtful of the emperor's hospitality and plotted to conquer the city from within. While he was away on a brief expedition to quell a controversy and potential rebellion in Veracruz, the commander in charge in Tenochtitlan, Pedro de Alvarado, murdered several Aztec nobles, causing a violent backlash against the Spanish. Cortés returned against the advice of his peers to claim back the city, but when the Emperor Montezuma II was stoned to death in the confusion of the violence, he feared that the Spanish were too greatly outnumbered to achieve a victory. Attempting to escape the city in the middle of the night carrying the treasure in gold and silver they had acquired during their stay, the Spanish troops and their native allies were discovered and suffered devastating losses in a brutal battle on July 1, 1520. Despite this, the small remaining troops defeated the Aztecs the following day in a feat of military strategy. Over the course of the next year, Cortés methodically sought to reconquer the city, succeeding in May 1521 to claim Tenochtitlan for the Spanish Crown and establish it as the capital of the new Viceroyalty of New Spain, a significant event in the course of the **Spanish Conquest** that also signaled a dramatic and abrupt end to an important and once great pre-Columbian culture and society.

Despite his achievements in conquering the territories of Mexico, Cortés faced constant political controversy due to the opposition of Governor Velazquez, who sought to tarnish his reputation and initiate proceedings against him among the members of the Council of the Indies. In response, Cortés wrote a series of powerful letters to Emperor Charles V defending his case and reminding the emperor of all that he had obtained for Spain. Following an expedition to Honduras in 1524, Cortés returned to find his status had worsened and embarked in 1528 to appeal personally to Charles V, with whom he had an audience at Toledo, where he was successful in appealing for confirmation of his titles and assets as captain general and was also conferred with the title of Marqués de Valle de Oaxaca; however, he did not receive the administrative post he had sought. Cortés returned to New Spain in 1530 only to find an atmosphere of political chaos in which his authority had eroded even further. He assisted in restoring order to the government and, when a viceroy was appointed, moved to his country estate in Cuernavaca. In 1540 he returned to Spain to continue his attempts to restore his claims to financial and royal privileges as well as property relating to his conquests in the name of the crown. After Charles V ignored his appeals for several years, he set out to return to Mexico but died on December 2, 1547, in the small village of Castilleja de la Cuesta, near Seville, following a serious case of dysentery, never having resolved his circumstances of misfortune.

See also: Pizarro, Francisco; Spanish Conquest; Tenochtitlan.

Further Reading

Cortés, Hernán. 2001. *Letters from Mexico*, ed. Anthony Pagden. New Haven, CT: Yale Univ. Press.
Diaz Del Castillo, Bernal. 2004. *The Discovery and Conquest of Mexico, 1817–21*. Cambridge, MA: Da Capo Press.
Leon-Portillo, Miguel, and Lysander Kemp, trans. 2006. *The Broken Spears: The Aztec Account of the Conquest of Mexico*. Boston: Beacon Press.
Prescott, William H. 2010. *History of the Conquest of Mexico*. New York: The Modern Library. (Orig. pub. 1843).
Thomas, Hugh. 1995. *Conquest: Montezuma, Cortés, and the Fall of Old Mexico*. New York: Simon and Schuster.

COX, JAMES

James Cox (ca. 1723–1800) was a London goldsmith and inventor renowned in the international watchmaking industry through his innovations in both design and function, as well as for charting new territory for luxury goods in overseas trade. During the mid-1760s, Cox established a far-reaching reputation for his extravagant novelty timepieces and other creations, many of which were automated. Cox actively pursued clients in the luxury trade markets in India and China, where his pieces quickly gained popularity.

Despite his swift path to success, however, his failure to obtain powerful, wealthy patrons among the European aristocracy and the British government's 1772 ban on Cox's delivery of luxury items to China resulted in an escalation of financial troubles for the entrepreneurial designer. That same year, in an effort to salvage his enterprise, Cox established a museum in Spring Gardens, Charing Cross, London, to generate revenue by charging admittance to see his collections. He promoted the museum aggressively in print publications and catalogues, such as "A Descriptive Catalogue of the Several Superb and Magnificent Pieces of Mechanism and Jewellry, Exhibited in Mr. Cox's Museum" (1772).

Although the museum caught the attention of London elites who buzzed about the original works on display, especially the cutting-edge automata, Cox attempted to dispose of his collection by lottery the following year, seeking written permission through an Act of Parliament to proceed with the lottery, which may have also been a marketing ploy to attract attention to the sale by means of the parliamentary handbill.

By 1778, Britain's weakened foreign trade in the wake of the American Revolution, especially the loss of his primary trade with Asia, doomed Cox's enterprise to bankruptcy for the second time in 1778 and the company's stock in London and in Canton, China, was fully liquidated by 1792.

Two of the most famous of Cox's pieces, still extant today, are the Peacock Clock, a large automaton timepiece conserved at the Hermitage Museum in St. Petersburg, Russia, and the clock that has become known as "Cox's Timepiece." The Peacock Clock features fully mechanized birds, a peacock, a cockerel, and an owl, and was crafted of gold, agate, and precious jewels. It was reputedly a gift from Gregory Potemkin to Catherine II in 1781, possibly by means of one of Cox's own suppliers who had the aim of securing the empress's patronage. The clock was displayed at the Hermitage, where it gained notoriety for Cox's work among the public. Equally well-known as this enchanting peacock, Cox's Timepiece was allegedly the first clock powered by atmospheric pressure, born out of Cox's larger experimentations with perpetual motion machines. The clock is conserved at the Victoria and Albert Museum in London. Upon witnessing an exhibit of Cox's creations in London, one visitor observed the spectacle with wonder:

> A peacock screeched and spread its tail when the hour struck, while a cock crowed and a cage with an owl inside revolved and twelve bells rang. A silver swan with an articulated neck glided across a surface of artificial waters . . . sixteen elephants supported a pair of seven-foot high temples adorned with 1,700 pieces of **jewelry**. . . a chronoscope inlaid with 100,000 precious stones evidently needed no animal guise. (Altick 1978, 69)

See also: Goldsmithing; Jewelry.

Further Reading

Altick, Richard. 1978. *The Shows of London*. Cambridge, MA: Harvard Univ. Press.
Brepohl, Erhard, and Tim McCreight. 2001. *The Theory and Practice of Goldsmithing*. Brunswick, ME: Brynmorgen Press.
Pointon. 1999. "Dealer in Magic: James Cox's Jewelry Museum and the Economics of Luxurious Spectacle." *History of Political Economy* 31:423–451.

Smith, Roger. 2000. "James Cox (c. 1723–1800): A Revised Biography." *Burlington Magazine* 142, no. 1167 (June):353–361.
Vincent, Clare, and J. H. Leopold. 2000. "James Cox (ca. 1723–1800): Goldsmith and Entrepreneur." In *Heilbrunn Timeline of Art History*. New York: The Metropolitan Museum of Art. http://www.metmuseum.org/toah/hd/jcox/hd_jcox.htm.

CROESUS

Croesus (ca. 595–547 BC) reigned as King of Lydia, a kingdom in Anatolia, from 560 to ca. 547 BC, a period in which his legendary wealth achieved a mythical status in historians' chronicles for centuries to come. The historical association between Croesus and wealth that would come to be the basis for the English proverb "as rich as Croesus," can be traced to Herodotus's accounts of the Lydian king's deeds, especially King Croesus's infamous dialogue with the Athenian philosopher and statesman Solon on the question of the nature of happiness. According to Herodotus, in response to Croesus's question "Who is the happiest of men?" (the king likely anticipated the response would be himself), Solon replied that three other common men were the happiest for reasons in direct contrast to tragic incidents in the life of Croesus, a reply intended to emphasize the fickle fate of fortune and fleeting impact of wealth. Solon concluded his remarks with the reminder that it is not possible to gauge the happiness of one's life relative to another until the end of a life, for fortune and prosperity once achieved are not guaranteed to last.

According to the ancient chronicles, the source of Croesus's wealth lay in his kingdom's rich gold mines and advantageous geographic position on the primary trade routes with Asia, in a territory spanning from the Aegean Coast of Asia Minor and to the west as far as the River Halys, this latter region comprised of the remaining Greek Ionian states subjugated to Lydia under Croesus, the tributes of which accounted for a significant portion of his wealth. The kingdom also benefited from the rich natural resources of alluvial gold that flowed from the Pactolus River near the Lydian capital of Sardis.

In addition to the good fortune of his kingdom's resources, Croesus consolidated his wealth even further through a series of economic reforms. Although Herodotus credits Croesus as the first ruler ever to mint coinage, there is evidence that Croesus's father, Alyattes, was the first to issue a form of gold coinage cast from electrum, an alloy of silver and gold formed naturally in rivers. The stability of these coins contributed to the expansion of Lydian trade relations. Archeological evidence and the written record indicate, however, that Croesus had all of the electrum coins melted down and reminted as bimetallic coins of purer levels of gold and silver at a 10:1 ratio through a method in which the gold was purified by heating it with lead and salt. These new coins were imprinted on one side with marks to indicate their value and on the other side with a stamp of the Sardis coat of arms depicting a lion and a bull. The coins were particularly noteworthy in the history of currency for the care Croesus took in maintaining their uniformity of weight and value, on account of which they were widely accepted throughout Greece and Asia Minor.

After having built up his kingdom to such heights, Croesus grew increasingly fearful of the imposing threat of innovation from the Persians under the rule of King Cyrus and began preparations to meet the Persian king in battle. Croesus sought the council of the oracle at Delphi as well as other oracles on the question of which course to pursue, to which the Delphic oracle replied that a great empire would be destroyed if he sent his army to attack the Persians and that he should ally with another great power in the region. Assuming that this prophecy referred to victory for Lydia, Croesus is alleged to have rewarded the oracle at Delphi with generous gifts of gold and other treasures.

In response to the news from Delphi, Croesus promptly allied himself with Sparta and launched a campaign against Persia in 547. During the winter, however, Cyrus launched a surprise attack on Sardis and captured the king, testimony to the fickleness of fate that the oracle had been referring to the defeat of the kingdom of Lydia rather than Persia.

Upon his capture, Croesus was set upon a funeral pyre in Sardis, where he is said to have been overheard by Cyrus exclaiming "Solon" three times. Out of curiosity, the Persian king ordered that he be cut down so that he could discover the reason behind this strange outburst. According to some sources, Cyrus was so inspired by the tale of Solon's prophetic words that he chose to spare Croesus's life after this event, making the Lydian king one of his most trusted advisors. There is contrary evidence, however, that Croesus's life ended with this battle.

See also: Alloying; Mining.

Further Reading

Davies, Glyn. 1995. *A History of Money from Ancient Times to the Present Day*. Cardiff: Univ. of Wales Press.
Head, Barclay Vincent. 1877. *The Coinage of Lydia*. London: Trubner.
Ramage, Andrew, and Paul Craddock. 2000. *King Croesus' Gold: Excavations at Sardis and the History of Gold Refining*. London: British Museum.
Tassel, Janet. 1998. "The Search for Sardis." *Harvard Magazine* (April):51–60, 95–96.

CROWNS, ROYAL

Royal crowns are a form of ornamental headdress typically worn by emperors, kings, and queens as well as religious rulers as a sign of distinction and as symbols of their power and authority. Crowns have existed in this capacity in human societies throughout the world from prehistory to the present, with cultural variations in the style and materials used to create crowns serving as reflections of their ritual, ceremonial, or political function. Gold has been used almost universally as the predominant precious metal used by goldsmiths to design crowns befitting the glory and, in many cases, perceived immortality of members of the royalty. Crowns have been worn for a variety of occasions, for example, coronation ceremonies, government functions, or

> "A crown, golden in show, is but a wreath of thorns."
>
> —JOHN MILTON

St. Edward's Crown was copied in the reign of Charles II (r. 1660–1685) from the ancient crown worn by Edward the Confessor (1004–1066), the Anglo-Saxon king who founded Westminster Abbey. Edward was canonized in 1161 and given the shrine in the Abbey. (AP/Wide World Photos)

religious rituals; and, in the past, rulers were often buried with their crowns.

The earliest versions of crowns exist from archeological excavations in ancient Mesopotamia, where the royalty and nobility wore a form of headdress that tied in the back with ribbons or strands of beading, a style that evolved into the early type of crown known as a diadem. Excavations of the royal burial complex of the ancient third-century BC Sumerian city of **Ur** revealed the tomb of Queen Puabi, in which the queen's body was adorned with a gold crown consisting of two intertwined wreaths of poplar leaves, encircled by another wreath of willow leaves and encrusted with ornate **filigree**, inlaid precious stones such as lapis lazuli and carnelian, and strings of gold beads. The evolution of this style of crown can be seen in the royal diadems worn by the Egyptian pharaohs, for whom different crowns were made to represent their rule over Egypt's Upper and Lower Kingdoms, as well as for certain functions such as war. Excavations of the tomb of the 14th-century BC pharaoh **Tutankhamun** revealed the first intact Egyptian royal crown ever discovered. Tutankhamun's crown consisted of an ornate gold fillet encircling the head, with the pharaoh's personal totem animals, the falcon and the cobra, symbols of the kingdoms he ruled, protruding from the front, and ornate streamers of gold extending down from the back. The crown was decorated with inlay of colored glass, obsidian, and precious stones such as carnelian. Ancient crowns such as these also served a practical function in fastening the ruler's hair back or keeping a wig in place. References to gold crowns in ancient Hebrew culture are abundant in Old Testament sources. Written sources and ancient Egyptian tablets provide important clues to the political significance of crowns in early antiquity.

Gold crowns were also worn by rulers and religious leaders in early antiquity in Asia, India, and pre-Columbian Latin America. Archeological excavations of the **Royal Tombs of Silla**, an ancient kingdom once located in present-day Gyeongju, South Korea, uncovered six exquisitely crafted crowns of pure gold dating from the fifth and sixth centuries. These crowns exhibit such a high level of metalworking techniques in intricate forms of **granulation** or early **filigree** that they are thought to reveal cultural exposure to Scythian or Etruscan goldsmiths' work. Crowns in China, India, Eurasia, and Arabia (for example, those of Arab sultans, Hindu rajah, Persian shahs, and the Chinese emperors) were typically more ornate than those in the Mediterranean regions, with elaborate **filigree** of foliage or animals, and often took the form of a type of cap, rather than a base consisting of a simple band.

Crowns were not as common in ancient Greece until the reign of Alexander the Great in the fourth century BC, who may have adopted the custom due to the influence of the Near Eastern cultures he encountered during his conquests. The significance of crowns of pure gold to Mediterranean cultures of classical antiquity is evident in the legend of **Archimedes** and the golden crown, in which the third-century BC King of Syracuse, Hieron II, is said to have asked the Greek mathematician **Archimedes** to determine whether the goldsmith who made his new crown, consisting of a wreath of gold laurel leaves, had indeed made it out of the pure gold provided to him or had stolen some of the gold and added silver as an alloy to the crown. Archimedes's experiments to determine the fineness of the crown's gold are said to have led to his discovery of a hydrostatic balance system for weighing metals.

In Imperial Rome, emperors wore simple bands or wreath-like crowns of gold that varied according to the occasion being celebrated or the office of state being performed. With the advent of Christianity in the Roman Empire, Byzantine crowns originally maintained the diadem shape and assumed a more ornate style around the 11th century, with strings of gold adorned with precious stones or pools dangling from the gold fillet, often encrusted with precious stones. Gold crowns entered the iconography during this era as well, with the emperor often depicted in art as receiving the crown from Jesus or a holy saint, a symbol of the ruler's divine ordination. Iconography also included depictions of saints with a golden crown as a symbol of their sanctity. One of the oldest extant gold crowns from early Christian Europe, the Iron Crown of Lombardy, thought to be of late Roman Imperial or Langobard origins and dating to around the 6th or 7th century, was said to contain an iron nail from the Holy Cross. Preserved at the Cathedral of Monza in Italy, the crown was used to crown those royals laying claim to the title "King of Italy," including many holy Roman emperors, from the 11th century to the early 19th century.

During the High Middle Ages, royal crowns evolved from open bands to include enclosures across the top, typically adorned with a cross and precious jewels. The Holy Crown of St. Stephen of Hungary, conserved today at the Hungarian Parliament, is an example of this later Byzantine style. Versions of these styles prevailed in Europe until the rise of absolutism in the 18th century, when crowns became slightly more stylized and elevated by curved arches on top of a banded base, sometimes encircling sumptuous fabrics or furs, with additional insignia indicating the royal line added to the crown's design; for example, the fleur-de-lis of the French monarchy.

Although many crowns are conserved in religious or government treasuries, offices, or museums as symbols of power and legitimacy or as sacred relics, few crowns are worn today with the same significance that they once held. Several coronation ceremonies still exist in which the new ruler or religious leader receives the crown from the legitimating authority, such as in the coronation of the kings or queens of Great Britain and the domains of the United Kingdom, the coronation of new popes with the papal tiara, and the coronation of new kings in the South Pacific Kingdom of Tonga. Among the most famous historical crowns still in official use today are the Imperial State Crown and St. Edward's Crown, housed with the Crown Jewels of the British Monarchy in the Jewel House at the Tower of London. Created in 1661 for Charles II after the Oliver Cromwell's destruction of the previous crowns, the crowns have been used continuously in coronation ceremonies and official state occasions, with St. Edward's Crown used for coronation and the Imperial State Crown used for official state ceremonies such as the opening of Parliament. The St. Edward's Crown was last used to crown Queen Elizabeth II in Westminster Abbey on June 2, 1953. After the coronation, the newly vested sovereign processes out of the church wearing the Imperial State Crown. For lesser state occasions, Queen Elizabeth and her predecessors have typically worn a selection of tiaras from the vast royal collection.

See also: Archimedes; Filigree; Granulation; Silla, Royal Tombs of; Tutankhamun; Ur Royal Tombs of.

Further Reading

The British Monarchy. "The Crown Jewels." http://www.royal.gov.uk/The%20Royal%20Collection%20and%20other%20collections/TheCrownJewels/Overview.aspx.

Bunker, Emma C. 1993. "Gold in the Ancient Chinese World: A Cultural Puzzle." *Artibus Asiae* 53, no. 1–2:27–50.

Clark, Grahame. 1986. *Symbols of Excellence*. Cambridge: Cambridge Univ. Press.

Historic Royal Places. *Tower of London, Crown Jewels*. http://www.hrp.org.uk/TowerOfLondon/stories/crownjewels.aspx.

Jones, William. 1902. *Crowns and Coronations: A History of Regalia*. London: Chatto and Windus.

Lee, Soyoung. (2003, October). "Golden Treasures: The Royal Tombs of Silla." In *Heilbrunn Timeline of Art History*. New York: Metropolitan Museum of Art. http://www.metmuseum.org/toah/hd/sila/hd_sila.htm.

Pollen, John Hungerford. 1878. *Gold and Silversmiths' Work*. London: R. Clay and Sons.

Rousseau, Vanessa. 2004. "Emblem of an Empire: The Development of the Byzantine Empress's Crown." *Al-Masaq* 16, no. 1 (March):5–15.

Scarisbrick, Diana, Christophe Vachaudez, and Jan Walgrave, eds. 2008. *Royal Jewels: From Charlemagne to the Romanovs*. New York: Vendome Press.

CYANIDE

Cyanide is the most commonly used complexant in the chemical reaction required to extract gold from low-grade ore deposits employed by modern gold **mining** operations. As a noble metal, gold is not soluble in water and therefore a chemical reaction is required to convert the gold into a water-soluble metallic complex from which the gold dust can then be extracted and refined. This can be done using a variety of chemical solutions, such as mercury, chloride, or bromide, yet cyanide is the least expensive and most effective. Due to its high levels of toxicity, however, the use of cyanide in hydrometallurgical **mining** operations can have potentially disastrous environmental effects and is monitored carefully by local and national governments and organizations.

Gold cyanidation is also known as the "MacArthur-Forrest Process," a reference to industrialist brothers Robert and William Forrest and John Steward MacArthur, the Scottish chemist who discovered how to obtain the correct chemical reaction for extracting gold in 1887, when he applied the stoichiometrical equation discovered in 1846 by L. Elsner for dissolving gold ore in an alkaline cyanide solution, an equation known as the Elsner reaction, in which sodium cyanide is used as the complexant in combination with atmospheric oxygen as an oxidant to transform the gold into an aurocyanide compound comprised of sodium cyanoaurite and sodium hydroxide.

After the gold is leached from the ore in the process, the metal is then extracted from the residue through various methods for separating the gold solution from the other compounds, or by a process of adsorption of the gold pulp onto an activated granulated carbon in which the extracted gold is then separated from the carbon compound by means of either precipitation or electrodisposition.

The cyanidation process was crucial to the development of the gold **mining** industry in the Transvaal region of South Africa during the late 19th century in an area where gold sources were found in low-grade ore in deep gold reefs, which required significant utilization of such technologies to extract a more profitable percentage of pure gold from the ore relative to the large capital outlay needed to mine and process the gold.

Due to the highly toxic properties of cyanide, strict regulations monitor its transport, storage, and use in hydrometallurgical **mining** operations, and cyanide is banned altogether by various local, state, or national governments throughout the world.

See also: Mining; Witwatersrand Gold Rush.

Further Reading

Fivaz, C. E. 1988. "How the MacArthur-Forrest Cyanidation Process Ensured South Africa's Golden Future." *Journal of the South African Institute of Mining and Metallurgy* (September):309–318.

Habashi, Fathi. 1987. "One Hundred Years of Cyanidation." *Historical Metallurgy Notes* 80, no. 905:108–114.

International Cyanide Management Code. "The Use of Cyanide in the Gold Industry." http://www.cyanidecode.org/cyanide_use.php.

Marsden, John. 2006. *The Chemistry of Gold Extraction*. Littleton, CO: Society for Mining, Metallurgy, and Exploration.

Rose, Thomas Kirke, and W. A. C. Newman. 1896. *The Metallurgy of Gold*. London: C. Griffin.

Schnabel, Carl. 1907. *A Handbook of Metallurgy*. New York: MacMillan.

DE GAULLE, CHARLES (1890–1970)

Charles de Gaulle was a French politician, military leader, and political theorist from the interwar period through the 1960s who is acknowledged as the architect of France's Fifth Republic, of which he was president from 1958 to 1969, and commemorated as the controversial champion of the reestablishment of French preeminence in the European sphere through vigorous policies of national sovereignty that were often viewed as isolationist and anti-American. Among these policies was his advocacy of a restoration of the **gold standard** as a means of stabilizing international monetary policy and preventing the further development of the U.S. dollar as the primary exchange currency in international markets.

De Gaulle was born into an upper-middle-class Roman Catholic family in Lille, France, on November 22, 1890. The son of philosophy and literature professor Henri de Gaulle and his wife Jeanne Maillot, both of whom came from educated, aristocratic families with traditionalist, nationalist viewpoints, young de Gaulle attended military school at the academy in Saint-Cyr and subsequently enlisted in 1912 in the French Army infantry regiment led by Colonel Philippe Pétain, who promoted him to 2d lieutenant in 1913. During the latter half of World War I, he was captured during combat at the Battle of Verdun in 1916 and spent over two years as a prisoner of war. In 1921 he married Yvonne Vendroux, with whom he had three children, Philippe, Elisabeth, and Anne.

During the interwar period de Gaulle taught military theory in Poland and France and developed his theories of military strategy and engagement in a series of written studies. In 1925 he was promoted by Marshal Pétain to serve on the Supreme War Council as an army major, and during 1927–1929 he oversaw the occupations in the Rhineland and the Middle East. During the 1930s he achieved the rank of lieutenant colonel and a position on the National Defense Council.

With the onset of World War II in 1940, de Gaulle commanded a tank brigade in the French 5th Army, and after a series of tactical victories in which he put his military strategies of armored warfare into action, he was promoted to the rank of brigadier general, a position he would retain for the remainder of his career. During this period, however, his insistence on certain military strategies and political philosophies had provoked the ire of Pétain, who sought to obtain an armistice with Germany and from whom de Gaulle was now disaffected. On June 6, 1940, he joined the short-lived government of Prime Minister Paul Reynaud as undersecretary of state for defense and war and liaison to the United Kingdom. Only 10 days later, however, de Gaulle fled to Britain upon learning of the resignation of Reynaud and the appointment of Pétain as prime minister with extraordinary powers enabling him to negotiate an armistice with Germany on June 22 and create the Vichy government. From London, de Gaulle appealed to the French people in a series of radio addresses to continue their fight against the Germans and established the Free French Forces as a means of organizing the resistance movement and collaborating with the Allied Special Forces. In a series of court martials and trials in the subsequent months, de Gaulle was tried and convicted by the Vichy regime in absentia of treason. In 1943 he relocated to Algiers where he continued his resistance activities and established the French Committee of National Liberation.

With the liberation of France following the Allied invasion of Normandy in the summer of 1944, De Gaulle swiftly returned to Paris to lead a provisional government and ensure the preservation of the autonomy of the French state following liberation.

Yet his frustrations with the partisan politics and ineffective constitution of France's Fourth Republic led him to resign on January 20, 1946, after which he focused his efforts on creating a movement for political reform that emerged as the Rally of the French People, a popular reform movement that functioned effectively as a political party by 1951, yet, which lost steam a few years later. Disillusioned, de Gaulle retired to his home in Colombey-les-deux-Églises to write his war memoirs. In May 1958, the weakness of the Fourth Republic government in the face of the potentially devastating conflict in Algeria prompted him to come out of retirement. On June 1 he was presented as prime minister of France to the National Assembly, and on the following day, he obtained parliamentary approval of constitutional reforms that bestowed his office with greater powers. Thus the Fifth Republic was born on the political foundations that de Gaulle had vehemently advocated for, namely, a strong state that enabled its leader to act decisively.

In December 1958, de Gaulle was elected as president of the Fifth Republic, an office that he held for two consecutive terms until 1969. During the initial years of his presidency, he faced the challenges of severe civil unrest and organized domestic terrorism resulting from the Algerian conflict and the continued postwar reconstruction of the state. After negotiating Algerian independence in 1962, however, De Gaulle turned his attention to his broader political vision for reestablishing France as an international power by pursuing a series of policies intended to strengthen politics and society in France that were perceived by many Western nations and the United States as isolationist and anti-American. Among such policies were his argument for neutrality in Vietnam, his limiting of French involvement with NATO, and, perhaps most significantly, his vehement advocacy of a return to the **gold standard** based on the argument that the continued U.S. deficit and global dependence on the dollar as the benchmark currency only functioned to perpetuate American imperialism and threatened to undermine the establishment of a healthy international monetary system.

Despite differences in the motivations behind his viewpoints on the dismantling of the Bretton Woods agreement and the progressive development of the dollar as the international reserve currency, which were largely political, de Gaulle's position on this matter rested largely on the work of economist Jacques Rueff, whose concern for the potential price instability and **inflation** as a result of the shift away from the **gold standard** led him to appeal vigorously to de Gaulle for monetary reform. De Gaulle appointed Rueff in 1958 to lead a committee dedicated to ensuring economic stability in France through prudent monetary policy. De Gaulle argued for the reestablishment of the **gold standard** by raising the price of gold from the threshold that had been set in the 1930s, and in doing so, he forced the United States to deal with its deficit and restrict its liberal extension of credit, which de Gaulle viewed as an instrument of economic imperialism. De Gaulle expressed his views in a controversial press conference on February 4, 1965, in which he claimed that the dollar as the unit of international exchange allowed the United States to "indebt itself freely to foreign countries" and "expropriate" business from abroad while also imposing and expanding its military power. According to de Gaulle, only gold had a real value that was acknowledged globally and could not be manipulated by individual governments. Later that same year, he addressed a grievance at the November 1966 meeting of the **International Monetary Fund** and G-10 ministers, demanding to know why an increase in the price of gold and a recommendation that the United States be obliged to international oversight over its balance of payments in the same manner as European Economic Community countries had not been included in the summit's agenda.

By 1966, however, de Gaulle's demands for monetary reform that would restore the **gold standard** were increasingly criticized in foreign policy circles as antagonistic and anti-American. Over the course of the next year, the French government developed a more conciliatory policy with respect to economic relations with the United States based in part on concerns that a return to the **gold standard** would be of great benefit to rogue nations such as the Soviet Union and South Africa. Shortly after the end of his second term as president in 1969, de Gaulle retired once again to Colombey-les-deux-Églises, where he died in his sleep of a heart attack on November 9, 1970.

Throughout the 1960s and 1970s, however, Rueff continued to argue for the fragility of the dollar, warning that a system based on economic cooperation and consensus alone was vulnerable to "unforeseeable events such as the continuation of the U.S. balance of payments deficit, a deterioration of economic circumstances, or even some banking or financial incident, a shift in the balance of power, a possible reversal of some alliances, or simply the evolution of thought and feeling" (Chivvis, 705), which threatened to lead to a collapse of the entire political and economic system of the democratic, Western world. According to historian Christopher Chivvis, Rueff continued to fear the predictions that he had advised de Gaulle of so passionately, namely, that "there could be no lasting international economy, indeed no lasting free society, without gold" (Chivvis 2006, 720).

See also: Bretton Woods System; Gold Standard; International Monetary Fund.

Further Reading

Calleo, David P. 1994. "De Gaulle and the Monetary System: The Golden Rule." In *De Gaulle and the United States. A Centennial Reappraisal*, ed. Robert O. Paxton and Nicholas Wahl, 239–255. Oxford: Oxford Univ. Press.

Chivvis, Christopher S. 2006. "Charles de Gaulle, Jacques Rueff, and French International Monetary Policy under Bretton Woods." *Journal of Contemporary History* 41, no. 4 (October):701–720.
Crawley, Aidan. 1969. *De Gaulle : A Biography*. London: Collins.
de Gaulle, Charles. 1993. "Press Conference February 4, 1965." *Discours et Messages, vol. 4*. Paris: Omnibus/Plon.
de Gaulle, Charles. 1975. *Discours et Messages*. Paris: Plon.
Gough, Hugh, and John Horne, eds. 1994. *De Gaulle and Twentieth-Century France*. Oxford: A Hodder Arnold Publications.
Shennan, Andrew. 1993. *De Gaulle*. London: Longman Group.

DENTAL CROWNS

Dental crowns function as a form of restorative or cosmetic dentistry in which a mold cast from a dental impression is used to create a cap that is adhered to the tooth with a bonding agent to cover the entire tooth to the gum line. This procedure is typically prescribed by dentists to restore a tooth to its original shape or strengthen a tooth that is broken, has become worn, or is damaged from tooth decay. It is also used as a form of cosmetic improvement. Dental crowns are made from a variety of substances, such as porcelain and gold. Gold has been used for dental crowns as early as antiquity due to its strength and durability.

What is known as a "full gold crown" by the American Association of Dentistry is actually a crown comprised of a metal alloy that is recommended to contain a minimum of 60 percent noble metal (gold, platinum, or palladium), of which a minimum of 40 percent must be gold. This alloyed substance may also contain a variety of other types of elements such as copper, silver, or tin. During the 1990s–2000s, a series of studies raised concerns about a potential link between sustained exposure to gold and gold alloys in dental implants such as crowns and the manifestation of allergic responses to gold in the form of skin rashes or stomatitis (an inflammation of the mucous lining in the mouth).

The earliest evidence of the use of gold crowns in dental treatment is among the Etruscans during 166–201. Excavations at an archaeological site in Satricum, an ancient town about 40 miles southeast of Rome, uncovered specimens consisting of two plates of gold that were stamped to form the impression of either side of the tooth and then soldered together and adhered to the tooth by means of a gold apparatus that slipped between adjacent teeth to keep the crown in place. There is scarce evidence of gold crowns after Roman antiquity, however, with the first subsequent references to this form of treatment appearing in dental medical books during the 18th century. The 20th century saw significant improvements in the effectiveness of gold crowns due to advancements in metal **casting**.

See also: Allergies; Alloying; Casting.

Further Reading

American Dental Association. "History of Crowns." http:www.ada.org.
Taylor, James Anderson. 1922. *History of Dentistry*. Philadelphia: Lea & Febiger.

DRAKE, SIR FRANCIS

Sir Francis Drake (ca. 1545–1595) was a legendary English sea captain who served as vice admiral in the navy under Queen Elizabeth I, during which he achieved international renown as a rogue for plundering Spanish sailing vessels for their precious cargoes of gold and other booty from the New World. With his 1577–1580 journey, he became the first Englishman to circumnavigate the globe in the name of the English Crown, and in 1588, he led the English victory against the Spanish Armada.

Drake's exact date of birth is unknown; however, it is thought to be between 1539 and 1542 in Crowndale, near the town of Tavistock, in Devon, where his grandparents cultivated a sizeable plot of land, around 150 acres, as leaseholders. Likewise little is known about Drake's parentage and early childhood. Although some historians have suggested that his father, Edmund, may have also been a sailor who later fled persecution as a Protestant during the religious persecution of the Catholic Queen Mary by moving the family to Kent, where he was a chaplain to the Royal Navy and the family initially lived in an old, grounded ship, this story has been proven false. In fact, Edmund fled Tavistock in 1548 after his arrest and arraignment for robbery, and Francis was raised in Plymouth by his distant relative Sir John Hawkins, who attended to his education and apprenticeship as a crewmember of a local merchant ship. By the age of 18, he served as the officer in charge on a trading ship serving important Spanish ports in the Bay of Biscay. Drake married his first wife, Mary Newman, in 1569 and remarried after her death 12 years later to Elizabeth Sydenham, of a significantly higher stature family. Neither marriage yielded any children.

In his early twenties, Drake accompanied Hawkins on a series of voyages to the Spanish Americas. On one such journey in 1568, the fleet commanded by Drake and Hawkins was ambushed by the Spanish in the Gulf of Mexico port of San Juan de Ulúa. Despite having suffered heavy material losses in the affair, upon his return to England, Drake was heralded for his bravery in the skirmish. This event is often seen as critical to the development of the legendary antagonism between Drake and the Spanish.

During 1570–1571, Drake led two expeditions to the West Indies with the support of Elizabeth I. In 1572 he set out on yet another voyage to Spanish America, intent on avenging his losses from the San Juan de Ulúa defeat, targeting the port settlement of Nombre de Dios on the Atlantic coast of Panama, an area that was heavily trafficked with Spanish ships shuttling precious cargo of gold and silver from Peru. Drake waged a successful initial raid on the Spanish treasure in July of 1572, yet suffered an injury and was forced to retreat. He continued to pursue successful raids inland in 1573 and returned to England in August of that year with boats full of riches. Legend has it that during his expedition to Panama, Drake peered from a tree on the isthmus to catch a glimpse of the Pacific Ocean, which he then resolved to navigate one day.

In 1577, Queen Elizabeth once again provided financial support for his proposed expedition to explore the Pacific Coast of the Americas by sailing through the Straits of Magellan. Drake set sail in December 1577 with a fleet of five ships manned by under 200 sailors, reaching South America by the spring of 1578 and passing through the Straits of Magellan on August 21, 1578, with a fleet reduced to three ships manned by significantly fewer men due to the hardships of the journey and an alleged mutiny among the crew led by Thomas Doughty. Drake then infamously made his way up the Pacific Coast of New Spain on a voyage of piracy, pillaging the unsuspecting Spanish forts and merchant ships along the way. Drake claims in his ship logs to have reached the 48th parallel of the West Coast, approximately the area of Vancouver, British Columbia, in his attempt to find a Northwest passage through to the Atlantic, but was forced to turn back due to stormy weather and anchor in the San Francisco Bay, where he repaired the ship and restocked its provisions. Despite the fact that the area had already been explored and claimed by the Spanish Crown, Drake staked a claim for the English Crown, naming the supposed new territory New Albion.

In the summer of 1579, Drake set out west from San Francisco across the Pacific on a route that would complete his circumnavigation of the globe, passing through several islands of the South Pacific, the Philippines, the Malaku Islands of Indonesia, then across the Indian Ocean, and through the Cape of Good Hope to the Atlantic again. On September 26, 1580, Drake's ship the *Golden Hind* entered Plymouth Harbor full of treasure from the voyage. The now infamous sea captain was met with a hero's welcome by Queen Elizabeth herself, who, despite diplomatic concerns about English relations with the Spanish due to Drake's piracy, bestowed Drake with the honor of knighthood on his very own ship.

Drake was soon appointed mayor of Plymouth, and following his remarriage to Elizabeth Sydenham in 1585, purchased a country home known as Buckland Abbey. Later that year, he was called to the service of the queen to command a large naval fleet in defense of England against the Spanish navy, which he successfully defeated in the Cape Verde Islands and the New Spain ports in Colombia, Florida, Hispaniola, and, later, Cádiz, greatly weakening the Spanish in their attempts to build up their navy for an offensive against England. As captain of the ship *Revenge*, Drake was among the famous naval commanders credited with the English defeat of the **Spanish Armada** in the English Channel in July–September 1588. Although recent research cites luck and weather as the most significant factors in the English victory, many of the tactical and technological developments innovated by Drake and Hawkins during their expeditions against the Spanish likely played a key role in the defeat of the outdated Spanish fleet.

Following a failed attempt to thwart the Spanish in Lisbon in 1589, Drake continued to pursue the Spanish threat, departing in 1595 on an expedition with John Hawkins to attack Spanish settlements in the West Indies. Throughout this voyage, Drake suffered a series of disappointing setbacks. Struck by fever from a deadly bout of dysentery, Drake died on January 28, 1596, naming his nephew Thomas as his sole heir in his last will and testament. His body was buried at sea.

See also: Spanish Armada.

Further Reading

Kelsey, Harry. 2000. *Sir Francis Drake: The Queen's Pirate*. New Haven, CT: Yale Univ. Press.
Pretty, Francis. 1580. *Sir Francis Drake's Famous Voyage around the World*. Modern History Sourcebook. http://www.fordham.edu/halsall/mod/1580Pretty-drake.html.
Whitfield, Peter. 2004. *Sir Francis Drake*. New York: New York Univ. Press.

DUCAT

The first ducat was introduced as a silver coin ca. 1140 by Roger II of Sicily as currency for the Duchy of Apulia in commemoration of the centralization of his kingdom in southern Italy following a period of war with the Holy Roman Empire and regional uprisings. The original ducats depicted an image of Christ on one side and that of Roger II with his son, Roger, Duke of Apulia on the other side with the inscription "*Rogerus Dux Apulia*," with the term "ducat" deriving from the Latin root *dux*, or "duke."

In 1284, the first gold ducats were minted by Doge Giovanni Dandolo in Venice with a weight of around 3.5 grams of gold at a purity of 0.986 during a period when gold coins were uncommon in the West. Although the coin would later be minted in denominations, the specifications of the full ducat remained the same for 700 years, a remarkable feat in a world in which coinage was commonly debased. The first Venetian gold ducats depicted Dandolo kneeling in supplication to St. Mark, the patron saint of Venice, on the obverse and a bust of Christ on the reverse with the inscription, "SIT. T. XTE. D.Q.T.V. REG. ISTE. DUCAT.," a Latin abbreviation for *sit tibi, Christe, datus quem tu regis, iste ducatus*, or "Unto thee, O Christ, is dedicated this duchy, which thou rulest." Again the term ducat was assigned based on the abbreviation of the Latin term for duchy. The coin was also referred to as a *zecchino*, a derivative of *zecca*, the Italian term for "mint." Originally intended by the doge to serve as a standardized currency that could compete on the global market with that of other powerful mercantile states such as Florence and Genoa, the ducat was minted by every consecutive doge until the fall of the Venetian state to the French in 1797. Venetians continued to mint ducats, however, until the state was ceded to Austria in 1819, and the coin was redesigned in conformity with those of the Austrian Empire.

Ducat depicting the duke with Saint Mark the Evangelist, the patron saint of Venice. (Corbis)

Indeed, the ducat could be said to have replaced the once dominant **bezant** as the primary currency for trade in Europe and the Mediterranean during the Middle Ages and the Renaissance. From the 16th century onward, the ducat was imitated by several other national mints, such as the Holy Roman Empire and the Netherlands, on account of the reputation and convenience of the coinage as a standardized form of payment. The ascendancy of the ducat and its acceptance as a uniform standard of currency helped to solidify the power and influence of the Venetian Republic and fueled the expansion of trade and commerce that gave birth to modern European states. Commemorative ducats are still minted today in Austria and the Netherlands with the same specifications as the originals.

See also: Bezant; Florin.

Further Reading

Doty, Richard G. 1982. *Macmillan Encyclopedic Dictionary of Numismatics*. New York: Macmillan.

Hobson, Burton, and Robert Obojski. 1970. *Illustrated Encyclopedia of World Coins*. Garden City, NY: Doubleday & Co.

Grierson, P. 1991. *Coins of Medieval Europe*. London: Seaby.

Lane, Frederic, and Reinhold Mueller. 1985. *Money and Banking in Medieval and Renaissance Venice, vol. 1: Coins and Money of Account*. Baltimore, MD: Johns Hopkins Univ. Press.

MacDonald, George. 1916. *The Evolution of Coinage*. Cambridge: Cambridge Univ. Press.

Room, Adrian. 1987. *Dictionary of Coin Names*. London & NY: Routledge & Kegan Paul.

EL DORADO

El Dorado, also known as "The Golden Man," or "The Gilded Man," is a Spanish reference to a native ritual in the region of modern-day Colombia in which a hereditary successor chief is gilded with a coating of gold dust and set out upon a lake as a spiritual offering in a ceremony officially acknowledging him as chief. Upon contact with the local native populations, Spanish conquistadors heard stories of the ritual that compelled them to explore the region in search of what they assumed would be the tribe's wealth in gold. Over time, the legend of El Dorado assumed mythical proportions as it evolved to signify a City of Gold, which many explorers of various nationalities sought on adventurous, dangerous expeditions to the Colombian highlands. Their failure to find the elusive city resulted in the legend evolving even further to refer to the possibility of a City of Gold anywhere in the New World and subsequently became symbolic of a type of quest for the unobtainable in memoirs and literature along the lines of the legend of the **Holy Grail**. Despite this, explorers' pursuit of the City of Gold known as El Dorado led to the rapid conquest and colonization of the region.

The earliest known references to the Gilded Man exist from around 1529–1530, when accounts of several explorers' journeys refer to the story of a native tribe whose leader is gilded in gold. In fact such a ritual did exist in the territory around Lake Guatavita, near present-day Bogotá, Colombia, at that time occupied by the Muisca tribe, although there is evidence that the El Dorado ritual may have been practiced by an independent tribe that was later conquered by the Muisca. Similar to other pre-Columbian cultures, the local natives cherished gold as a sacred vessel representative of the creative energy of their gods. The precious metal was thus used in a ceremony anointing the new hereditary tribal chief, the *uzaque*, in which the designated successor chief emerged from a period of seclusion to attend the lakefront ritual. With assistance from four other chiefs inferior to his rank who were adorned sumptuously in gold, the new chief was undressed, covered in a type of adhesive resin, and gilded with gold dust. Serenaded by song and musical accompaniment from flutes and trumpets on the shore, the leaders sailed out onto the lake on a type of raft elaborately decorated with gold and precious gems, including vessels burning holy incense. When the raft arrived at the center of the lake, the spectators were silenced and the designated chief and his attendants tossed his gold and precious gems into the lake as a ritual offering to the gods. The *uzaque* then cleansed his body of the gold dust in the lake and returned to the shore to be officially acknowledged as the new chief. Although Spaniards encountered the story of this ritual upon contact with the local natives in the early

16th century, it is widely believed among archeologists and historians that it had ceased in the late 16th century when the tribe came under the rule of the neighboring tribe.

The Spanish conquistadors mistakenly assumed that the abundance of gold employed in this elaborate ritual signaled an equal abundance of gold to be discovered in the region; however, the local natives obtained their gold via trade with tribes far to their north, with whom they exchanged other precious items such as emeralds, salt, and cotton linen for gold. In fact, the region around the lake contains scarce evidence of gold sources. Nevertheless, throughout the 16th and 17th centuries, explorers planned daring journeys to discover the legendary El Dorado. In 1540, Gonzalo Pizarro, half-brother of Francisco Pizarro, led a deadly expedition to the Amazon River basin in search of the legend. Additional searches were launched by German Explorer Philipp von Hutten during 1541–1545 after landing at the Venezuelan port of Coro. Spanish conquistador Gonzalo Jiménez de Quesada, founder of Bogotá, explored Colombia's interior in 1536 and 1569 in the hopes of discovering the treasure of El Dorado. In 1595, Sir Walter Raleigh claimed to have discovered the location of El Dorado at Lake Parime, up the Orinoco River, in Guyana. This position is even found on later maps that identify it as "El Dorado." Alternate locations in the region of Brazil were also included in contemporary maps. During a journey tracing the paths of these explorers who sought El Dorado, Prussian scientist and adventurer Alexander von Humboldt disproved Raleigh's claims as to the city's location.

A pre-Columbian, Muisca votive figurine (*Tunjo*) depicting the ceremony of El Dorado (copper & gold alloy). (AFP/Getty Images)

The location of the reputed golden city of El Dorado continued to elude explorers, yet the desire and adventure of the quest for the famed treasures drove them to conquer and settle the regions of Latin America along the way. By the 17th century, El Dorado had already come to signify a type of unobtainable wealth, a meaning evident in literary references such as Miguel de Cervantes' *Don Quixote* (1605–1615), John Milton's *Paradise Lost* (1667), and Voltaire's *Candide* (1759), in which it is depicted as a utopian kingdom, or ideal society. During the modern era, El

Dorado was also used in various place names, particularly in certain gold rush settlements such as El Dorado County, in California's gold country. Edgar Allen Poe featured El Dorado as the title and subject of a poem, with more ominous tones about why such riches and perfection are unobtainable:

> Gaily bedight,
> A gallant knight,
> In sunshine and in shadow,
> Had journeyed long,
> Singing a song,
> In search of Eldorado.
>
> But he grew old—
> This knight so bold—
> And o'er his heart a shadow
> Fell as he found
> No spot of ground
> That looked like Eldorado.
>
> And, as his strength
> Failed him at length,
> He met a pilgrim shadow—
> "Shadow," said he,
> "Where can it be—
> This land of Eldorado?"
>
> "Over the Mountains
> Of the Moon,
> Down the Valley of the Shadow,
> Ride, boldly ride,"
> The shade replied—
> "If you seek for Eldorado!" (Poe 1849)

Since the mid-16th century, a range of attempts have been made to drain or dredge portions of Lake Guatavita to uncover the treasures thought to rest beneath the water. A significant portion of gold and emeralds have been uncovered.

See also: Gilding; Holy Grail; Pre-Columbian Gold; Spanish Conquest.

Further Reading

Bandelier, Adolph. 1893. *The Gilded Man: El Dorado and Other Pictures of the Spanish Occupancy of South America.* New York: D. Appleton.

Bottiglia, W. F. 1958. "The Eldorado Episode in *Candide*." PMLA 73, no. 4 (September):339–347.

Bray, Warwick. 1980. *The Gold of El Dorado.* New York: Abrams.

Hagen, Victor Wolfgang von. 1968. *The Gold of El Dorado: The Quest for the Golden Man.* Farnborough, UK: Saxon House.

Hemming, John. 1978. *The Search for El Dorado.* London: Michael Joseph.

Van Heuvel, Jacob A., and Sir Walter Raleigh. 1844. *El Dorado: Being a Narrative of the Circumstances which Gave Rise to Reports, in the Sixteenth Century, of the Existence of a Rich and Splendid City in South America.* New York: New World Press.

ELECTROPLATING

Electroplating, also known as electrodeposition, is a chemical process in which a thin metallic coating, often of gold, silver, or copper, is deposited onto a surface of another type of metal by means of an electrolyte chemical compound such as salt, which converts into ions when exposed to a solvent such as **cyanide** and creates an electrically charged current. Gold is used as the metal of choice in electroplating technologies in an increasingly wide capacity to fulfill many purposes ranging from gold plating on silverware, **jewelry**, and appliances for aesthetic purposes; as an optimal conductor of electricity on circuit boards; as the metallic base for **bonding wire**, and in semiconductor technology in many common electronics; in optical electronics and sensors; and in automobile and other types of motor engines to protect against corrosion and optimize electric charges.

The technology first appeared ca. 1800 as a result of the experiments of Italian professor of natural philosophy at the University of Pavia Luigi Brugnatelli, a colleague of physicist Alessandro Volta, inventor of the chemical process that induced electronic cells, known today as "voltaic cells." After Brugnatelli successfully electroplated with gold, he utilized this method to experiment with various technical applications of electroplating in the creation of voltaic electricity, in a manner along the lines of the use of gold-plated voltaic cells in modern fuel cell technology. In an 1803 letter to a colleague in Brussels, Brugnatelli described his discovery, noting that he had "recently gilt in a complete manner two large silver medals, by bringing them into communication by means of a steel wire, with a negative pole of a voltaic pile, and keeping them one after the other immersed in ammoniuret of gold newly made and well saturated" and citing the formula for the electrolyte solution as consisting of "one part of the saturated solution of gold in nitromuriatic acid," to which he added "six parts of solution of ammonia, by which the solution is decomposed and oxide of gold is precipitated, and a portion is set free, forming ammoniuret of gold" (Hunt, 2).

News of Brugnatelli's work did not spread beyond Italy initially for largely political reasons, and in the course of the early 19th century, several other researchers throughout Europe developed similar technologies for electroplating with metals such as copper and gold. On March 25, 1840, two cousins from Birmingham, England, in the gilt and bronze casting manufacturing industry, Henry and George Elkington, registered British patent number 8447 for a type of electroplating process for use in the commercial manufacturing industry as a means of applying electricity to metals by means of a galvanic current. In 1841, the Elkingtons purchased and patented a technology invented by Birmingham surgeon John Wright, who had determined the formulas for creating potassium **cyanide** plating baths as the ion-producing electrolyte for electroplating gold and silver. The technology grew rapidly throughout Europe and the United States in the course of the 19th century in a variety of commercial applications. As the Elkingtons proceeded to register their patents in other countries, however, a series of controversies ensued over their claim to the rights for the processes as other scientists had experimented successfully with similar formulas. The dispute over who was in fact the first to invent electroplating technology continued through the 19th century.

Upon the initial spread of the technology within the industrial manufacturing industry, electroplating was predominantly used to create luxurious looking objects inexpensively, such as gold-plated religious icons, art reproductions, eye glasses, and flatware. In the course of the 20th century, commercial use of electroplating with various metals spread rapidly to include applications in the auto industry, low voltage connectors in telecommunications technology and computers, semiconductor and circuitry manufacture, and magnetic recording devices and sensors, among many other applications.

With this increase in manufacture processes utilizing electroplating techniques, the industry increasingly acknowledged, during the latter half of the 20th century, the need to maintain safe industrial standards and minimize toxic waste water emissions. On June 28, 1988, a serious industrial accident occurred at the Bastian Plating Company plant in Auburn, Indiana, in which four workers died by asphyxiation resulting from exposure to hydrogen cyanide gas when an employee accidentally mixed muriatic acid with zinc cyanide, which prompted even greater regulation of the electroplating manufacturing industry. Along with safer production, more effective methods for electroplating are continually being researched, particularly with a view to the use of electroplating in the rapidly expanding telecommunications and small electronics industries, as well as for use as an infrared reflective coating in the **aerospace industry**.

See also: Aerospace Industry; Bonding Wire; Cyanide; Fuel Cells; Jewelry.

Further Reading

Associated Press. 1988. "4 Killed in Cyanide Accident at Indiana Plant." *New York Times*, June 29.

Hunt, L. B. "The Early History of Gold Plating: A Tangled Tale of Disputed Properties." *Gold Bulletin*. http://www.goldbulletin.org/assets/file/goldbulletin/downloads/Hunt_1_6.pdf.

Schlesinger, Mordechay. "Electroplating." *Electrochemistry Encyclopedia*. http://electrochem.cwru.edu/encycl/art-e01-electroplat.htm.

Schlesinger, Mordechay, and M. Paunovic. 1998. *Fundamentals of Electrochemical Deposition*. New York: Wiley.

Schlesinger, Mordechay, and M. Paunovic, eds. 2000. *Modern Electroplating*, 4th ed. New York: Wiley.

EMERGENCY BANKING RELIEF ACT OF 1933

The Emergency Banking Relief Act of 1933 was a bill passed by the House of Representatives and the U.S. Senate on March 9, 1933, under President Franklin D. Roosevelt to avert further economic disaster in the face of an extreme banking crisis resulting from the financial instability of the Great Depression. The bill in effect legalized *ex post facto* the president's emergency closure of all U.S. banks during March 6–9 and conferred a wide range of powers on the Secretary of the Treasury to regulate the banks once they reopened and to prohibit the use of gold in financial transactions. The stated intent of the bill was "to provide relief in the existing national emergency in banking, and for other purposes."

Franklin D. Roosevelt was inaugurated as President of the United States on March 4, 1933, during the Great Depression, the worst economic crisis in American history. At the time of his inauguration, the financial mess precipitated by the stock market crash in 1929 was worsening as the national monetary system became strained to its limit and frightened consumers continued to withdraw their money from banks out of fear that they would fail, which further weakened the already fragile banking system. One day later, Roosevelt ordered a Special Session of the 73rd Congress, to be convened on March 9, to address the urgent situation. To prevent further erosion of bank deposits, however, in the meantime, by means of executive order, he declared a bank holiday from March 6 to 9, during which all financial institutions were required to be closed for business and all trading in gold halted.

On March 9, Roosevelt submitted the Emergency Banking Relief Act to Congress and urged them to pass it swiftly with a view to authorizing the government to oversee the reopening and possible reorganization of national banks and to prevent further hoarding or export of gold from the United States. After only 38 minutes of debate, the House passed the bill by acclamation, with many representatives not having seen the bill. The Senate passed the bill later that day by a vote of 73 to 7. On March 12, Roosevelt announced that banks would reopen on March 13, and, indeed, despite their questionable legal justification, the president's emergency actions succeeded initially in restoring Americans confidence in the banking system, preventing the plausible possibility of an entire collapse.

The new legislation contained five provisions that were reinforced and amended by additional amendments and proclamations in the immediate aftermath of its passage. The act itself was constructed as follows:

- Title I addresses the problem of hoarding or exporting gold and legalizes the bank holiday declared by the president by amending the 1917 Trading with the Enemy Act, which allowed for the regulation or suspension of transactions deemed by the president as a national threat during a time of war. The Emergency Banking Relief Act amended this power to include the provision "During time of war *or during any other period of national emergency declared by the President*, the President may, through any agency that he may designate, or otherwise, investigate, regulate, or prohibit, under such rules and regulations as he may prescribe, by means of licenses or otherwise, any transactions in foreign exchange, transfers of credit between or payments by banking institutions as defined by the President, and export, hoarding, melting, or earmarking of gold or silver coin or **bullion** or currency, by any person within the United States or any place subject to the jurisdiction thereof." Failure to comply with the law is deemed as punishable by a possible fine of up to $10,000 and/or imprisonment for up to 10 years.

    Title I also requires citizens and institutions to furnish all gold coins, gold bullions, and gold certificates to the U.S. Treasury in exchange for equivalent U.S. currency in coins or paper money. Failure to comply is punishable by fines of up to twice the value of the gold held by the individual or institution. Furthermore, all Federal Reserve member banks were under the custodianship of the Secretary of the Treasury for the duration of the emergency period as declared by the president.

- Title II, cited as the "Bank Conservation Act," states that the Comptroller of the Currency may appoint a conservator to any banking institution whose assets are deemed to be at risk of insolvency. The legislation entrusts the Comptroller with a range of

powers to assess, regulate, and reorganize banking institutions as needed to ensure they reopen and meet the needs of depositors in a timely, secure manner, or, if need be, liquidate insolvent banks. These duties are not to supersede in any way the powers of the presidents, the Treasury, or the Federal Reserve Bank.
- Title III permits national banks to issue preferred stocks at a rate and quantity approved by the Comptroller of the Currency, with cumulative dividends not to exceed 6 percent per year. With the approval of the Secretary of the Treasury, these stocks could be purchased by the Reconstruction Finance Corporation or used as collateral for a loan for the corporation. Such measures assisted with strengthening the capital holdings of banks whose assets were reduced to a level that threatened their capital structure.
- Title IV amends the Federal Reserve Act to allow the Federal Reserve Bank to use all direct financial obligations (such as bonds) of the United States acceptable as collateral for federal reserve bank notes issued at an equivalent value, and it authorized the purchase of additional U.S. government bonds and applied much broader lending powers to the Federal Reserve Bank in an effort to stimulate economic growth.
- Title V appropriates $2 million for the expenses related to the implementation of the act and validates the legislation by limiting the ability to amend or invalidate its measures.

In the initial few weeks following the enactment of the legislation, Roosevelt issued a series of presidential proclamations that further clarified aspects of the bill by further prohibiting U.S. citizens to export or in a manner remove gold from the United States, requiring banks to submit all gold to the Federal Reserve Bank and the U.S. Treasury, prohibiting foreign exchange and other transactions in gold, and canceling all U.S. "gold clause contracts," in which payments were agreed to be made in gold, substituting currency or securities instead.

See also: Banking and Credit; Bullion; Exchange Rates; Federal Reserve System, Gold Standard.

Further Reading

Bayoumi, Tamim, Barry Eigengree, and Mark Taylor, eds. 1996. *Modern Perspectives on the Gold Standard.* Cambridge: Cambridge Univ. Press.
Emergency Banking Relief Act of 1933. http://tucnak.fsv.cuni.cz/~calda/Documents/1930s/EmergBank_1933.html.
Holzer, Henry Mark. 1981. "How Americans Lost Their Right to Own Gold and Became Criminals in the Process." *Committee for Monetary Research and Education.* http://users.rcn.com/mgfree/Economics/goldHistory.html#note17.
Preston, Howard H. 1933. "The Banking Act of 1933." *American Economic Review* 23, no. 4 (December):585–607.

EXCHANGE RATES

Exchange rates reflect the value of one nation's domestic currency relative to that of another nation and can be determined at either fixed or floating rates. A fixed exchange rate is one in which the price of the currency is pegged at a determined, stable rate to the price of a precious metal such as gold or silver or to another currency or index of currencies. A floating exchange rate is determined by market conditions and is subject to the forces of supply and demand. Hybrid exchange rates that are pegged but allow for floating adjustments according to market conditions

> "As far as gold is concerned . . . the dollar remains a dominant currency in reserves."
> —BEN BERNANKE, INTERVIEW (JANUARY 4, 2010, AMERICAN ECONOMIC ASSOCIATION)

also exist in contemporary national markets.

From the mid-19th century to the early 20th century, the majority of developed countries adhered to a fixed exchange rate set according to either the **gold standard** or a bimetallic standard assessed on the relative values of gold and silver. As the **gold standard** was abandoned in response to the economic pressures of the Great Depression and the socioeconomic effects of war in the 1930s, a secondary form of exchange that emerged from the Bretton Woods Agreement of July 1944 became the predominant measure of international foreign currency exchange. In the **Bretton Woods system**, the U.S. dollar was tied to **gold reserves**, and the majority of other developed nations pegged their currencies to the dollar. After this system was abandoned due to the depreciation of the U.S. dollar and increasing U.S. balance-of-payments deficits, most developed countries adopted a floating exchange rate system.

Establishing the value of domestic currencies according to a floating exchange rate is considered to be the most advantageous option for developed countries with healthy, established economies because floating exchange rates typically correct according to market conditions to compensate for trends such as **inflation**, shifts in foreign investment, or levels of imports versus exports. Monetary policy is also used to strategically guide a healthy and stable exchange rate in international currency markets. Less-developed nations in emerging markets generally value their currency at a fixed exchange rate tied to another foreign currency, most commonly the U.S. dollar. This requires that the government maintain adequate reserves in the foreign currency to which the domestic currency is pegged in order to protect against the possibility of **inflation**.

Under the **Bretton Woods System**, the flow of imports versus exports in merchandise and commodities was one of the biggest indicators affecting fluctuations in foreign exchange rates, which typically remained fairly stable. The period since the dissolution of Bretton Woods has seen a significant expansion of capital that is more mobile along with an unprecedented global integration of financial markets, changes that have made financial transactions the dominant supply and demand indicator of foreign exchange rates. Such changes in the global economy have contributed to a higher level of volatility in exchange rate levels since the 1990s, along with significant uncertainty about how to predict exchange rate patterns. A wide variety of econometric models exist to measure and predict future exchange-rate patterns. In the face of so many unknowns, many contemporary economists point to the ultimate significance of macroeconomic trends over the long term when analyzing the course of foreign exchange rate values.

In the present, the U.S. dollar continues to prevail as the benchmark currency in most global capital markets. Current data indicate that 50 percent of the total currency exchanged in the world is comprised of transactions in dollars and euros.

See also: Banking and Credit; Bretton Woods System; Gold Reserves; Gold Standard; Inflation; International Monetary Fund.

Further Reading

Dornbusch, Rudiger. 1976. "Expectations and Exchange Rate Dynamics." *Journal of Political Economy* 84:1161–1176.

Friedman, Milton. 1953. "The Case for Flexible Exchange Rates." In *Essays in Positive Economics*, ed. M. Friedman, 157–203. Chicago: University of Chicago Press.

Garrett, Geoffrey, 2000. "Capital Mobility, Exchange Rates, and Fiscal Policy in the Global Economy." *Review of International Political Economy* 7, no. 1 (Spring):153–170.

Isard, Peter. 1995. *Exchange Rate Economics*. Cambridge: Cambridge Univ. Press.

Taylor, Mark. 1995. "The Economics of Exchange Rates." *Journal of Economic Literature* 33, no. 1:13–47.

FABERGÉ, HOUSE OF

The House of Fabergé was a retail **jewelry** firm established in St. Petersburg in 1842 by Russian jeweler Gustav Fabergé. The firm assumed notoriety in 1885 under Gustav's son, master goldsmith Peter Carl Fabergé (1846–1920), who received the distinction of Goldsmith by Special Appointment to the Imperial Crown following the overwhelming success of his first commission from the imperial family for an Easter egg crafted of gold and precious jewels.

In commemoration of his 20th wedding anniversary in 1885, which also coincided with the celebration of Easter in the Orthodox calendar, Tsar Alexander III commissioned Fabergé to design and create the first imperial Easter egg for his wife, Tsarina Maria Fedorovna, who was already an admirer of Fabergé's work. Known as the "Hen Egg," this first piece in what would become an annual series of imperial Easter eggs was crafted entirely in gold and then coated in white enamel to resemble a real egg. Inside the egg was a gold yoke, within which was concealed a golden hen with tiny ruby eyes. The hen's tail was hinged, and opened to reveal two final surprises: a miniature replica of the imperial crown crafted in gold and diamonds and a tiny ruby pendant in the shape of an egg, which the Tsarina would wear on a chain as a necklace. The Tsarina was so delighted with her gift

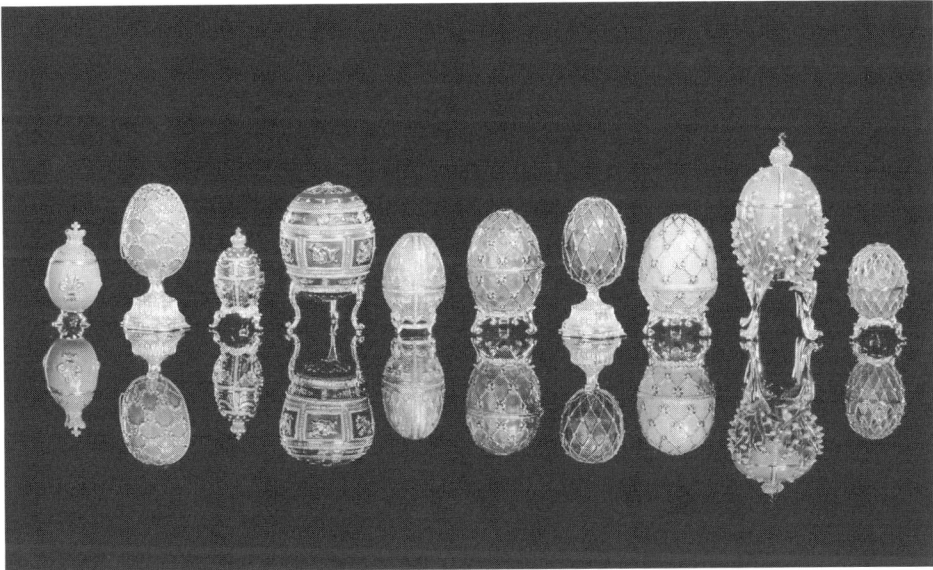

Fabergé eggs. (Alexander Makarov/Dreamstime.com)

that Alexander engaged the House of Fabergé to create an original Easter egg for the imperial dynasty every year.

After 1887, Carl Fabergé was free to design the eggs as he pleased, and the creations remained a surprise until they were completed and presented to the Romanovs. He used his knowledge of the lives and personal interests of the members of the imperial family to design increasingly extravagant works of art that incorporated symbolic and representational elements of the Romanov dynasty.

Upon assuming the crown in 1894, Alexander's son and successor Nicholas II continued with the tradition until his abdication in 1917. The production of the Fabergé eggs during this turbulent period in Russia's history can be seen at times as an ironic reflection of the looming fate of the imperial dynasty. Following an international debut at the World Exhibition in Paris, where the innovative and opulent jewels caught the attention of nobility and elites throughout Europe, production of the eggs ceased for a year during the Russo-Japanese War in 1904–1905. The "15th-Anniversary Egg" created in 1911 is a poignant family album depicting portraits, crests, and symbols of Tsar Nicholas and his family just years before their arrest and execution during the Revolution in 1917–1918. Nicholas's mother, the Dowager Empress Maria Fedorovna, was the only family member to escape, and as she fled, she carried with her the "Order of St. George Egg," which her son had given her in 1916.

Following the Russian Revolution, Fabergé fled to Switzerland, where he remained in exile until his death in 1920. The Communist government assumed control of the House of Fabergé firm and all of the imperial eggs that remained in the Romanov palaces were confiscated and housed at the Kremlin Armory. Of the 50 imperial Fabergé eggs created during 1885–1916, 42 are extant today, along with 15 other Fabergé eggs created for other elite clients such as wealthy industrialist Alexander Kelch, the Rothschilds, the Nobels, the Yusupovs, and the Duchess of Marlborough.

After Carl Fabergé's death, his sons Alexander and Eugene continued to produce works in the Fabergé style in Paris, where he opened the company Fabergé et Cie and established the trademark Fabergé. In the mid-20th century the Fabergé heirs sold the rights to the trademark to an American company that established itself as Fabergé, Inc., to produce a wide range of beauty products and accessories. In 1989, Fabergé, Inc., appointed Victor Meyer, a German **jewelry** company based in Pforzheim, Germany, to produce the official, limited-edition Fabergé designs, including the renowned Fabergé eggs. In 1991, a Victor Meyer–manufactured Fabergé egg was presented as a gift to former Soviet leader Mikhail Gorbachev on the occasion of his receipt of the Nobel Peace Prize. Subsequent limited-edition eggs have honored likewise distinguished historical moments; for example, the Millennium Egg produced in 1999.

See also: Goldsmithing; Jewelry.

Further Reading

Faber, Toby. 2008. *Fabergé's Eggs: The Extraordinary Story of the Masterpieces that Outlived an Empire*. New York: Random House.

Habsburg, Geza von. 2005. *Fabergé Then and Now*. Munich: Hirmer Verlag.

Hill, Gerald. 2007. *Fabergé and the Russian Master Goldsmiths*. New York: Universe.

FEDERAL RESERVE SYSTEM

The Federal Reserve System is the central banking organization established by the Federal Reserve Act of December 23, 1913, to guide national monetary policy and credit regulation in the aftermath of the 1907 banking panic that shook Wall Street and undermined Americans' confidence in the commercial banking system.

Following a period of fierce debate between progressives and conservatives over the nature and scope of proposed banking reforms, President Woodrow Wilson appointed a committee of finance experts led by Virginia congressman Carter Glass and economist H. Parker Willis to analyze the issue upon his election to the presidency in 1912. The committee presented the president with an act that would pass in December 1913 with minor modifications as the Federal Reserve Act, establishing a compromise solution with a system that contained elements of both a centralized and decentralized bank. After almost a year of planning for the implementation of the legislation, 12 regional branches of the Federal Reserve Bank were opened in Boston, New York, Philadelphia, Cleveland, Richmond (Virginia), Atlanta, Chicago, St. Louis, Minneapolis, Kansas City, Dallas, and San Francisco, all governed by the board of governors based in Washington, D.C., that is comprised of seven members appointed by the president of the United States and confirmed by the Senate for terms of 14 years. The appointment of the seven board members is scheduled to have one term expire every other year, so that appointments are filled in a rotating manner. The president further designates two of the board members as chairman and vice chairman of the board for terms of four years, positions that also must be confirmed by the Senate. The organization of the districts was designed to ensure that the financial interests across every industry and region are served indiscriminately.

One of the most significant tools utilized by the Federal Reserve Bank since the 1920s is the "Open Market System," an instrument of monetary policy established to abate recession in 1923 that was not directly tied to the **gold standard** but had a significant effect on the availability of credit. The Federal Open Market Committee (FOMC) was thus established to monitor the purchase and sale of government and some private securities traded through the Federal Reserve Bank of New York as a means of either stimulating or curbing the flow of credit and pace of economic growth. The FOMC board consists of the seven Federal Reserve Bank board members, the president of the Federal Reserve Bank of New York, and four presidents of regional Federal Reserve Bank branches, the latter of which serve one-year terms on a rotating schedule that ensures regional interests are represented by voting board members.

In addition to guiding monetary policy through the Open Market System, the Federal Reserve Bank also influences the supply and demand of money by setting the discount rate, the interest commercial banks and other financial institutions pay on money borrowed from the Federal Reserve Bank, and the reserve requirements of depository financial institutions as a safety net intended to offset banks' liabilities with holdings in assets.

Still in its infancy in 1929, the Federal Reserve Bank was blamed by many for not predicting the economic depression that resulted from stock market speculation. As part of his Depression-era financial reforms, President Franklin D.

Roosevelt passed the Banking Act of 1933, also referred to as the Glass-Steagall Act, which required much greater collateral against the transactions of commercial and investment banks and created the Federal Deposit Insurance Corporation (FDIC) and tighter government regulation of financial institutions in general. That same year, Roosevelt also effectively ended the **gold standard** by executive order as many economists argued that the Depression was prolonged by the ineffectiveness of the Federal Reserve System to implement monetary policy based only on **gold reserves**.

Since the 1950s, the scope and responsibilities of the Federal Reserve Bank have evolved to meet the needs of the economy, maintaining its decentralized position while working closely with the U.S. Treasury to develop and implement monetary policy and assess market conditions. Since 2003, the Federal Reserve board has faced significant challenges and threat of economic decline due to deregulation in the banking and finance industries in the context of an expanding global economy in which market conditions are more complex and variable. The current mission of the Federal Reserve Board is stated as consisting of the following duties:

- conducting the nation's monetary policy by influencing the monetary and credit conditions in the economy in pursuit of maximum employment, stable prices, and moderate long-term interest rates
- supervising and regulating banking institutions to ensure the safety and soundness of the nation's banking and financial system and to protect the credit rights of consumers
- maintaining the stability of the financial system and containing systemic risk that may arise in financial markets
- providing financial services to depository institutions, the U.S. government, and foreign official institutions, including playing a major role in operating the nation's payments system (http://www.federalreserve.gov/aboutthefed/mission.htm)

See also: Banking and Credit; Exchange Rates; Gold Reserves; Gold Standard.

Further Reading

Federal Reserve System Official Web site. http://www.federalreserve.gov/aboutthefed/mission.htm.

Gerdesmeier, Dieter, Francesco Paolo Mongelli, and Barbara Roffia. 2007. "The Eurosystem, the U.S. Federal Reserve, and the Bank of Japan: Similarities and Differences." *Journal of Money, Credit and Banking* 39, no. 7 (October):1785–1819.

McGregor, Rob Roy. 2007. "Federal Reserve Transparency: The More Things Change, the More They Stay the Same." *Public Choice* 133, no. 3–4 (September):269–273.

Meltzer, Allan H. 2003. *A History of the Federal Reserve, 1913–51*. Chicago: Univ. of Chicago Press.

Sprague, O.M.W. 1914. "The Federal Reserve Act of 1913." *The Quarterly Journal of Economics* 28, no. 2 (February):213–254.

Wells, Donald R. 2004. *The Federal Reserve System: A History*. Jefferson, NC: McFarland.

FERDINAND II OF ARAGON

Ferdinand II (1452–1516) of Aragon reigned as King of Aragon from 1479 to 1516 and as king consort by rights of marriage over Castile from 1479 to the death of his wife **Isabella I of Castile** in 1504, after which he remained regent over Castile. He also ruled as the King of Sicily during 1468–1516 and King of Naples during 1504–1516. Under the joint rule of Ferdinand and his queen Isabella, the various kingdoms of Spain were united to form the foundations of an empire that swiftly rose to supremacy during the early modern era as it was strengthened by the wealth and prestige of Ferdinand and Isabella's patronage of **Christopher Columbus**'s voyage of discovery to the New World and the wealth in gold and other natural resources that poured into the Spanish royal treasury from conquistadors' gold-seeking expeditions.

> "Get gold, humanely if possible—but at all hazards get gold."
> —FERDINAND II OF ARAGON

Ferdinand was born on March 10, 1452, in the Aragonese town of Sos, son of John II of Aragon and Navarre and John's second wife, Johanna Henriquez of Castile. In 1469, Ferdinand married **Isabella I of Castile**, sister of Henry IV of Castile and heir to the throne. The strategic marriage alliance between Ferdinand and Isabella would prove to be a significant watershed event in the history of Spain as their consolidation of the kingdoms of the realm and administrative

Royal Couple, a painting of King Ferdinand (1452–1516) and Queen Isabella (1451–1504) of Spain, supporters of Christopher Columbus's voyage to discover America. Illustration from 1754. (MPI/Getty Images)

reforms created the foundations of what would become one of the greatest empires in Europe during the 16th century.

Among Ferdinand's most notable accomplishments were the successful union of the kingdoms of Aragon and Castile and a series of effective domestic policies to improve the functioning of the royal state; the issuance of a Papal Bull by Pope Sixtus IV endorsing the establishment of the Spanish Inquisition to persecute heresy; tribunals that grew to a level at which they greatly contributed to the consolidation of state power and increased the wealth of the royal treasury; the Reconquest of the Kingdom of Granada; the expulsion or forced conversion of the Jews and the Moors in 1492, a campaign in which significant property was seized, which increased the empire's wealth; engagement in the Italian Wars with France during 1494–1504 over disputed territories in southern Italy; and, famously, support of **Christopher Columbus**'s 1492 voyage in search of westward trade routes to the Indies, which resulted in the discovery of the Americas, an event that rapidly filled Spain's coffers even further as the conquistadors of the early 16th century sent back vast quantities of gold riches from the explorations and conquests of Latin America.

The reforms and expansion initiatives undertaken by Ferdinand and Isabella increased the wealth of the Spanish royal treasury by nearly sixfold. From the onset of his rule, Ferdinand pursued his goals for the consolidation of royal power and authority shrewdly and ambitiously. These traits cast him in a ruthless position at times: most notoriously in the cruel persecutions of the Spanish Inquisition; the forced conversions and expulsions of the Reconquest in violation of the treaty with the King of Granada; and his betrayal of **Christopher Columbus**'s contract terms, which stipulated that the Genoese explorer was to receive the office of Governor of the Indies for life and 10 percent of any profits accumulated by the Spanish Crown from the New World. Because he had been stripped of his governorship in 1500 amid accusations that he was not fit to rule, Columbus's contract was deemed by the king to be null and void.

With the death of Isabella in 1504, their daughter Joanna of Castile was recognized as heiress to the crown of Castile, while Ferdinand sought to exclude her husband, Archduke Philip of Austria, from his claim to the throne by establishing himself as regent. Amid the objections of the Castilian nobles to Ferdinand's claims to the regency, Philip took his place as Philip I of Castile until his death in 1506, at which point Joanna was deemed to be mentally unfit to rule and Ferdinand once again assumed the crown of Castile. In 1505 Ferdinand solidified his ties to French King Louis XII by marrying his niece, Germaine of Foix.

Ferdinand died on January 23, 1516, at Madrigalejo in Estremadura, upon which Joanna and Philip's son ascended to the throne as Charles I of Spain and later assumed the title Holy Roman Emperor Charles V in 1519, poised to oversee the expansion of the magnificent empire that Ferdinand and Isabella had set out to build.

See also: Columbus, Christopher; Isabella I of Castile; Spanish Conquest.

Further Reading

Edwards, John. 2004. *Ferdinand and Isabella.* New York: Longman.
Prescott, William H. 2003. *History of the Reign of Ferdinand and Isabella the Catholic.* Honolulu, HI: University Press of the Pacific.
U.S. Library of Congress. "Spain, the Golden Years: Ferdinand and Isabella." Country Studies Series. http://countrystudies.us/spain/7.htm.

FIELD OF THE CLOTH OF GOLD

During the initial years of the reign of King of England Henry VIII in the early 16th century, power and dominance of Europe was held largely in the hands of King Francis I of France and Holy Roman Emperor Charles V. It was essential that Henry VIII position himself strategically in an alliance with one of these monarchs in order to maintain England's power and influence on the continent. On the advice of his royal councilor Cardinal Thomas Wolsey, in 1520 Henry arranged for a diplomatic meeting with Francis I across the Strait of Dover, the narrowest part of the English Channel, in the Calais region of northern France in the villages of Guines and Ardres. Because the castles in the respective villages were considered unworthy of housing the distinguished guests, instead the royal retinues established lavish, gold-embellished encampments in the fields in between the towns, Francis in Ardres and Henry in Guines. To observers, the respective kings' encampments and attendants appeared so resplendent with gold that they were said to appear to be fields filled with cloth of gold.

From their youth both kings had been bitter rivals, personally and politically. In setting the stage for their encounter, they each sought to outdo one another by remarkable levels of lavish displays of wealth, erecting silken tents and pavilions made of cloth woven with golden thread and decorated with precious gems, the inspiration for the historical term for the meeting as that of "The Field of the Cloth of Gold." Accompanied by his queen, Catherine of Aragon, King **Henry VIII** spent three weeks in Calais during June 7–24, 1520, attempting to impress his rival. Both kings hosted such an abundance of festivities and pageantry, ranging from jousts, tournaments, fireworks displays, balls, and masquerades, that the affair created a serious strain on the treasuries of both kingdoms. The meeting ended abruptly and bitterly when Henry challenged Francis to a wrestling match and humiliated the French king with his victory.

Although it might be argued that an implicit agreement on an alliance had been formed, evidenced by Cardinal Wolsey's favor of the French and the betrothal of Henry and Catherine's daughter Mary to the Dauphin, heir to the French throne, the meeting on the Field of the Cloth of Gold had no concrete political outcome. In fact, despite these ties, Catherine herself favored Emperor Charles V, her nephew. Two weeks later Charles V met with Henry in England, and the two signed a treaty creating an alliance that forbade Henry from signing any new treaties from that point forward. Mary's betrothal to the Dauphin was nullified, and the princess was subsequently promised in marriage to Charles V himself, an arrangement that was also broken a short time later. Such a sequence of political maneuvering among the three monarchs would characterize European diplomatic relations throughout their reigns as each sought the most advantageous and timely alliance from which to establish their own dominance.

See also: Henry VIII.

Further Reading

Richardson, Glenn.2002. *Renaissance Monarchy: The Reigns of Henry VIII, Francis I, and Charles V.* London: Arnold.

Russell, Joycelyne Gledhill. 1969. *The Field of Cloth of Gold: Men and Manners in 1520.* London: Routledge & Kegan Paul.

FILIGREE

Gold filigree is a **goldsmithing** technique employed since earliest antiquity for creating intricate **jewelry** and ornamentation. The process of filigree uses fine threads or wire of gold that are twisted, curled, or bent into intricate designs, often against a background of precious metal, then soldered together using a filler metal and a blowtorch to join the two metallic surfaces into one at a low melting point.

The earliest extant examples of gold filigree were found during excavations at the **Royal Tombs of Ur**, the capital of ancient Sumeria during the mid-third century BC. Many other ancient cultures of early antiquity demonstrated advanced filigree techniques, including Egypt, Greece, India, pre-Columbian South America, and, perhaps most famously, the Etruscans, whose early **granulation** techniques of the seventh century BC evolved by the fourth century BC into a

Pair of gold filigree earrings from Greater Iran, 11th–12th century. (The Metropolitan Museum of Art/Art Resource, NY)

renowned tradition of filigree gold **jewelry**. These techniques continued among ancient Romans and developed further in the Byzantine Empire. During the Middle Ages, gold filigree was popular for other uses as well, such as decorative elements of precious manuscript covers, armor, reliquaries, and **royal crowns** and scepters. The British Isles Saxons, Britons, and especially the Celts were renowned for their filigree work during the early medieval period. Filigree designs became even more intricate at the hands of the master goldsmiths of the Renaissance.

The art of filigree in jewelry making remains popular today, particularly in the Mediterranean region, Mexico, and India. Many artifacts illustrating the history of this fine art are concentrated in some of the most famous collections around the world, primarily at the Metropolitan Museum of Art in New York, the Louvre in Paris, the British Museum in London, and the Vatican Museum in Rome.

See also: Crowns, Royal; Granulation; Jewelry; Ur, Royal Tombs of.

Further Reading

Evans, Joan. 1982. *A History of Jewelry, 1100–1870.* London: British Museum Publications.
Jones, Julie. 2002. "Gold of the Indies." *The Metropolitan Museum of Art Bulletin* 59, no. 4 (Spring):3–4.
Lee, Soyoung. 2003, October. "Golden Treasures: The Royal Tombs of Silla." In *Heilbrunn Timeline of Art History*. New York: Metropolitan Museum of Art. http://www.metmuseum.org/toah/hd/sila/hd_sila.htm.
Maryon, Herbert. 1954. *Metalwork and Enamelling: A Practical Treatise on Gold and Silversmiths' Work and Their Allied Craft.* London: Chapman & Hall.
Philips, Clare. 1996. *Jewelry: From Antiquity to the Present.* London: Thames and Hudson.
Tait, H. 1986. *Seven Thousand Years of Jewelry.* London: British Museum Publications.
Walters Art Gallery. 1979. *Jewelry Ancient to Modern.* Baltimore: Studio.

FLORIN

The florin was a gold coin introduced by the Republic of Florence in 1252. Known originally as the *fiorino d'oro*, the coin depicted the fleur-de-lis of the city's coat of arms on one side with the inscription "FIORENTIA" and on the reverse side an image of the city's patron saint, John the Baptist. The coin was minted continuously until 1523 at the same standard weight and grade of 3.5 grams of 24-**karat** gold. The florin was originally minted at this weight as the equivalent in value to 240 Florentine denari, or 1 lira, in order to provide a currency standard that could be used as a money of account in international markets.

The coin was created by the ruling class in Florence to consolidate the merchant republic's supremacy as one of Europe's most significant exporters of a variety of products, and the nascent banking empire fueling this expansion of trade. The state coffers were full of gold due to the fact that gold was the most common form of payment required for large-scale trade of goods. Florence thus had a reserve supply of gold substantial enough to mint the coins consistently at a standardized weight and measure, along with the financial infrastructure to facilitate their circulation. The coins were struck largely using gold originating from **mining** operations in Africa.

By the 14th century, a number of other European countries had begun to mint their own versions of the florin, although these tended to vary in measure and weight when balanced against the **gold standard** maintained by the original Florentine coin. By the early 16th century, however, debasement of the lira led to an exchange rate of 7 lira per 1 florin, a fact that may have led to the demise of the Italian version.

See also: Banking and Credit; Bezant; Ducat; Gold Standard; Karat; Mining.

Further Reading

Doty, Richard G. 1982. *Macmillan Encyclopedic Dictionary of Numismatics*. New York: Macmillan.

Goldthwaite, Richard. 2009. *Economy of Renaissance Florence*. Baltimore, MD: Johns Hopkins Univ. Press.

Hobson, Burton, and Robert Obojski. 1970. *Illustrated Encyclopedia of World Coins*. Garden City, NY: Doubleday & Co.

Room, Adrian. 1987. *Dictionary of Coin Names*. London & New York: Routledge & Kegan Paul.

FORT KNOX. *See* United States Bullion Depository.

FUEL CELLS

Fuel cells are diverse types of devices that produce electricity by converting chemical energy into electrical energy in a noncombustive environment in which a fuel source, typically hydrogen, is exposed to an oxidant, typically atmospheric oxygen, to produce a chemical reaction that is triggered by an electrolyte. Fuel cell systems consist of three distinct segments that trigger the reaction: the reactant (fuel) flows into the cell through the anode level, which contains two identical electrodes that furnish electrons to the electrolyte level where ions are conducted between the electrodes. When fuels such as hydrogen are channeled through the anode, a process of **catalysis** occurs by means of the oxidant in order to oxidize the fuel, which produces hydrogen ions and electrons. The ions are then absorbed in the third level, the cathode, in which a process of catalyst emits the ions as a byproduct in the form of water or carbon dioxide. Since the 1990s, innovations in **nanotechnology** research have utilized nanoparticularite gold as a highly efficient and cost-effective catalyst agent in the chemical reactions that generate the power emitted by the fuel cells.

Fuel cell technology originated in the mid-19th century with the discovery by British physicist William Grove in 1839 that hydrogen fuel could by catalyzed with an oxidant on platinum electrodes to produce energy. Yet a variety of roadblocks to the development of such an energy source due to the complex requirements of the configuration of the fuel cells prevented such research from reaching its fruition until advancements in physics and chemistry in the late 20th century prompted renewed interest in developing commercially viable fuel cell systems as a source of alternative energy that has the potential to be more efficient, more compact; utilize a variety of fuel sources with a much higher output of energy per fuel input; and reduce toxic emissions and noise.

Since the 1980s, fuel cells have been utilized in the aerospace industry as an optimal power source in limited environments such as spacecraft, for satellites and other remote telecommunications operations, and in certain military applications. During the 1990s, however, acknowledgment of the potential efficiency of fuel cells as an alternative fuel source with lower harmful environmental emissions spurred increased research into the development of various fuel cell technologies for broader commercial markets such as automobile industries and public transportation and domestic electricity usage. Yet such advances were limited due to the prohibitive cost and stringent temperature requirements of platinum as the catalytic substance used to trigger the chemical reaction between the elements of the fuel cells. A much more cost-effective alternative to platinum is gold in nanoparticle form, which can facilitate the chemical reaction at a wider range of temperatures than platinum, thereby ensuring the longevity of the fuel cell systems by reducing corrosion and wear.

In coordination with general advancements in the use of fuel cell technology, gold as a catalyst has received increasing attention in industrial research and development due to its potential application in more commercial environments and broader markets. In 2003, President George W. Bush instituted a five-year research program known as the Hydrogen Fuel Initiative to fund research on the use of hydrogen fuel cells as a more efficient alternative energy source. In 2007, scientists at the U.S. Department of Energy's Brookhaven Lab made significant advancements during a research project funded by this initiative in which the use of gold nanoparticle clusters to stabilize platinum electrocatalysts in fuel cell systems powering electric cars, greatly minimizing the oxidation of the platinum in the process of **catalysis**. In the course of the research project, the team also gained more detailed knowledge of the precise nature and function of the gold particles in the catalytic reaction of the fuel cells. According to Radoslav Adzic, coauthor of the team's report, "Fuel cells are expected to become a major source of clean energy, with particularly important applications in transportation. Despite many advances, however, existing fuel-cell technology still has drawbacks, including loss of platinum cathode electrocatalysts, which can be as much as 45 percent over five days, as shown in our accelerated stability test under potential cycling conditions. Using a new technique that we developed to deposit gold atoms on platinum, our team was able to show promise in helping to resolve this problem. The next step is to duplicate results in real fuel cells" (Brookhaven 2007).

See also: Catalysis; Nanotechnology.

Further Reading

Brookhaven National Laboratories. 2007, January 11. "Brookhaven Lab Scientists Discover Gold Clusters Stabilize Platinum Electrocatalysts for Use in Fuel Cells." http://www.bnl.gov/cfn/news/news.asp?a=07-04&t=pr.

Cameron, D., R. Holliday, and D. Thompson. 2003. "Gold's Future Role in Fuel Cell Systems." *Journal of Power Sources*:118, 298.

Corti, Christopher W., Richard J. Holliday, and David T. Thompson. "Potential of Catalysis by Gold for Fuel Cell and Pollution Control Applications." *World Gold Council*.

http://www.nacatsoc.org/18nam/Posters/P020-Potential%20of%20Catalysis%20by%20Gold%20for%20Fuel%20Cell.pdf

DOE/Brookhaven National Laboratory. 2007, December 14. Unexpected Activity Of Fuel Cell Catalysts Revealed. *ScienceDaily.* http://www.sciencedaily.com/releases/2007/12/071213140335.htm.

Keel, Trevor, Richard Holliday, and Tim Harper. 2010, January. "Gold for Good: Gold and Nanotechnology in the Age of Innovation." *World Gold Council Bulletin.*

Schmidbaur, Hubert, ed. 1999. *Gold: Progress in Chemistry, Biochemistry, and Technology.* Chichester, UK: Wiley and Sons.

Vielstich, Wolf et al., eds. 2009. *Handbook of Fuel Cells: Advances in Electrocatalysis, materials, diagnostics, and durability.* 6 vols. Hoboken, NJ: Wiley.

GILDED AGE

The Gilded Age is a period of United States history from the 1870s to 1900 that saw tremendous economic development, social change, and political activism in the wake of the Reconstruction era, as well as significant expansion of the nation's population and infrastructure and the consolidation of extreme wealth among a few exceptional industrialists. The term "Gilded Age" was first coined in an 1873 novel of the same name by Mark Twain and Charles Dudley Warner that parodies the political corruption and greedy, unscrupulous industrialists known as the "robber barons," who were perceived to have consolidated their wealth at the expense of the common worker and to have engaged in unethical business tactics.

The combination of population growth, westward expansion, the establishment of national transportation and communication systems, rapid industrialization and a subsequent expansion of production, and urbanization contributed to

Caricature of J. P. Morgan as a bull blowing bubbles labeled "inflated values" for which people are reaching, from a 1901 issue of *Puck*. Morgan helped to create some of the largest trusts and monopolies during his lifetime, which allowed companies great control over prices and competition. Political cartoon by Joseph Keppler for *Puck*. (Library of Congress)

the emergence of a modern industrial economy and social and economic changes stemming from the consolidation of wealth among a select group of elites and the growth of a working-class identity among laborers.

Such changes brought a shift in business structure toward large corporations and a new managerial class in industrial society. The extent of these new modes of business operation and labor identity also contributed to the intense political partisanship between the Republicans and the Democrats and a wider political participation during the era. The concerns of the working class led to the creation of strong labor union organization, which was pitted against the wealthy industrialists in the courts and at the polls. Farm workers also united to form the Populist Party, a significant third-party political interest in the hotly contested elections of the period. The political environment was also rife with corruption, and a series of political scandals and the antics of local political machines such as Tammany Hall tainted the democratic process at various levels.

> And so we are told this is the golden age. And gold is the reason for the wars we wage.
> —U2, "New Year's Day"

Although legendary industrialists such as Andrew Carnegie, John D. Rockefeller, and Cornelius Vanderbilt were parodied in the media for their perceived greed and shrewd conduct in financial affairs, they also established the foundations of major philanthropy in the United States, creating legacies that have driven powerful philanthropic organizations to the present day.

In a 2008 article in the *Nation*, Doug Henwood suggested that the early 2000s experienced a new kind of Gilded Age, in which the consolidation of outrageous wealth among select few entrepreneurs and chief executive officers in banking and finance sectors led to decisions that ultimately proved detrimental to American society and contributed to subsequent financial decline and recession at the end of the decade. In particular, he compares a costume ball hosted in February 1897 by high-society lawyer Bradley Martin and his wife Cornelia at the Waldorf in Manhattan to a birthday party given in February 2007 by Blackstone CEO Steve Schwarzman, who spent a million dollars to have singer Rod Stewart perform. Such comparisons of periodization are common in contemporary American historiography, which continues to analyze the circumstances in which Mark Twain implied that all that glitters is not gold in characterizing the decades of the late 19th century as glittering with a gilded effect on the surface to conceal the decay underneath.

See also: Banking and Credit.

Further Reading

Buenker, John D., and Joseph Buenker, eds. 2005. *Encyclopedia of the Gilded Age and Progressive Era*. Armonk, NY: M. E. Sharpe.

Edwards, Rebecca. 2009. "Politics, Social Movements, and the Periodization of U.S. History." *Journal of the Gilded Age and Progressive Era* 8, no. 4:461–473.

Henwood, Doug. 2008. "Crisis of a Gilded Age." *Nation*, October 13.
Henwood, Doug. 2008. "Our Gilded Age." *Nation*, June 30.

GILDING

Gilding is a decorative arts technique in which gold leaf or powder is applied to a prepared surface such as wood, metal, ivory, leather, fabric, glass, and architectural features of buildings by hand, by mechanical or chemical means, or, in modern times, through the use of **electroplating**.

Ancient evidence for the practice of spreading or hammering gold leaf on a surface for ornamentation exists from a variety of cultures. In Egypt as early as 2300 BC, there are examples of gilding on mummies' caskets and personal effects buried in royal tombs. There are also examples of gilded precious objects from Asia as early as the eighth century BC and among the pre-Columbian people of South America. The first known written references to gilding appear in the Old Testament and in Homer's *Odyssey*. Gilding appears to have been adopted in ancient Rome following the destruction of Carthage in 146 BC, after which the technique was used to embellish public temples and private palaces in a manner modeled after the techniques common among the Greeks.

During the Middle Ages, advancements in technology led to the spread of gilding applications to a wider range of mediums, such as fabric, wooden frames, and **illuminated manuscripts**. The modern period saw the development of processes such as chemical and mechanical gilding, which alloyed for a smoother application of the gold. Italian chemist Luigi Brugnatelli's invention of the technique of **electroplating** in 1805 soon came to replace the more ancient methods.

Two of the most famous examples of gilding in decorative and fine arts include Lorenzo Ghiberti's (1425–1452) *Gates of Paradise*, a pair of gilded bronze doors originally designed for the Baptistry of Florence's Duomo in the mid-15th century, and Gustav Klimt's painting *The Kiss* (1907), created using oil on canvas with gold leaf applied for effect.

> ### Klimt's Golden Phase
>
> During what is referred to as his "Golden Period" from ca. 1901–1908 Austrian symbolist painter Gustav Klimt (1862–1918) used applications of gold leaf so prominently that gold appears as the dominant color on famous Klimt paintings such as the 1907 *Portrait of Adele Bloch-Bauer I* and *The Kiss* (1907–1908). Klimt's eclectic style was influenced by artistic techniques from ancient Greece and Egypt and from Byzantine-era mosaics, which he traveled to Venice and Ravenna to view. In 2006, Klimt's *Portrait of Adele Bloch-Bauer I* earned the highest recorded price for a work of art auctioned publicly when it was purchased by Ronald Lauder for the Neue Galerie in New York City.

See also: Electroplating; Illuminated Manuscripts; Pre-Columbian Gold; Versailles, Palace of.

Further Reading

Beuster, Kirsten. 2007. *Gold: Gilding History and Techniques*. Atglen, PA: Schiffer Publishing.

Bigelow, Debora, ed. 1980. *Gilded Wood, Conservation, and History*. Madison, CT: Sound View Press.
Oddy, A. 1981. "Gilding Through the Ages." *Gold Bulletin* 14:75–79.

GOLD ALBUM CERTIFICATION

Gold Album Certification, also commonly known as a "gold record," is a popular music recording industry accolade awarded by the Record Industry Association of America (RIAA) for high-level sales of original record albums and singles, and in the contemporary era compact discs, digital downloads, and ringtone downloads, based on the net totals of wholesale distribution of the music to retail and other resale markets.

The phenomenon of the "gold album" began informally in 1942 as a gimmick when RCA Records gilded a master LP of Glenn Miller's single "Chattanooga Choo Choo" with gold lacquer and presented it to him during a live radio appearance to publicize the fact that his single had sold 1.2 million copies. Unofficial gold albums were similarly awarded by the respective record labels to Elvis Presley in 1956 for his single "Don't Be Cruel" and to Harry Belafonte for sales of his album *Calypso*. In 1958, the RIAA capitalized on the success of such marketing efforts to establish official criteria for gold album certification and trademarked the term gold record as their proprietary phrase. Record companies were required to submit a request that a particular album or single be certified and pay for an external audit of their domestic sales distribution numbers. Although similar awards exist around the world, for example, that given by the International Federation of the Phonographic Industry (IFPI), an organization with members from 70 countries that does not include the United States, the award is restricted to U.S. companies and export sales do not apply. To obtain recognition, a single must sell 500,000 units and a full-length album must sell the same number with a minimum manufacturer's market dollar value of one million dollars.

The first official gold record award was given on March 14, 1958, to Perry Como for the RCA Records–produced hit single "Catch a Falling Star." Later that year, the recorded soundtrack of the musical *Oklahoma!* was the first full-length record album to receive gold album certification. As a result of the rapid expansion of the music recording industry and increased record sales during the 1970s, the RIAA added additional designations of Platinum and Multi-Platinum certifications for singles and albums selling over one and two million units, respectively. Further growth in the mid-1980s tied to the phenomenon of MTV and music videos led to the addition of more awards in other related formats. In 2004, the Digital Sales Award was added in response to a consumer shift in purchases of digital downloads online, which occur at the expense of compact disc purchases. In 2006, the RIAA recognized the significance of another digital format with the addition of the Master Ringtone Sales Award. Among the top artists who have achieved record high numbers of total gold albums awarded since the establishment of the certification are Elvis Presley, The Beatles, the Rolling Stones, and Barbara Streisand.

Whereas the achievement of multiple gold and platinum albums or singles is acknowledged worldwide in the popular music industry as an indication of extraordinary artistic success, music critics often debate the degree to which the award is the reflection of industry or publicity factors pertaining to the popular market versus a reflection of true musical talent.

See also: Gilding.

Further Reading

Chung, Kee H., and Raymond Cox. 1994. "A Stochastic Model of Superstardom: An Application of the Yule Distribution." *Review of Economics and Statistics* 76, no. 4 (November):771–775.
Denisoff, R. Serge, and William Schurk. 1986. *Tarnished Gold: The Record Industry Revisited*. New Brunswick, NJ: Transaction Books.
Recording Industry Association of America. "History of the Awards." http://www.riaa.com/goldandplatinum.php?content_selector=historyx.
Stambler, Irwin. 1989. *The Encyclopedia of Pop Rock and Soul*. New York: St. Martin's Press.

Elvis's Gold Cadillac

During the late 1960s, Elvis enlisted the help of famous car customizer George Barris to design his "Dream Cadillac," a 1965 Eldorado convertible decorated entirely with gold, yet Elvis died tragically in 1977 before the car's completion. In 1986, the Elvis Presley Cosmetics Company commissioned Barris to complete the car according to Elvis's designs in order to tour with the car as a promotion for the company. The car was unveiled in 1987 customized with an exterior painted in gold pearlescent paint and accented with stripes of gold leaf; guitar-shaped sun visors; gold-plated steering wheel, television, and telephone; three gold RCA Elvis 45's mounted on the cloth convertible cover; and a 14-karat gold hood ornament. The completed car was insured for an estimated value of $250,000. After touring with Elvis Presley Cosmetics in 1987, the car was given away as a grand prize in a sweepstakes. It was then sold several times among collectors and was last sold in 2008 by a St. Louis, Missouri, specialty auto dealer to a private collector for $119,000.

GOLD NUGGETS

Gold nuggets are fragments of gold that occur naturally within the bedrock or alluvium and are most prevalently discovered in alluvial gold sources by means of placer **mining**, although gold nuggets are less commonly found in vein sources in plains and ravines. Once extracted, the purity level of gold retrieved from nuggets is typically between 20 and 22 **karats**, with a significant variance according to regional geological differences. The percentage of fineness of gold nuggets is affected by the degree of the presence of other metallic elements such as copper or silver.

The scarcity of gold nuggets in lodes of ore versus alluvial deposits can be attributed partly to the accumulation of mineral deposits in a lode source over thousands of years versus the significantly lesser accumulation of a river's gravel deposit in comparison. Because gold nuggets found in alluvial sources were altered by the transport process of their passage, the two types of nuggets diverge significantly. Gold nuggets discovered in veins of ore have a greater concentrated accretion of crystalline substances compared to gold nuggets found in alluvial sources and also have a lesser fineness due to a higher incidence of alloys such as silver. Nuggets mined from bedrock have protrusions and a rough shape from the build-up of such minerals, whereas gold nuggets from the river silt look like smooth stones or pebbles, polished on the outside from their passage in the water.

The current record for the largest gold nugget ever discovered is held by the gold nugget known as the "Welcome Stranger," uncovered by English prospectors John Deason and Richard Oates in Moliagul, Victoria, Australia, on February 5, 1869, just below the surface at the entrance to a gully. The famous nugget measured 24.0157 x 12.2047 inches and weighed 172.8 pounds, yielding a net 2.284 troy ounces of gold.

See also: Karat; Mining.

Further Reading

Egleston, T. 1881. "The Formation of Gold Nuggets and Placer Deposits." *American Institute of Mining Engineers*.

Maslowski, Andy. 1989. "Legendary Gold Nuggets." *Rock & Gem* 19, no.7 (July):40–41, 44–45.

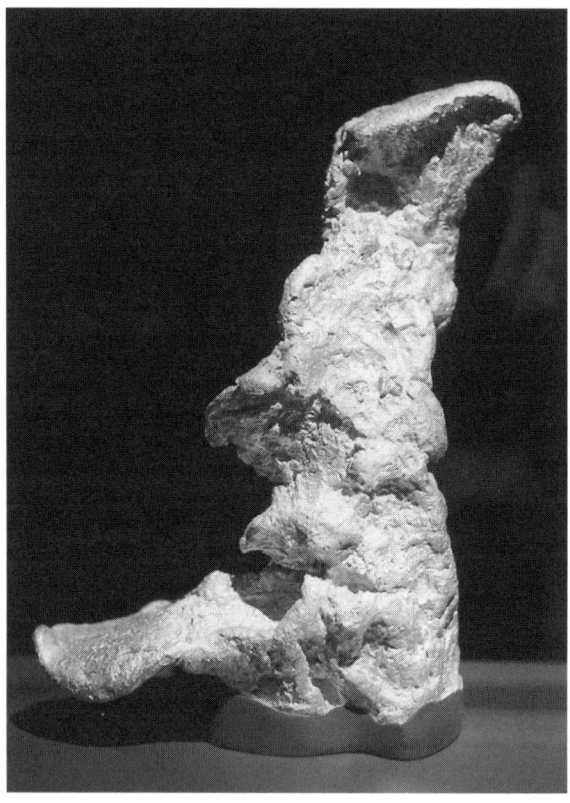

A large gold nugget is on display at the *Gold* exhibit at the American Museum of Natural History in New York, 2006. (AP/Wide World Photos)

GOLD RESERVES

Gold reserves are the balance of gold held in the form of **bullion** or coins by central banks and national treasuries as a guarantee of redemption for depositors and bank notes. Whereas issuance of currency was tied in the past to an obligation to exchange bank notes for their face value in gold, after the decline of the **gold standard** in the **banking and credit** markets, gold reserves functioned as a measure of credibility and reassurance for national currencies and bond issues.

Throughout history, rulers and political leaders have recognized the importance of acquiring and maintaining a reserve of gold and treasure as a measure of security and authority and an essential safeguard in the event of the need to wage a costly war or overcome famine or economic crisis. With the tenfold increase in the production of gold throughout the world following the discovery of gold in California, Australia, the Klondike, and South Africa during the mid-19th century, a period when annual output of gold increased from 35 metric tons in 1847 to 265 metric tons in 1852, global banking institutions increasingly adopted the international **gold standard** from the 1870s to the 1900s. These trends occurred amid larger changes in **banking and credit** practices involving a greater movement toward gold reserves as guarantees in international currency systems following the lead of the Bank of England in the early 19th century.

Thus in the course of the 19th century, banks emerged as the primary holders of gold reserves, and as the **gold standard** was gradually adopted, commercial banks were required to maintain a level of gold reserves consistent with their obligation to repay gold on demand to their depositors, although the reserve of gold **bullion** and coin was not intended to match deposits equally, but rather represented a percentage of total deposits. During the 1870s alone, stocks held in gold by commercial banks around the world increased by 70 percent. As commercial banking became increasingly consolidated and centralized and currency issue and monetary policy was supported by nationalized or semi-nationalized central banking institutions such as the Federal Reserve Bank in the United States, commercial banks were no longer obligated to maintain individual gold reserves to guarantee the holdings of their depositors, and gold reserves were largely held by national or central banks that in turn issued notes of guarantee backed by gold reserves to the individual commercial banks. The international **gold standard** and levels of national gold reserves also functioned as a method of guaranteeing balance of payments and financial obligations among nations.

This trend accelerated rapidly in the interwar period during the early 20th century amid general policies of economic nationalization and the concentration of international gold holdings among a few great powers. During World War I and World War II, the policy of concentrating gold reserves among central banking institutions increased dramatically in part due to the urgency to remove gold specie and **bullion** from circulation as a precautionary measure to protect national treasuries under duress from the financial pressures of war on such a grand scale. During this period, the export of gold was severely restricted among Western nations, and the **gold standard** was, in effect, suspended out of necessity, a measure perceived as temporary at the time. By the end of World War II, the United States gold reserves had tripled from a level of 6,000 tons in 1925 to a reported 18,000 tons, a total that comprised 60 percent of the total reported gold reserves in the world at that time. This concentration of gold reserves among government-regulated or state-run financial institutions was also the result of a decline in the **gold standard** during the 1920s–1930s, a period when economists such as **John Maynard Keynes** began to perceive the **gold standard** as an antiquated measure of currency value that was potentially restrictive to effective monetary policy. In the interwar period, Keynes and other economic advisors advocated for various ways to maintain the stability offered by gold reserves through a system of foreign exchange managed by international institutions that would guarantee transactions among foreign currencies while requiring less direct circulation of gold as currency or notes backed by gold.

Keynes warnings against the potential pitfalls of any monetary policy that might attempt to legislate a return to the **gold standard** at parity with prewar values initially went unheeded, as the British Parliament passed the **Gold Standard Act of 1925**, which restored the British currency to the **gold standard** with disastrous economic consequences that caused Great Britain to go off the **gold**

standard permanently in 1931. The global effects of the Great Depression made the **gold standard** unsustainable on a wider level. Keynes also advised U.S. President Franklin D. Roosevelt, who sought to stabilize U.S. monetary conditions with such measures as the Banking Act of 1933, which created the Federal Deposit Insurance Corporation and an executive order that effectively ended the **gold standard** in the United States. Roosevelt's policies devalued the dollar relative to gold by increasing the price of gold by over 50 percent, which was one of the factors that contributed to the dramatic increase in gold holdings by the United States during this period as gold poured in from other national gold holding institutions seeking to take advantage of this higher price. By the 1950s, U.S. depositories held 75 percent of the world's total reported monetary gold stores. Roosevelt also decreed that private holdings of gold be turned in to the U.S. Treasury for storage at national facilities in exchange for face value in currency or bonds.

Today gold reserves are maintained in national mints and central banks throughout the world, and transactions in gold as monetary specie are limited to settlement of international exchange on rare occasions. In the United States, gold reserves are stored at the **United States Bullion Depository** in Fort Knox, Kentucky; the Philadelphia Mint; the Denver Mint; and the West Point Bullion Depository, with additional reserves at the Federal Reserve Bank in New York. Although certain restrictions from Roosevelt's legislation forbidding the use or export of gold coin and **bullion** as a means of currency have been lifted, U.S. financial institutions are still prohibited from fulfilling any transactions in gold coin.

Gold to Go

In 2009, German luxury manufacturer TG Gold-Super-Markt, a division of Ex Oriente Lux AG, introduced Gold to Go, a gold-dispensing ATM machine, at the Frankfurt Airport. Developed by German inventor Thomas Geissler, the large vending machine, coated in a thin layer of 24-karat gold, dispenses a choice of 320 solid gold items, including gold bars with a weight of 1, 5, or 10 grams, and gold coins that can be customized with logos or selected engravings. The machine calculates the market price of gold every 10 minutes. In May 2010 another Gold to Go machine was installed in the lobby of the Emirates Palace Hotel in Abu Dhabi.

Source: Behler, Liz. 2010. "Abu Dhabi Gets Gold Vending Machine." *AOL Travel*, May 13. http://news.travel.aol.com/2010/05/13/emirates-palace-gold-dispensing-atm/.

Despite the fact that gold holdings no longer have any technical correlation to national currency levels, gold remains an essential component of international monetary policy as a the primary reserve asset supporting the currency issued by national treasuries. During the mid-20th century, the U.S. dollar emerged as the international benchmark for currency exchange as international currencies tied to the dollar were theoretically convertible to gold at their official dollar value. Yet even this system was difficult for the United States to maintain and was abandoned under President Nixon in 1971 due to a drop in international confidence resulting from prolonged U.S. balance-of-payments deficits and converted into a floating exchange rate system in which currency values fluctuate in alignment with foreign exchange markets tied by virtue of market conditions to certain core currencies, which since the 1970s has largely been the U.S. dollar.

During the 1980s and 1990s, due to cooperation among central banks in Europe and non-Western countries through financial institutions such as the European Community and the **International Monetary Fund**, official gold reserves continue to be maintained at varying levels, with the United States as the largest holder of gold reserves with reported holdings in 2010 at 8,965.65 tons with a value of $358.63 billion; Germany, the second largest with 2006 levels at 3,754.29 tons valued at $150.17 billion; and the **International Monetary Fund**, the third largest gold holdings of 3,311.84 tons valued at $132.4 billion.

See also: Banking and Credit; Bullion; California Gold Rush; Federal Reserve System; Gold Standard; Gold Standard Act of 1925; Hargraves, Edward Hammond; International Monetary Fund; Keynes, John Maynard; Klondike Gold Rush; United States Bullion Depository; Witwatersrand Gold Rush.

Further Reading

Bordo, Michael, and Barry Eichengreen. 1998. "The Rise and Fall of a Barbarous Relic: The Role of Gold in the International Monetary System." Cambridge, MA: National Bureau of Economic Research, Working Paper 6436.

Green, Timothy. 1993. *The World of Gold*. London: Rosendale Press.

Green, Timothy. 1999. "Central Bank Gold Reserves: An Historical Perspective since 1845." *World Gold Council* Research Study no. 23 (November). http://www.gold.org/assets/file/pub_archive/pdf/Rs23.pdf.

Harmston, Steven. 1998. "Gold as a Store of Value." *World Gold Council* Research Study no. 22 (November). http://www.goldbullion.com.au/pdf/research_study_22.pdf.

World Gold Council. 2010. "Gold as a Reserve Asset." http://www.reserveasset.gold.org/.

World Gold Council. 2010. "World Gold Holdings—Volume and Value." March. http://www.reserveasset.gold.org/.

GOLD STANDARD

The gold standard was a practice in which governments committed to assigning a fixed value to their national currencies that was tied to a determined quantity of gold, often at a set price. Paper currency and other forms of bank issue were consequently required to be fully convertible into their gold equivalent at the official price. The gold standard as a measure of valuing and securing domestic currencies flourished in international markets from the 1870s to the 1930s.

Early examples of a form of gold standard existed in England during **Sir Isaac Newton**'s tenure as Mint Master at the Royal **Tower Mint** in London, when Newton issued a report in September 1717 to the Lords Commissioners of His Majesty's Treasury recommending that the price relationship between gold and silver be managed by the treasury, a form of bimetallic standard. Yet the Royal Proclamation by King George I in December in response to Newton's recommendation had the unintended result of shifting British guineas from a bimetallic standard to a gold standard when the exchange of gold guineas was capped at 21 silver shillings.

During the late 18th century, however, the financial pressures of war with France constrained the Bank of England to temporarily suspend the convertibility of the currency to a set price of gold. Following the recommendations of economist **David Ricardo** to the House of Commons **Bullion Committee of 1810** to convert bank notes into gold as a solution to the rampant inflation caused by the devaluation of the pound, Great Britain officially adopted the gold standard during 1816–1819 in a process of legislation that set the official currency as the sovereign tied to an equivalent price of "one standard ounce of gold" valued at the prewar price of .3.17s.10 1/2d and fully convertible into this regulated gold equivalent by 1821.

> "You shall not press down upon the brow of labour this crown of thorns, you shall not crucify mankind upon a cross of gold."
> —WILLIAM JENNINGS BRYAN, SPEECH AT DEMOCRATIC NATIONAL CONVENTION, CHICAGO, 1896

The spread of the use of the gold standard in domestic currencies did not occur until the later 19th century, however, following the influx of gold pouring in from the gold rushes of the mid-19th century in California, Australia, the Klondike, and South Africa. In the decades immediately following the production of gold from mines and alluvial gold sources in these regions, the world output of gold increased tenfold, generating a supply in gold that contributed to increased gold reserves and coin production and a gradual adoption of the international gold standard. The earliest currencies tied to the gold standard were that of Great Britain and Australia, while other countries maintained the traditional silver standard or bimetallic standard. Yet as total world monetary stocks shifted increasingly toward gold held by governments, rather than private firms or individuals, alignment with an international gold standard became increasingly perceived as advantageous, with further incentive in the 1890s as countries like Japan, India, Russia, and the United States joined at the turn of the century (although the United States had adhered to a de facto gold standard since 1834 by fixing the price of the convertibility of currency to gold at $20.76 per ounce). Yet conversion to the gold standard was not without its critics, and the pros and cons of bimetallism versus the classical gold standard became a divisive political issue during the U.S. presidential campaign of 1896, when democratic candidate Williams Jennings Bryan famously delivered his "Cross of Gold" speech at the Democratic National Convention in which he railed against the monied class for their support of the gold standard to the detriment of the working class, who he argued benefited from a bimetallic standard as he exclaimed, "You shall not crucify mankind on a cross of gold." The Populist Democratic Party's support of bimetallism during this era is considered to be an underlying allegory for Frank L. Baum's novel *The Wizard of Oz* (1900), in which the **Yellow Brick Road** signified the gold standard, essentially, a fancy path to nowhere. Despite such opposition, the prominence of the gold standard continued to escalate, with 59 countries adopting either a gold standard or gold exchange standard during 1880–1914 as joining assumed an international level of prestige and privileged membership in the network of developed nations that participated. Such attitudes also had elements of patriotism that were reinforced by the rise of nationalism in the early 20th century.

As the pressures of impending war mounted on some of the biggest holders of gold reserves, such as Great Britain, Germany, and the United States, national treasuries acquired and held on to gold with a frenzy. During World War I, Great Britain was forced to suspend the convertibility of the currency into gold, and other nations such as the United States in effect did the same by limiting the monetary circulation of gold, with the assumption among policy makers and the citizenry alike that these measures were only temporary. Despite the vehement opposition of economic advisor **John Maynard Keynes**, William Churchill ceded to popular support of a return to the gold standard at its prewar parity of $4.86 at all costs in his support of the **Gold Standard Act of 1925**, a policy that Churchill himself would acknowledge in hindsight as misguided and unsustainable given the economic conditions at that time. Ensuing deflationary pressures and an alarming outflow of gold and capital forced Great Britain to abandon the gold standard just a short time later in 1931. Shortly after that, the economic crisis of the Great Depression constrained the United States to alter its monetary policy dramatically. Through a series of executive orders and the Banking Act of 1933, President Franklin D. Roosevelt abandoned the classical gold standard and nationalized all privately owned gold **bullion** and coinage, yet maintained an international gold exchange standard and stimulated monetary supply by raising the fixed price of gold to $35.00 per troy ounce.

In the post–World War II era, international policymakers still attached a high degree of credibility to the theory of international gold exchange as a means of monetary stability. With the creation of the **International Monetary Fund** and the World Bank following the Bretton Woods Agreement crafted at the international economic summit in Bretton Woods, New Hampshire, in June 1944, these new multilateral institutions would regulate sovereign nations' currency exchange values tied to the dollar's fixed ratio relationship to **gold reserves** with a conversion ratio of dollars to gold at $35.00 per ounce. The **Bretton Woods System** supplanted the gold standard until 1971, when U.S. President Richard Nixon charted the path toward a floating exchange rate system when persistent U.S. deficits in international balance-of-payments undermined the world's faith in the dollar's convertibility to gold.

For a relatively brief period, mankind's faith in the value of gold as a unit of currency assumed the ultimate proportions of power and influence with the institution of the gold standard on a worldwide scale. In effect, the benefits of the gold standard relied in large part on its credibility and the cooperation among participating governments and institutions. By upholding fixed rates of exchange among member nations, the classical period of the gold standard supported the expansion of global capital markets and broadened international trade, supported a theoretically stable money supply, and minimized the risk of **inflation**. Yet the pitfalls of the gold standard were difficult to control as this form of monetary policy was extraordinarily vulnerable to external, unpredictable factors such as war, gold rushes, or individual nations' attempts to "sterilize" their money supply by increasing or lowering the levels of circulation through the sale or purchase of national securities. The policy Keynes once derided as a "barbarous relic" has thus given way to floating exchange rates in the present era, although a certain nostalgia for the perceived stability of the gold standard persists, especially in times of crisis and instability.

See also: Banking and Credit; Bretton Woods System; Bullion; Bullion Committee of 1810; California Gold Rush; Federal Reserve System; Gold Standard Act of 1925; Hargraves, Edward Hammond; International Monetary Fund; Keynes, John Maynard; Newton, Sir Isaac; Ricardo, David; Tower Mint; United States Bullion Depository; Witwatersrand Gold Rush; Yellow Brick Road.

Further Reading

Bayoumi, Tamim, Barry Eichengreen, and Mark Taylor, eds. 1996. *Modern Perspectives on the Gold Standard*. Cambridge: Cambridge Univ. Press.
Bordo, Michael, and Barry Eichengreen. 1998. "The Rise and Fall of a Barbarous Relic: The Role of Gold in the International Monetary System." Cambridge, MA: National Bureau of Economic Research, Working Paper 6436.
Bordo, Michael, and Sanna J. Schwarts, eds. 1984. *A Retrospective on the Classical Gold Standard, 1821–1931*. Chicago: Univ. of Chicago Press.
DeCecco, Marcello. 1974. *Money and Empire: The International Gold Standard, 1890–1914*. London: Oxford Univ. Press.
Eichengreen, Barry J., Jorge Bragade Macedo, and Jaime Reis, eds. 1996. *Currency Convertibility: The Gold Standard and Beyond*. London: Routledge.
Ford, A. 1962. *The Gold Standard, 1880–1914: Britain and Argentina*. Oxford: Clarendon Press.
Friedman, Milton, and Anna Schwartz. 1971. *A Monetary History of the United States, 1867–1960*. Princeton, NJ: National Bureau of Economic Research & Princeton Univ. Press.
Gallarotti, Giulio M. 1995. *The Anatomy of an International Monetary Regime: The Classical Gold Standard, 1880–1914*. New York: Oxford Univ. Press.
Green, Timothy. 1993. *The World of Gold*. London: Rosendale Press.
Green, Timothy. 1999. "Central Bank Gold Reserves: An Historical Perspective since 1845." World Gold Council Research Study No. 23, November. http://www.gold.org/assets/file/pub_archive/pdf/Rs23.pdf.
Harmston, Steven. 1998. "Gold as a Store of Value." World Gold Council Research Study No. 22, November. http://www.goldbullion.com.au/pdf/research_study_22.pdf.
Kenwood, A. G., and A. L. Longheed. 1999. *The Growth of the International Economy, 1820–2000*. London: Routledge.
Keynes, John Maynard. 1925. *The Economic Consequences of Mr. Churchill*. London: L. and V. Woolf.
Redish, Angela. 1990. "The Evolution of the Gold Standard in England." *Journal of Economic History* 50:789–806.
Rockoff, Hugh. 1990. "The *Wizard of Oz* as a Monetary Allegory." *Journal of Political Economy* 98:739–760.
World Gold Council. "Gold as a Reserve Asset." http://www.reserveasset.gold.org/.

GOLD STANDARD ACT OF 1925

The Gold Standard Act of 1925 was passed by the British Parliament on April 28, 1925, with the stated goal of facilitating a return to the **gold standard**, thus restoring the convertibility of bank notes issued by the Bank of England into the established fixed exchange rate equivalency in gold at the prewar parity of $4.86. The bank no longer produced redeemable gold coins, however, so any requests to exchange bank notes into gold were to be rendered only in 400-ounce gold bullion bars.

During World War I, financial strain due to military expenses and fears that gold would be hoarded or exported led many European governments engaged in the war to suspend the convertibility of their currencies to gold, including Great Britain, which took this measure in 1914. In the aftermath of the war, faced with rising unemployment and deflation, Winston Churchill convened a commission in the summer of 1924 to study the possibility of reinstating the **gold standard** through a variety options, including maintaining the current floating rate established for the British pound, identifying a timeline for restoring the **gold standard** at the prewar rate at a future date, returning to the **gold standard** at a reduced rate, or immediately restoring the **gold standard** at prewar parity. Emphasis on the need to reestablish credibility in the global economy largely influenced the decision to implement the latter solution. Economists such as **John Maynard Keynes** opposed this decision based on the fact that it would very likely result in deflation and the devaluation of the British pound given the contemporary market conditions in the United States and Europe. In a speech to Parliament addressing the debate over the pros and cons of the Gold Standard Act amid criticisms in the immediate aftermath of its passage from among the proponents of the policy, who accused the Churchill cabinet of "undue precipitancy" in the act's implementation, Churchill acknowledged Keynes's expertise in matters of monetary policy, noting that in response to Parliament's general majority opinion that the **gold standard** was the best option for Britain: "Mr. Keynes was writing in the *Nation* a series of searching and brilliant articles, formidable and instructive, in favour of a managed currency" (Churchill 1925).

Great Britain's efforts to restore the classical gold exchange standard were thwarted by a series of external economic and political factors that led to its collapse only six years later, in September 1931, when Britain once again abandoned the **gold standard** due to unsustainable capital outflows resulting from the devaluation of the currency. This event is widely credited with precipitating the overall collapse of the international **gold standard** as a means of foreign exchange. Churchill himself would reflect soon after the return to gold parity that such a policy had been misguided and largely politically motivated.

See also: Banking and Credit; Gold Standard; Keynes, John Maynard.

Further Reading

Bayoumi, Tamim, Barry Eichengreen, and Mark Taylor, eds. 1996. *Modern Perspectives on the Gold Standard*. Cambridge: Cambridge Univ. Press.

Bayoumi, Tamim, and Michael D. Bordo. 1999. "Getting Pegged: Comparing the 1879 and 1925 Gold Resumptions." *Oxford Economic Papers* 50, no. 1:122–149.

Churchill, Winston. 1925, May 4. Address to the House of Commons on the Proposed Gold Standard Bill. http://www.winstonchurchill.org/learn/speeches/speeches-of-winston-churchill/115-gold-standard-bill.

"The Gold Standard Act, 1925." 1925. *The Economic Journal* 35, no. 138 (June):311–313.

Keynes, John Maynard. 1971–1989. *The Collected Writings of John Maynard Keynes*, ed. Austin Robinson and Donald Moggridge. 30 vol. London: Macmillan.

Moggridge, D. E. 1968. *The Return to Gold, 1925.* Cambridge: Cambridge Univ. Press.
Moggridge, D. E. 1972. *British Monetary Policy, 1924–31: The Norman Conquest of $4.86.* Cambridge: Cambridge Univ. Press.

GOLDEN CALF

The story of the Golden Calf was originally recounted in the Hebrew Bible in Exodus 32, which describes how the Israelites became doubtful and restless when Moses was absent for 40 days and 40 nights on Mount Sinai, where he had been summoned by God to receive the Ten Commandments after he led the Israelites out of their captivity in Egypt. Fearing that Moses might never return, the people asked Moses's brother, Aaron, to make gods for them to worship, to which Aaron replied by asking the women and children of the camp to collect all the gold earrings among their family members and deliver them to him. Aaron then used this gold to fashion an idol in the image of a calf for the people to worship.

Aaron also built an altar for the Golden Calf and proclaimed that a symbolic sacrifice and feast would be held on the following day to honor Yahweh (God). Upon witnessing the feast and the transgression of the people, God ordered Moses to descend from the mountain and threatened to set his wrath upon the people for their betrayal. Moses successfully pleaded with God to "repent of this evil against thy people" (Exodus 32:11) and descended from the mountain with the tablets containing the Ten Commandments.

Upon returning to the Israelites' camp, Moses saw the Golden Calf on the altar and the people celebrating. In anger he threw the tablets to the ground, breaking them. He then threw the molten idol into the fire and once it was incinerated, tossed its ashes into the river and made the people drink from the water. In response to Moses's demand that whoever remained loyal to God identify themselves, the men of the tribe of Levi came forth. Biblical exegesis (commentary)

Moses Destroying the Golden Calf (oil on canvas), by Italian painter Andrea Celesti (1637–1712). (Cameraphoto Arte Venezia / The Bridgeman Art Library International)

claims that the tribe of Levi did not participate in the worship of the calf. Moses ordered them to "go in and out from gate to gate throughout the camp, and slay every man his brother, and every man his companion, and every man his neighbor" (Exodus 32:27), which they did, killing 3,000 Israelites.

Moses then appealed to God once again, citing this purge as atonement for the sins of the people who had made "gods of gold" (Exodus 32:31 [King James]). God replied that Moses was free to lead the tribes of Israel to the Promised Land, but would nevertheless punish them for their sin in the future, as the passage states, "the Lord plagued the people, because they made the calf, which Aaron made" (Exodus 32:35).

The story of the Golden Calf was also refashioned in the Koran in Taha 20:83, with a slight variation in which a man of an evil nature named Samiri suggests to the people that they should seek another leader in the face of Moses's prolonged absence and offers to create a new god for them. He collected the **jewelry** from among the people and used it to create the Golden Calf. Aaron spoke against Samiri's actions and in defense of Moses, beseeching them to be obedient to the Lord. Upon his return, Moses scolded his brother in anger for failing to avert this transgression and sent Samiri into exile, promising to burn the golden idol and toss its remnants into the sea.

In biblical exegesis, the incident of the Golden Calf represents the second worst sin of the people after the Fall of Man in the Garden of Eden. The most straightforward interpretation of these passages reads the episode as a warning against the practice of idolatry, forbidden by the Second Commandment, and also in some interpretations the First Commandment, which prohibits worship of other gods. This reading is based on the assumption that having lived in Egypt for so long, the Israelites' faith may have been corrupted and they were vulnerable to reverting to worshipping traditional pagan gods or ancient Semitic religions, many of which were represented as various bovine deities, such as El, the supreme god of the Canaanites known as the "bull god." Bull calf sacrifices and the fashioning of altars with horns on the corners were also common practices among ancient Semitic cultures. Additional analysis focuses on the prominence of gold in the story, which may have served as a warning against worldly materialism. Astrological interpretations read the story as a rejection of the previous era, the Age of Taurus (ca. 3800–1658 BC), and the prominence of bull deities, as well as Egypt by extension during that period. The fact that the gold melted down to create the Golden Calf was gold that the Israelites had collected for self-adornment during their time in Egypt is seen as significant in this latter interpretation. Practical questions also remain as to exactly how the gold was melted down and ground into a powder, as the biblical description is contrary to the realities of the chemical process in certain respects.

There are many probable political layers to this passage, as the primary themes of disobedience, rebellion, and atonement were reinforced in centuries of rabbinical literature. The incident of the Golden Calf is ultimately viewed as significant for the evolution of the authoritative structure of the religious practices of Judaism due to the fact that the people faltered in the absence of Moses as their leader and intercessor between the people and God. Furthermore, the loyalty of the tribe of

Levi earned their descendants exclusive rights to perform certain rituals as mandated by the Torah, which led to the ascendancy of the Levite rabbis.

The story of the Golden Calf continues to hold a high degree of significance in contemporary Judeo-Christian religious practices and remains an essential aspect of Jewish and Christian teachings on man's relationship to God and fellow man as well as to greater society and political realms through traditions of law and covenant.

See also: Jewelry.

Further Reading

Exodus. *King James Version*.

Rubin, Uri. 2001. "Traditions in Transformation. The Ark of the Covenant and the Golden Calf in Biblical and Islamic Historiography." *Oriens* 36:196–214.

GOLDEN FLEECE

The legend of the Golden Fleece, which originated in ancient Greek mythology as early as the eighth century BC, tells of the fleece of the winged ram Chrysomallos, who was born of the nymph Theophane, granddaughter of the sun titan Helios, and the god Poseidon, who, in the form of a ram, seduced the nymph.

The significance of the Golden Fleece begins with its appearance in the tale of brother and sister Phrixius and Hella, children of the cloud goddess Nephele and Athamas the Minyan of Halos. Upon learning that her children's lives were gravely threatened by the jealous plotting of their stepmother Ino, Nephele enlisted the winged ram with a fleece of pure gold in their rescue from Thessaly. As Phrixius and Hella escaped across the sea on the ram's back heading north, Hella looked down and fell to her death in the strait later known as the Hellespont. Her brother continued on to the eastern shore of the Black Sea, landing in Colchis (in modern-day Georgia), where he sacrificed the ram in honor of Poseidon and then slung the fleece in a sacred oak grove that was said to have been protected by a dragon.

The Golden Fleece achieved further renown in the ancient Greek myth of Jason and the Argonauts. Before Jason's birth, his father Aeson, rightful heir to the throne of Iolcus, was murdered by his half-brother Pelias, who then usurped the throne. Worried about the fate of her newborn infant son, Jason's mother sent him to be raised by the centaur Chiron. As a young man, Jason returned to Iolcus to claim his right to the throne, upon which Pelias sent him on a quest to find the Golden Fleece and bring it back in order to assume his title.

Jason set out on a great adventure on his ship the *Argo* with a crew of loyal sailors known as the "Argonauts." Finally arriving in Colchis after a treacherous journey, Jason had to proceed through another series of challenges ordered by King Aeetes of Colchis before he could collect his prize. Third-century-BC poet Apollonius of Rhodes describes Jason's feat eloquently in his epic *Argonautica*:

> Lord Jason held up the great fleece in his arms. The shimmering wool threw a fiery glow on his fair cheeks and forehead; and he rejoiced in it, glad as a girl who catches on her silken gown the lovely light of the full moon as it climbs the sky and looks into her attic room. The ram's skin with its golden covering was as large as the hide

of a yearling heifer on a brocket, as a young stag is called by hunting folk. The long flocks weighed it down and the very ground before him as he walked was bright with gold. When he slung it on his left shoulder, as he did at times, it reached his feet. But now and again he made a bundle of it in his arms. He was mortally afraid that some god or man might rob him on the way. (Book IV)

Upon their return, Jason's wife, the sorceress Medea, tricked Pelias's daughters into brutally murdering their father by chopping him to pieces and boiling him in a cauldron. Pelias's son Acastus had Jason and Medea exiled for their treachery, and they finally settled in Corinth. Jason later betrayed Medea when he left her to become betrothed to Creusa, daughter of the King of Corinth. Medea's wrath led to Creusa's cruel death, and Jason ended his life alone and insignificant.

From antiquity to the present, scholars and poets have proposed a variety of explanations for the symbolic meaning of the Golden Fleece. Primary among them is the theory that the Golden Fleece represented kingship and legitimacy, which is reinforced by Jason's symbolic quest. A predominance of biblical references, including the story of the magi, and an abundance of archeological evidence indicates that ancient cultures in the region considered gold as a symbolic representation of the power and authority of kings. Others have pointed to more technical explanations, suggesting the myth of the fleece derived from the placer **mining** practice in the eastern Black Sea region of submerging sheepskins in streams to collect flecks of gold, then hanging them on trees to dry and shaking or combing off the gold. Fleeces were also used on washing tables in the region's gold mines since antiquity. Greek scholar Palaephatus's fourth-century-BC compendium on the Greek myths, *On Incredible Tales*, relates the Golden Fleece to a known book on alchemy. Finally, a variety of scholars have proposed that the myth of the fleece of pure gold was tied to legends of the wealth of gold found in the East.

See also: Biblical Magi; Mining.

Order of the Golden Fleece

The Order of the Golden Fleece is a chivalric order founded by Duke Philip III of Burgundy in 1430 on the occasion of his marriage to Isabel of Aviz to commemorate the union of the kingdoms of Flanders and Burgundy. Only the sovereign with the title of Duke of Burgundy is permitted to appoint the knights of the order, one of the highest honors of the realm awarded to select individuals for their service and loyalty to the kingdom and their Catholic piety. The symbolic badge of the order is a pendant of a Golden Fleece dangling from a jewel-encrusted gold collar with ornate *b*'s on each link and inscriptions containing the order's motto and a statement of allegiance. The significance of Philip's choice of the Golden Fleece for the badge is unknown; however, it has been suggested that it was chosen due to either the importance of the wool trade in Flanders to Philip's political success, or to his idealization of the East and the chivalric traditions of the Crusades. The right to confer the title passed to the Habsburgs after they absorbed the lands of Burgundy into the Holy Roman Empire, and later the order split into an Austrian and Spanish branch following the War of the Spanish Succession during 1701–1714. Both branches continue to the present day.

Further Reading

Appollonius Rhodius. 2008. *The Argonautica*. Charleston, SC: Biblio Bazaar.
Lordkipanidze, Otar. 2001. "The Golden Fleece: Myth, Euhemeristic Explanation and Archaeology." *Oxford Journal of Archaeology* 20:1–38.
Newman, John Kevin. 2001. "The Golden Fleece. Imperial Dream." In *A Companion to Apollonius Rhodius* (*Mnemosyne Supplement* 217), ed. Theodore Papanghelis and Antonios Rengakos, 309–340. Leiden: Brill.
Palaephatus [fourth-century-BC author]. 1902. "On the Incredible." *Mythographi Graeca* III, 2, ed. N. Festa, 89. Lipsiae.

GOLDEN GOOSE

The tale of the Golden Goose is a fairy tale that first appeared in print in the 1812 collection by the Brothers Grimm, *Children's and Household Tales*. The below story of a simpleton's reward of a golden goose for his acts of generosity is perhaps most significant because it employs a variety of folkloric archetypes found in many other Indo-European fairytales.

Illustration from "The Golden Goose," in *Household Tales by the Brothers Grimm*, 1906. (*Household Tales by the Brothers Grimm*. London: J. M. Dent & Co., 1906.)

There was a man who had three sons, the youngest of whom was called Dummling, and was despised, mocked, and sneered at on every occasion.

It happened that the eldest wanted to go into the forest to hew wood, and before he went his mother gave him a beautiful sweet cake and a bottle of wine in order that he might not suffer from hunger or thirst.

When he entered the forest he met a little grey-haired old man who bade him good day, and said: "Do give me a piece of cake out of your pocket, and let me have a draught of your wine; I am so hungry and thirsty." But the clever son answered: "If I give you my cake and wine, I shall have none for myself; be off with you," and he left the little man standing and went on.

But when he began to hew down a tree, it was not long before he made a false stroke, and the axe cut him in the arm, so that he had to go home and have it bound up. And this was the little grey man's doing.

After this the second son went into the forest, and his mother gave him, like the eldest, a cake and a bottle of wine. The little old grey man met him likewise, and asked him for a piece of cake and a drink of wine. But the second son, too, said sensibly enough:

"What I give you will be taken away from myself; be off!" and he left the little man standing and went on. His punishment, however, was not delayed; when he had made a few blows at the tree he struck himself in the leg, so that he had to be carried home.

Then Dummling said: "Father, do let me go and cut wood." The father answered: "Your brothers have hurt themselves with it, leave it alone, you do not understand anything about it." But Dummling begged so long that at last he said: "Just go then, you will get wiser by hurting yourself." His mother gave him a cake made with water and baked in the cinders, and with it a bottle of sour beer.

When he came to the forest the little old grey man met him likewise, and greeting him, said: "Give me a piece of your cake and a drink out of your bottle; I am so hungry and thirsty." Dummling answered: "I have only cinder-cake and sour beer; if that pleases you, we will sit down and eat." So they sat down, and when Dummling pulled out his cinder-cake, it was a fine sweet cake, and the sour beer had become good wine. So they ate and drank, and after that the little man said: "Since you have a good heart, and are willing to divide what you have, I will give you good luck. There stands an old tree, cut it down, and you will find something at the roots." Then the little man took leave of him.

Dummling went and cut down the tree, and when it fell there was a goose sitting in the roots with feathers of pure gold. He lifted her up, and taking her with him, went to an inn where he thought he would stay the night. Now the host had three daughters, who saw the goose and were curious to know what such a wonderful bird might be, and would have liked to have one of its golden feathers.

The eldest thought: "I shall soon find an opportunity of pulling out a feather," and as soon as Dummling had gone out she seized the goose by the wing, but her finger and hand remained sticking fast to it.

The second came soon afterwards, thinking only of how she might get a feather for herself, but she had scarcely touched her sister than she was held fast.

At last the third also came with the like intent, and the others screamed out: "Keep away; for goodness' sake keep away!" But she did not understand why she was to keep away. "The others are there," she thought, "I may as well be there too," and ran to them; but as soon as she had touched her sister, she remained sticking fast to her. So they had to spend the night with the goose.

The next morning Dummling took the goose under his arm and set out, without troubling himself about the three girls who were hanging on to it. They were obliged to run after him continually, now left, now right, wherever his legs took him.

In the middle of the fields the parson met them, and when he saw the procession he said: "For shame, you good-for-nothing girls, why are you running across the fields after this young man? Is that seemly?" At the same time he seized the youngest by the hand in order to pull her away, but as soon as he touched her he likewise stuck fast, and was himself obliged to run behind.

Before long the sexton came by and saw his master, the parson, running behind three girls. He was astonished at this and called out: "Hi! your reverence, whither away so quickly? Do not forget that we have a christening today!" and running after him he took him by the sleeve, but was also held fast to it.

Whilst the five were trotting thus one behind the other, two labourers came with their hoes from the fields; the parson called out to them and begged that they would set him *and the sexton free. But they had scarcely touched the sexton when they were held fast, and now there were seven of them running behind Dummling and the goose.*

Soon afterwards he came to a city, where a king ruled who had a daughter who was so serious that no one could make her laugh. So he had put forth a decree that whosoever should be able to make her laugh should marry her. When Dummling heard this, he went with his goose and all her train before the king's daughter, and as soon as she saw the seven people running on and on, one behind the other, she began to laugh quite loudly, and as if she would never stop. Thereupon Dummling asked to have her for his wife; but the king did not like the son-in-law, and made all manner of excuses and said he must first produce a man who could drink a cellarful of wine. Dummling thought of the little grey man, who could certainly help him; so he went into the forest, and in the same place where he had felled the tree, he saw a man sitting, who had a very sorrowful face. Dummling asked him what he was taking to heart so sorely, and he answered: "I have such a great thirst and cannot quench it; cold water I cannot stand, a barrel of wine I have just emptied, but that to me is like a drop on a hot stone!"

"There, I can help you," said Dummling, "just come with me and you shall be satisfied."

He led him into the king's cellar, and the man bent over the huge barrels, and drank and drank till his loins hurt, and before the day was out he had emptied all the barrels. Then Dummling asked once more for his bride, but the king was vexed that such an ugly fellow, whom everyone called Dummling, should take away his daughter, and he made a new condition; he must first find a man who could eat a whole mountain of bread. Dummling did not think long, but went straight into the forest, where in the same place there sat a man who was tying up his body with a strap, and making an awful face, and saying: "I have eaten a whole ovenful of rolls, but what good is that when one has such a hunger as I? My stomach remains empty, and I must tie myself up if I am not to die of hunger."

At this Dummling was glad, and said: "Get up and come with me; you shall eat yourself full." He led him to the king's palace where all the flour in the whole Kingdom was collected, and from it he caused a huge mountain of bread to be baked. The man from the forest stood before it, began to eat, and by the end of one day the whole mountain had vanished. Then Dummling for the third time asked for his bride; but the king again sought a way out, and ordered a ship which could sail on land and on water. "As soon as you come sailing back in it," said he, "you shall have my daughter for wife."

Dummling went straight into the forest, and there sat the little grey man to whom he had given his cake. When he heard what Dummling wanted, he said: "Since you have given me to eat and to drink, I will give you the ship; and I do all this because you once were kind to me." Then he gave him the ship which could sail on land and water, and when the king saw that, he could no longer prevent him from having his daughter. The wedding was celebrated, and after the king's death, Dummling inherited his kingdom and lived for a long time contentedly with his wife.

Scholars and folklorists have widely interpreted the little old man as a personification of wisdom and the goose's feathers of pure gold as a symbol of supreme enlightenment. The feathers play a key role in another common folkloric element employed in the story, namely, the use of incremental repetition to reinforce a theme, in this instance the majority of the characters repeatedly fall victim to their own greed, while Dummling is consistently rewarded for his acts of kindness.

See also: "The Goose and the Golden Egg"; Rumpelstiltskin.

Further Reading

Brothers Grimm. *Grimm's Fairy Tales*. http://www.gutenberg.org/files/2591/2591-h/2591-h.htm.

Cirlot, Juan Eduardo. 2002. *A Dictionary of Symbols*. Mineola, NY: Courier Dover.

GOLDEN RULE

The Golden Rule is a moral and ethical principle found almost universally among the world's religions that refers to many versions of the maxim reported in the New Testament's Book of Matthew to have been said by Jesus, namely, "Do unto others as you wish them to do unto you." Also commonly known as the "ethic of reciprocity," the Golden Rule has been stated in myriad ways by history's most esteemed philosophers and religious leaders, appearing in the earliest written records of ancient Egypt, in the writings of ancient Greek philosophers, in Confucius's *Analects*, and in the holy texts of Hinduism, Buddhism, Judaism, Islam, and Christianity.

The overwhelming prevalence of the Golden Rule throughout the history of mankind and in widely divergent cultures has led modern historians of philosophy and religion to debate the question as to whether this proverb might be considered as an innate, or supreme, moral principle upon which human relations are founded—the original source for our modern understanding of human rights and a universal ethical basis for social relations reinforced by a common association with a blessing from God if one conformed.

> "Get learning with a great sum of money, and get much gold by her."
> —ECCLESIASTICUS 51:28

The phrase was first referred to in written sources as the "Golden Rule" in 17th-century England in *The Comprehensive Rule of Righteousness, Do as You Would Be Done By* (1679), by the Reverend Father William Lord Bishop of St. David's of the Church of England. A subsequent reference followed shortly after in George Boraston's *The Royal Law, or the Golden Rule of Justice and Charity* (1683), although 17th-century editions of the *Oxford English Dictionary* note the earliest use of the phrase as occurring in the 14th century. These English references suggest that by the early modern era, the usage of the word "gold" in this context connoted something of the highest value, perhaps to signify that the ethic of reciprocity reigned supreme over all other principles governing the conduct of mankind.

Further Reading

Allinson, Robert E. 1985. "The Confucian Golden Rule: A Negative Formulation." *Journal of Chinese Philosophy* 12:305–315.

Bachmeyer, T. J. 1973. "The Golden Rule and Developing Moral Judgment." *Religious Education* 68 (May–June):348–365.

Wattles, Jeffrey. 1996. *The Golden Rule*. New York: Oxford Univ. Press

GOLDEN TICKET (*CHARLIE AND THE CHOCOLATE FACTORY*)

The Golden Ticket was a prominent feature of the plot in British children's book author Roald Dahl's classic novel *Charlie and the Chocolate Factory*, published in the United States in 1964. Dahl's tale centers on the fate of Charlie Bucket, a young boy from a poor family. Charlie is one of five lucky winners in a contest held by an eccentric chocolate factory owner, Willy Wonka, who grants the winners a tour of his factory and a free lifetime supply of chocolate. The prizewinners all had the good fortune to have received a golden ticket concealed in the wrapper of their Wonka Bar. Each ticket was printed with the mysterious message:

> Greetings to you the lucky finder of this Golden Ticket from Mr. Willy Wonka. Present this ticket at the factory gates at ten o'clock in the morning on the first day of October and do not be late. You may bring with you one member of your family . . . and only one . . . but no one else. In your wildest dreams you cannot imagine the surprises that await you.

When the contestants arrive at the factory, they are informed that at the end of the tour Wonka intends to select one of them to designate as his heir because he has no children himself. Through the course of the day, children are disqualified one-by-one from this honor on account of their bad behavior; only Charlie is left at the end of the tour and is therefore the candy maker's chosen successor. At the very end of the story, Willy Wonka enters a glass elevator with Charlie and his grandfather, at which point the story ends until it is continued in Dahl's sequel, *Charlie and the Great Glass Elevator* (1973).

In 1970–1971, Dahl adapted the novel for film, writing the screenplay for *Willy Wonka and the Chocolate Factory*, produced by David Wolper in connection with Warner Brothers and starring Gene Wilder as Willy Wonka, a role that earned him a Golden Globe award. In 2005 the book was adapted for a remake of the film now titled *Charlie and the Chocolate Factory*, produced by Tim Burton and starring Johnny Depp as Willy Wonka. The remake was nominated for an Oscar award in the Best Achievement in Costume Design category.

See also: Oscar Statuette.

Further Reading

Roald Dahl Museum and Story Centre Archives. http://www.roalddahlmuseum.org/discoverdahl/index.aspx.

GOLDFINGER (FILM)

Goldfinger is a 1964 James Bond spy film, the third in the series, a film adaptation of the 1959 novel of the same name by Ian Fleming. Directed by Guy Hamilton and produced by Albert R. Broccoli and Harry Saltzman, the film features Sean Connery as special M16 agent James Bond in pursuit of legendary gold thief Auric Goldfinger, played by Gert Fröbe.

As the plot unravels, Goldfinger learns that Bond has been assigned to uncover the details of his so-called Operation Grand Slam, a plan to taint the U.S. **gold**

Sean Connery as James Bond fights with Harold Sakata as Oddjob in the film *Goldfinger*, directed by Guy Hamilton and produced by United Artists. (Getty Images)

reserves at the depository in Fort Knox, Kentucky, with radiation using an atomic detonation in order to reduce the global gold supply as a means of increasing the value of his own holdings in gold. Goldfinger's obsession with gold is a recurrent theme throughout the movie: gold bars are used as spy gadgets, infamous bricks of gold **bullion** are at stake in a golf game between Bond and Goldfinger; and Goldfinger's pilot Pussy Galore (played by Honor Blackman) wields a gold-plated pistol. In perhaps the most iconic scene from the movie, Bond's love interest, former Goldfinger casino escort Jill Masterson (played by Shirley Eaton), is painted head-to-toe in gold paint by Goldfinger's henchman Oddjob as revenge for her betrayal. In the film, Bond discovers her face down on the bed, apparently dead from suffocation of the skin. In a 2003 episode of the Discovery Channel science show *MythBusters*, death by skin suffocation from gold paint was proven to be a myth.

The Fort Knox **United States Bullion Depository** is featured dramatically in the film's conclusion, as Bond enlists the help of the military garrison to derail Goldfinger's plot by trapping him along with his collaborators inside the vault complex

> "Mr. Bond, all my life I have been in love. I have been in love with gold. I love its colour, its brilliance, its divine heaviness . . . I have worked all my life for gold. I ask you, is there any other substance on earth that so rewards its owner?"
>
> —Auric Goldfinger, *Goldfinger*

to ensure they do not escape. Bond successfully aborts the plot and disables the atomic device just before detonation, yet Goldfinger escapes disguised as a U.S. soldier. While Bond is subsequently en route to the White House, where he is to be honored by the president for his heroics, his airplane is hijacked by Goldfinger and Pussy Galore. In defending himself, Bond inadvertently causes Goldfinger's gold-plated pistol to shoot through the window, and Goldfinger is sucked out of the cabin. In the finale, Bond and Pussy Galore parachute to safety.

See also: Bullion; United States Bullion Depository.

Further Reading

Bouzerau, Laurent. 2006. *The Art of Bond*. London: Macmillan.
"Larry's Lawn Chair Balloon, Poppy Seed Drug Test, *Goldfinger*." 2003, March 7. *Myth-Busters*. Discovery Channel.

GOLDSCHLÄGER

Goldschläger is a flavored schnapps liqueur with a taste of spicy cinnamon that has gold flake leaves added to it as a novelty. The fancy liqueur was originally produced in Switzerland until the 1990s, when the U.K. multinational wine, beer, and spirits distributor Diageo acquired the brand and transferred its manufacture to Italy. Originally developed with a 53.5 percent alcohol content equivalent to 107 proof, Goldschläger now contains 43.5 percent alcohol at 87 proof. Each 750 milliliter bottle contains less than 0.1 grams of gold leaf. The German name means "gold beater" in reference to gold leaf manufacturers who pound bricks of gold into thin sheets.

> **Breakfast of Champions**
>
> The Westin New York Hotel offers guests a $1,000.00 breakfast bagel spread with white truffle cream cheese, Riesling wine jelly, and a topping of edible gold leaf.

The addition of the gold to the clear liqueur originates from a long tradition of adding gold flakes to food and beverage for both prestige and presumed health benefits. There is evidence that the pharaohs of Ancient Egypt may have added gold to their wine. The nobility of medieval Europe added gold to their alcoholic beverages as a lavish display of wealth. Goldwasser ("gold water") is another gold-flecked beverage, manufactured in the Polish city of Danzig since its invention in the late 16th century. A variety of contemporary champagnes and sparkling wines also contain gold flakes for celebratory flare.

The Goldschläger brand has faced widespread unfounded rumors about the potential harmful effects of ingesting gold flakes. In response to concerns about metallic **allergies**, the quantity of gold was reduced. One popular urban legend suggests that the gold flakes lacerate the stomach or esophagus, enabling a much faster pace of intoxication as the alcohol enters the bloodstream more directly. This is a fallacy, however, as the edges of the gold leaf are soft and malleable, and gold as an approved food additive is an inert substance that is harmless to ingest.

See also: Allergies.

Further Reading

Diageo Corporate Site. http://www.diageo.com/Pages/default.aspx.

Stimmell, Gordon. 2009. "Blue Nun Sparkles with Bubbles and Bling." *Toronto Star*, June 3.

GOLDSMITHING

According to economic historian Wilbur Aldrich, "Goldsmiths appear to have been among the earliest artificers of the human race" (1903, 68). The craft of goldsmithing has indeed existed from earliest antiquity in the realm of metallurgy and metalworking in almost the full range that it exists to the contemporary period, encompassing the production of objects with a variety of design and purpose, from secular to sacred and from fine art to tableware. Throughout history, goldsmithing remained such a highly specialized and cherished art that goldsmiths had a significant level of influence in culture and society in artistic and patronage circles, religious ceremonial practices, **jewelry** making, and banking and finance.

> ### World's Most Expensive Ice Cream Sundae
>
> To commemorate their golden anniversary after 50 years of business, Serendipity 3 restaurant in New York City created the Golden Opulence Sunday, an edible gold-adorned luxurious desert that earned the distinction of "World's Most Expensive Sundae" by the Guinness Book of World Records for its outrageous price—$1,000.00. The dessert consists of five scoops of Tahitian vanilla bean ice cream infused with Madagascar vanilla, which is covered in edible gold leaf and topped with shavings of Amedei Porcelana and Chuao chocolate, two of the most expensive chocolates made; a small glass bowl inserted with a portion of the American Golden Grand Passion dessert caviar; passion fruit; orange; Armagnac brand liqueur; candied fruits imported from Paris; cherries made of marzipan from Italy; a sprinkling of Jordan almond confections covered in real edible gold; and a gilded sugar flower. The sundae is served in a Baccarat Harcourt crystal goblet with an 18-karat gold spoon. Diners are required to request the dessert with a reservation 48 hours in advance.
>
> *Source*: Morse, Jody. 2007. "$1,000 Ice Cream: The Golden Opulence Sundae." *Associated Content*, August 14. http://www.associatedcontent.com/article/340520/1000_ice_cream_the_golden_opulence.html.

The craft was passed on through apprenticeships in which the master goldsmith would instruct his apprentice in the goldsmithing methods required to create diverse objects, including techniques for testing the fineness of the metal through **assaying**, coloring the metal by **alloying** it with other metals, designing molds and **casting** gold objects, joining pieces together by means of **granulation** or **filigree**, creating settings for precious gems or cloisonné inlay, **gilding** and hammering, and finishing pieces by burnishing or polishing. These methods were used to create practical objects for the table such as utensils or goblets, sacred objects for use in religious rituals and ceremonies, **jewelry** and other pieces for personal adornment, and decorative works of art. Because gold was cherished and utilized in many of these capacities in societies throughout history from South America to India, Africa, the Middle East, and across Europe, archeological and historical records leave many traces of the history and evolution of goldsmithing in abundant gold artifacts from funeral and temple sites, coinage, written manuals and treatises, regalia, and fine art and **jewelry**. Due to its particular chemical properties as

> ### Saint Eligius
>
> Saint Eligius (590–660) is the Catholic patron saint of goldsmiths, blacksmiths, and metalworkers. Born in Limoges, France, Eligius was apprenticed to Master of the Mint of Limoges Abbo, a famous goldsmith. King Clotaire II later appointed him Master of the Mint at Marseilles, and in 629, Clotaire's successor Dagobert appointed him his chief councilor. In this position, Saint Eligius undertook many projects for the sake of the Church and on behalf of the poor, building monasteries, churches, and shrines and organizing alms for the poor. After his ordainment as a priest in 639, he was named Bishop of Noyon-Tournai. Upon his death in 660, Eligius was buried in Noyon. He is often depicted in Christian hagiographic art with a miniature gold church resting in his left upturned palm.

a noble metal, gold in effect lasts forever without tarnishing or oxidizing, and thus many gold objects exist from antiquity to the present in pristine condition.

One of the earliest archeological discoveries revealing ancient goldsmithing techniques was found in excavations at the site of the **Royal Tombs of Ur**, an ancient Sumerian civilization. The gold objects buried with members of the royal family and the nobility ca. 2900–2300 BC demonstrated an astoundingly high level of craftsmanship and encompassed almost the extent of goldsmithing as it was known in later periods. These techniques were developed to an even higher degree in ancient Egyptian society amid the wealth and sumptuous display of the pharaonic dynasties. Treasures discovered in excavations of burial chambers in the Valley of the Kings in Egypt, most notably the tomb of the **Tutankhamun**, contain an abundance of precious gold objects crafted with a variety of methods. Ancient Egyptian techniques were supplanted by early Greek and Etruscan metalworking practices around the sixth and seventh centuries BC, forms that would have a strong influence on Roman and Byzantine goldsmithing. During this same period, pre-Columbian civilizations in South America such as the **Inca** utilized advanced metalworking techniques to create gold **jewelry**, decorative objects, and other finery for royalty, nobility, and religious ceremony.

During the Middle Ages in Europe, metalworking techniques were transmitted among monks in monasteries for the production of religious objects and among guilds for the production of both religious and secular objects. Written sources from this period such as manuals, treatises, and account books provide valuable information on the methods they used, the people who the pieces were created for, and the function of goldsmiths in the economy and society. Among the earliest known treatises are the *De Diversis Artibus*, written by Benedictine monk Theophilus Presbyter in the early 12th century, who produced the work as a practical manual to a range of decorative arts with extensive commentary on goldsmithing techniques, and the 12th-century *Mappae Clavicula*, also produced in a Benedictine scriptorium, which contains several hundred recipes and chemical formulas for creating colors and textures in metalwork and other decorative arts.

Developments in the production of gold coinage, money changing, and banking and finance; and an increase in demand for sumptuous regalia among the rising nobility and monarchies of Europe and sacramental objects, decorative arts, and reliquaries for the growing Catholic Church fueled advancements in goldsmithing techniques and the social status of goldsmiths to the heights obtained by the craft in the Renaissance, when many fine artists were trained out of necessity in goldsmithing techniques, which were considered as core to artistic training. Written records and treatises from this period reveal the patronage and commercial environment in which legendary goldsmiths thrived, among them the 16th-century artist **Benvenuto Cellini**, author of a treatise on the craft and creator of the *Saliera*, an elaborate gold saltcellar produced for French king Francis I in the 1540s that is considered to be one of the finest pieces of goldwork every created.

Goldsmithing continued to flourish in the early modern period amid the influx of gold coming in to Europe from the Spanish settlement of territories in the New World and the growing need to meet the demand emanating from the great courts of the powerful monarchies of the 16th to 18th centuries; for example, in the ornate designs of the baroque and rococo styles found at such palaces as **Versailles**. During the 19th century, the neoclassical and romantic styles prevailed and the opening up of new markets in an unprecedented atmosphere of commercial consumption, along with vastly greater supplies of gold from the California Gold Rush and the discovery of gold in Australia, South Africa, and the Klondike, made the gold decorative arts and **jewelry** available to a wider population. These developments contributed to the growth of the goldsmithing industry in the United States and the rise of famous **jewelry** houses such as **Fabergé**, Cartier, and Bulgari in Europe.

In the contemporary period, many traditional **jewelry** making practices have been mechanized to be mass produced; however, the goldsmiths' craft continues to be passed on among artists and artisans who pursue highly specialized training in goldsmithing techniques that have existed largely unchanged since their origins in antiquity.

See also: Alloying; Asante Golden Stool; Assaying; Banking and Credit; Casting; Cellini, Benvenuto; Fabergé, House of; Filigree; Gilding; Granulation; Inca; Jewelry; Pre-Columbian Gold; Tutankhamun; Ur, Royal Tombs of; Versailles, Palace of.

Further Reading

Aldrich, Wilbur. 1903. *Money and Credit*. New York: Grafton Press.
Barsali, Isa Belli. 1969. *Medieval Goldsmiths Work*. London: Pauly Hamlyn.
Brepohl, Erhard, and Tim McCreight. 2001. *The Theory and Practice of Goldsmithing*. Brunswick, ME: Brynmorgen Press.
Bunt, Cyril G. E. 1926. *The Goldsmiths of Italy, Some Accounts of Their Guilds, Statutes, and Work Compiled from the Published Papers, Notes, and other Material Collected by Sidney J. A. Churchill*. London: Martin Hopkinson.
Cellini, Benvenuto. 2006. *The Treatises of Benvenuto Cellini on Goldsmithing and Culture*, trans. C. R. Ashbee. Whitefish, MT: Kessinger Publishing.
Cennini, Cennino d'Andrea. 1954. *The Craftsman's Handbook*, trans. Daniel Varney Thompson. New York: Thompson Courier Publications.
Hawthorne, J. G., and C. S. Smith. 1979. *Theophilius, On Divers Arts: The Foremost Medieval Treatise on Painting, Glassmaking, and Metalwork*. New York: Courier Dover Publications.
Pollen, John Hungerford. 1878. *Gold and Silversmiths' Work*. London: R. Clay and Sons.

"THE GOOSE AND THE GOLDEN EGG" (AESOP)

"The Goose and the Golden Egg" is a fable authored by the legendary storyteller Aesop, who lived in ancient Greece ca. 620–560 BC. Many details about Aesop's origins are unknown, yet it is widely believed that he was a slave from Phrygia who came to live in Samos. The corpus of moralistic fables attributed to Aesop has been passed on throughout the centuries from antiquity to the present. Although the fables have had varied significance according to the contexts in which the tales were read, Aesop's use of animal narrators and symbolic natural elements to subtly illustrate the ironies and lessons of the human condition have in many ways remained timeless.

"The Goose and the Golden Egg" tells of a man and his wife who had a goose that laid a single golden egg for them each day. As the man's wealth grew, he became dissatisfied with only a single egg each day and wished for more. Assuming the goose was filled with gold inside, the man foolishly killed it. Upon finding that the goose bore no treasure inside of her, he remarked to himself, "While chasing after hopes of a treasure, I lost the profit I held in my hands!" (adapted from Gibbs 2002). The moral of the story is that those who are greedy and seek to obtain more than they deserve risk losing everything they have.

The phrase "killing the goose for the golden egg" has evolved as a widespread idiomatic expression in the English language used in reference to events or actions that produce an immediate desirable result with potentially disastrous long-term consequences and is particularly common as a proverbial reference to shortsighted economic policies within the realm of banking and finance.

See also: Banking and Credit.

Further Reading

Gibbs, Laura. 2002. *Aesop's Fables*. Oxford: Oxford Univ. Press.
Lerer, Seth. 2007. "Aesop, Authorship, and the Aesthetic Imagination." *Journal of Medieval and Modern Studies* 37, no.3:579–594.

GOSPEL BOOKS OF SAINT MÉDARD DE SOISSONS

The Gospel Books of Saint Médard de Soissons is an early-ninth-century **illuminated manuscript** conserved today as MS lat. 8850 at the Bibliothéque Nationale in Paris. It is considered to be one of the most exquisite and advanced examples of a group of **illuminated manuscripts** produced at the scriptorium at Aachen in present-day Germany for the court of Charlemagne, known alternately as the Court School of the Carolingian Renaissance or the Ada School, due to the patronage of Charlemagne's sister Ada.

The style of the script, illustrations, and contents indicate that the book likely belonged to Charlemagne himself at one point, although it is first identified in the historical record in 827, when Frankish king Louis the Pious, Charlemagne's son and successor, presented it along with his wife Judith as a gift to the Church of Saint Médard in Soissons, in the present-day Aisne department in the Picardy region of northern France.

French illustration of St. Mark in *Gospel Book of St. Médard of Soissons*, from the court school of Charlemagne, early ninth century. (Bibliotheque Nationale, Paris, France /The Bridgeman Art Library International)

The precious manuscript consists of 239 vellum folios measuring 362 x 267 millimeters and contains the Vulgate text of the traditional four Gospels, the Eusebian canon tables, and several other exegetical works, all written entirely in gold using **chrysography** in Carolingian uncial calligraphic script, which is identified as the precursor to modern-day cursive. Twelve pages of the canon tables are ornately illuminated, as are six full-page miniatures, including the famous "Fountain of Life" illustration, and four decorative pages. The initial folio of each Gospel book is dyed a deep purple to brighten the effect of the gold writing. The richness and wealth of the colors used, the quantity of gold applied in writing and illustrations, and the ornate details evoke a regal sense of ceremony that is a hallmark of the group of 10 manuscripts identified together as the Ada School. These books also had sumptuous covers adorned with jewels set in gold, often **filigree**, with carved ivory panels. The style was further set apart from earlier Hiberno-Saxon or Merovingian **illuminated manuscripts** from the evident Italian influence, which accounts for the characterization of the Saint Médard Gospel Books as one of the first examples of a revival of Roman classicism. These manuscripts were also more ambitious in scope and conception, with advanced depictions of scale and form in human figures and the background in which they were portrayed.

The Saint Médard Gospel Books remained in Soissons until the French Revolution, after which the priceless book fell into private hands until the modern era when it was obtained by the Bibliothéque Nationale.

See also: Chrysography; Filigree; Illuminated Manuscripts.

Further Reading

Dodwell, C. R. 1993. *The Pictorial Arts of the West, 800–1200*. Pelican History of Art. New Haven, CT, and London: Yale Univ. Press.

Walker, Robert M. 1948. "Illustrations to the Priscillian Prologues in the Gospel Manuscripts of the Carolingian Ada School". *Art Bulletin* 30, no. 1 (March).

GRANULATION

Granulation is a decorative arts technique used by gold and silversmiths to apply miniscule "grains," or pieces, of silver or gold to a precious metal surface by means of a soldering process with an alloy metal or another type of adhesive to create an ornamental effect or design. Granulation was most commonly used to create **jewelry** or other items of personal adornment such as pins and head ornaments.

The term "granulation" derives from the Latin word for grain, *granum*, and is a relatively modern term. In ancient sources the technique is referred to through a variety of descriptions in reference to the general process, rather than a specific practice, due to the fact that many variations of the basic technique existed from early antiquity. In the basic practice of granulation, small, spherical pieces of gold or silver are created through a variety of methods, such as pouring the molten metal into a sieve and cooling the pieces in cold water or cutting tiny pieces of gold or silver and then melting the pieces to achieve the desired shape. These granules of precious metal are then affixed to a typically flat base piece of gold or silver to form a design using a metal alloy solder that has a lower melting point than the gold elements of the object along with another form of adhesive or by simply applying the pieces with an adhesive and then melting granules onto the base. Common adhesives used in the premodern era since antiquity included gum from the tragaganth bush, a plant native to modern-day Iran; a type of glue made from cattle hides; quince paste; or gum arabic, a hard Acacia tree sap.

The oldest extant examples of granulation were found in excavations at the **Royal Tombs of Ur**, the site of an ancient Sumerian city in modern-day Iraq, dating to ca. 2560–2500 BC. Artifacts from the mid-second century BC demonstrating granulation were also found in excavations of ancient Troy and Egypt. Perhaps the most famous ancient artisans of this method were the goldsmiths of the Etruscans in Italy during the eighth–sixth centuries BC. One of the most elaborate examples of granulation known to scholars is a gold Etruscan cup excavated from the site of the ancient city of Praeneste in modern-day Palestrina dating from the seventh century BC that is adorned with a pattern consisting of over 137,000 gold granules.

Although many scholars have considered granulation to be strictly an ancient practice that can be defined as a type of primitive **filigree**, written historical sources indicate that many variations on the ancient practice of granulation continued through to the modern era. Ancient references to granulation appeared in significant technical and historical treatises of classical antiquity, the Middle Ages, and the Renaissance such as Roman naturalist Pliny's *Naturalis Historiae*, German Benedictine Theophilus's *De Diversis Artibus* (1122), Italian metalworker Vannoccio Biringuccio's *Della Pirotechnia* (1540), German scientist Agricola's *De Re Metallica* (1556), and Benvenuto Cellini's *Treatise on Goldsmithing and Sculpture* (1568). The art of granulation is still found today and applied in largely the same manner as in antiquity in folk craft and jewelry making traditions in Central and Eastern Europe, Mongolia, Tibet, and Iran.

See also: Filigree; Goldsmithing; Jewelry; Ur, Royal Tombs of.

Further Reading

Evans, J. A. 1989. *A History of Jewellery 1100–1870.* London: British Museum Publications.
Maxwell-Hyslop, Rachel K. 1972. "The Art of Granulation in Early Iranian Gold Jewelry." *Proceedings of the 5th International Congress of Iranian Art and Archaeology.* http://www.cais-soas.com/CAIS/Art/granulation.htm.
Tait, H. 1986. *Seven Thousand Years of Jewellery.* London: British Museum Publications.
Wolters, Jochem. 1982. "Granulation: A Re-Assessment of an Ancient Craft." *Endeavor* 6, no. 1:2–9.

GREAT RECOINAGE ACT OF 1696

In 1696, the British Parliament passed the Great Recoinage Act as an emergency measure to address the worsening economic situation throughout the Kingdom of England in the wake of the Glorious Revolution of 1688 and the Nine Years' War of 1688–1697 on the European continent between English king William III at the head of the Grand Alliance against French king Louis XIV.

Prior to the recoinage, English currency was valued based on the silver standard. Upon his restoration to the throne following the English Civil War, King Charles II minted the **guinea** as a gold coin with an intended market value of 20 shillings, or a pound sterling. The wars of the late 1680s and early 1690s steadily devalued the silver coinage, however, and a large-scale capital outflow of both silver and gold from England to the Continent strained the treasury to its limit. By 1695, the value of silver had plummeted with the demand for gold, which led to an escalation of the relative value of the guinea to as high as 30 shillings that same year. To make matters even worse, the prevalence of **coin clipping**, in which the edges of silver coins were melted down as **bullion** and sold for a higher price on the Continent, and rampant counterfeiting undermined currency and trade throughout England and its empire, leading to riots of protest against the government's attempts at monetary control of clipped coins weighing well under their minted value.

In 1695, a parliamentary committee of the House of Commons was assembled to address the problem. During their deliberations, expert opinions were sought from among the nation's renowned political economists, most significantly John Locke, who favored recoinage, and **Sir Isaac Newton** for his metallurgical knowledge. Under the leadership of Chancellor of the Exchequer Charles Montagu, an act calling for the collection and reminting of all coinage using new, mechanized processes such as milling, rather than hammering by hand, that would help to prevent clipping and counterfeiting, and the establishment of a bimetallic valuation of the currency that favored gold and limited the value of the **guinea** to no more than 28 shillings was passed on January 17, 1696, as "An Act for Remedying the Ill State of the Coin of the Kingdom."

Montagu subsequently offered Newton the post of Warden of the Mint to oversee the implementation of the recoinage and to deal with the problem of counterfeiting. Newton moved from Cambridge to London in May 1696 to assume the post, later moving up to Master of the Mint in 1699, a position he held until his death in 1727. During his tenure at the **Tower Mint**, Newton produced a valuable series of reports on his progress to Parliament and also consulted with Locke about monetary policy. Newton was credited by his contemporaries as one of England's greatest civil servants for his diligent and successful reform of coinage and for his prosecution of counterfeiters, as well as being the originator of the **gold standard** that would come to prevail over modern currencies until the economic crises of the 1930s.

See also: Coin Clipping; Gold Standard; Newton, Sir Isaac; Tower Mint.

Further Reading

Davison, Lee. 1988. "John Locke, Edward Clarke, and the 1696 Guineas Legislation." *Parliamentary History* 7, no. 2:228–240.
Laslett, Peter. 1957. "John Locke, the Great Recoinage, and the Origins of the Board of Trade, 1695–99." *William and Mary Quarterly* no. 14:137–164.
Li, Ming-Hsun. 1963. *The Great Recoinage of 1696 to 1699*. Weidenfeld and Nicolson.
Redish, Angela. 1990. "The Evolution of the Gold Standard in England," *Journal of Economic History* 50, no. 4 (December):789–806.
Quinn, Stephen. "Gold, Silver, and the Glorious Revolution: International Bullion Arbitrage and the Origins of the English Gold Standard." Economic History Association. http://eh.net/Clio/Conferences/ASSA/Jan_95/Quinn.shtml.

GUINEA

The Guinea is a British gold coin that was produced at the **Tower Mint** from 1663 to 1813. During the Restoration of the British monarchy to the throne following the English Civil War, King Charles II devoted considerable attention to the reestablishment of a gold coinage that would rival Europe's most reputable currencies. In

George II guinea. (iStockPhoto)

1663, the mint began to use gold imported from West Africa by the Africa Company, who had received special permission to stamp the coins with their company mark, an elephant, placed below the bust of the king. The term "guinea" was used loosely from the 15th century on to refer to the gold-producing regions of West Africa.

The guinea soon replaced the common currency in circulation, the sovereign, by virtue of a royal proclamation from Charles II on March 27, 1663, calling for the **minting** of the guinea at a value equivalent to one silver pound sterling worth 20 shillings, with a weight of 8 grams, more than twice the weight of the famous **florin** gold coin of Renaissance Florence. The design of the coins was undertaken by master Dutch engraver John Roettier, who Charles II had brought back with him from exile to attend to the mint. Early guinea coins measured a diameter of 25 millimeters and were stamped on the obverse with a bust of the king accompanied by an elephant (and, sporadically after 1674, an elephant and a castle) and on the reverse with royal insignia that varied according to the current reigning monarchs. The elephant served as evidence that the coin was struck with gold imported by the Africa Company. The coins also included an edge inscription of the motto *Decus et Tutamen Anno Regni* ("Ornament and safeguard in the year of the reign of") followed by the year it was struck. Roettier also designed a 5 guinea coin that was produced during 1668–1753 with a weight of approximately 41 grams and measuring 37 millimeters in diameter. On account of Roettier's updating of the **Tower Mint** with new mechanization processes, the guinea was the first coin to be produced using mechanized rolling mills, edge markers, and blank-cutting machines. These manufacturing techniques greatly reduced the problem of devaluation by edge clipping and ultimately replaced all hammered gold coins in circulation.

The guinea maintained a value of 20 shillings in international and domestic trade until the economic crisis of the Revolution of 1688 resulted in a serious devaluation of silver currency used widely in domestic commerce, which in turn led to a progressive increase in the price of gold. These circumstances were further worsened by widespread **coin clipping** and counterfeiting, exportation of silver, and a dearth of currency due to costly foreign wars and revolution. During the early 1690s, concerns over debasement and wide variance in the actual weight of coinage led to confusion and popular unrest across Britain as coins were accepted at different values from place to place. In March 1694, the guinea was valued on the open market at 22 shillings; slightly over a year later, it had increased to 30 shillings. A parliamentary committee of the House of Commons was appointed to address the monetary crisis. In 1696, **Sir Isaac Newton** was appointed warden and then master of the **Tower Mint** to stop the counterfeiting and oversee the implementation of the **Great Recoinage Act of 1696**, intended to restore the value of the guinea to a uniform rate of 20 shillings. After collecting all of the devalued coinage in circulation, melting it down, and reissuing it, Newton implemented Great Britain's historic conversion to a form of **gold standard** in 1717, when the Law of Queen Anne shifted the value of the guinea from an equivalent in silver, to a bimetallic valuation based on silver and gold that fixed the value of the guinea at 21 shillings.

Between 1813 and 1816, the guinea was ultimately replaced the British pound, which from that point on remained the primary unit of currency in circulation. The term guinea is still commonly used today in reference to winning purses in the horse racing industry and at other forms of livestock auctions in Anglophone cultures.

See also: Florin; Gold Standard; Great Recoinage Act of 1696; Newton, Sir Isaac; Tower Mint.

Further Reading

Craig, John Herbert, Sir. 1946. *Newton at the Mint.* London: Univ. Press.

Dodd, Agnes F. 1911. *The History of Money in the British Empire and the United States.* Longmans, Green.

Hobson, Burton, and Robert Obojski. 1970. *Illustrated Encyclopedia of World Coins.* Garden City, NY: Doubleday.

Room, Adrian. 1987. *Dictionary of Coin Names.* London & New York: Routledge & Kegan Paul.

HAGIA SOPHIA

The Hagia Sophia is a Byzantine church in present-day Istanbul built by the Emperor **Justinian I** (ca. 482–565) in the early sixth century upon the site where two earlier churches had been built during the fourth and the fifth centuries, both of which were destroyed by fire in public riots. Following the destruction of the second basilica, **Justinian** proceeded with ambitious plans in 532 to build a glittering, gold-adorned basilica, which would be hailed as one of the greatest architectural achievements of all time and a powerful and enduring expression of Eastern Orthodox Christianity and the Byzantine style in art and architecture. Also contributing to its iconic status as a symbol of Byzantine power and influence is the Hagia Sophia's geographic position: it sits on a promontory extending out from the European side of the Bosphorus to face the direction of Asia at the intersection of the Sea of Marmara and the channel.

Gilded minbar at the Hagia Sophia in Istanbul. (Ahmet Ihsan Ariturk/Dreamstime.com)

The basilica was completed in only six years and first dedicated by the emperor in 537, who is said to have proclaimed upon viewing the awe-inspiring structure, "Glory to God who has thought me worthy to finish this work. Solomon, I have outdone you" (*Narratio*, quoted in Mainstone 1988). The dome covering the nave is one of the most renowned architectural feats of late antiquity. The approximate size of the Pantheon dome in Rome, the dome of the Hagia Sophia was the first cupola supported by pendentive, or triangular, stone segments constructed on the nave's rectangular base and reinforced with four large columns. For further effect, a row of 40 windows was built into the base of the dome, allowing light to stream in from the soaring dome to create an awe-inspiring visual sense of light streaming in from every direction. Yet the weight of the dome presented problems almost from its completion, as damage from several earthquakes led to substantial renovations later in the sixth century.

Equally as stunning as the scale of the large dome was the sumptuous decoration of the church's interior, which was adorned with gold mosaics throughout on the vaults and arches, gold leaf in decorative accents outlining other architectural features, polished slabs of colorful marble inlaid along the walls and lining the floor, an altar and crucifix gilded in gold and set with precious stones, gates gilded with silver, and gilded gold crosses adorning columns, podiums, and walls. Such lavish interior ornamentation inspired awe among contemporary observers such as Procopius, who remarked in his *On Buildings* that "the entire ceiling is covered in real gold, which is beautiful as well as ostentatious" (Swift 1940, 88). Historians estimate that the total square area covered in gold **mosaic** in the church is the equivalent of approximately four acres.

Many of the gold ornaments decorating the interior were looted by the Ottomans during the fall of Constantinople; however, the basic structure itself was left largely intact by Sultan Mehmet II, who rededicated it as the Mosque of Ayasofya. Under Mehmet, the interior was modestly reconfigured in accordance with the needs of Islamic religious practices, and a minaret was added to the exterior. His successors would undertake subsequent more extensive renovations to the building, including the addition of three more minarets in the remaining corners. Additionally, from the 15th to the early 19th century, most of the gold mosaics had been covered with whitewash. Perhaps the most famous restoration of the interior was undertaken by Sultan Abdul Medjid, who commissioned Swiss architects Gaspare and Giuseppe Fossati to restore parts of the interior in 1847, during which many of the original mosaics were uncovered.

In the 1930s, secular ruler Mustafa Kemal Ataturk, president of the Republic of Turkey, undertook further restorations to the building and established the building as a museum in 1934.

See also: Gilding; Justinian I; Mosaic.

Further Reading

Kleinbauer, W. Eugene, Antony White, and Henry Matthews. 2004. *Hagia Sophia*. London: Scala Publishers.
Mainstone, Rowland J. 1988. *Hagia Sophia: Architecture, Structure, and Liturgy of Justinian's Great Church*. New York: Thames & Hudson.
Mango, Cyril. 1962. *The Mosaics of St. Sophia at Istanbul*. Locust Valley, NY: J. J. Augustin.
Procopius. 2011. *Of the Buildings of Justinian*, trans. Aubrey Stuart. Port Chester, NY: Adegi Graphics.
Swainson, Harold. 2005. *The Church of Sancta Sophia Constantinople: A Study of Byzantine Building*. Boston, MA: Adamant Media Corp.
Swift, Emerson Howland. 1940. *Hagia Sophia*. New York: Columbia Univ. Press.

HAMILTON, ALEXANDER

Alexander Hamilton (ca. 1755–1804) was a Founding Father of the United States government who made fundamental contributions to the drafting of the first Constitution, its early legal interpretation, and the nascent state's centralized fiscal structure with respect to banking, particularly the establishment of a mint and the economic function of gold as a form of monetary exchange, credit, public debt, taxation, and trade policy.

Hamilton was born around 1755 in Nevis in the British West Indies, the illegitimate son of Scottish trader James Hamilton and Rachel Fawcett Lavine, who was at that time married to a German merchant, although she had left him several years before, relocating from St. Croix to Nevis, where she moved in with James, who later abandoned the family. With his mother's death in 1768, the young Alexander Hamilton was faced with poverty and went to work as a clerk in the local merchant office. In 1772, he was sent to New Jersey to study thanks to the financial support of local community benefactors. In 1773, he enrolled in King's College in New York.

Amid the tensions leading up to the American Revolution, Hamilton was an early yet measured supporter of the cause and enlisted in a New York militia in 1775. In 1776 he obtained a commission as artillery captain, and on March 1, 1777, was appointed as an aide-de-camp to George Washington, a position he served in its fullest capacity, in effect as chief of staff, for four years to become a close and trusted associate of Washington and many other important Revolutionary leaders. In 1780, Hamilton married the daughter of a prominent, wealthy New York merchant, Elizabeth Schuyler, with whom he had eight children.

After leaving his military post, Hamilton was appointed as a New York delegate to the Congress of the Confederation. In this capacity, he critiqued the weaknesses of the Articles of Confederation and presented his arguments for a strong, centralized state and the drafting of a new constitution. Many elements of his proposals were founded on his own experiences of the precariousness of government finances during his tenure as aide-de-camp to Washington and therefore argued that the federal government needed to be entrusted with the ability to create nationalist financial legislation in order to maintain a strong state. In 1782 he studied to become a lawyer and was admitted to the New York Bar in July 1783. The following year he founded the Bank of New York with a particular view to establishing a strong, stable currency.

Hamilton continued to advocate for a convention to draft a new constitution while serving as a representative to the New York state assembly in 1787. In this capacity he was appointed as a delegate to the Philadelphia Constitutional

> "To declare, that a less weight of gold or silver shall pass for the same sum, which before represented a greater weight, or to ordain that the same weight shall pass for a greater sum, which before represented a greater weight, or to ordain that the same weight shall pass for a greater sum, are things substantially of one nature. The consequence of either of these, if the change can be realized, is to degrade the money unit."
>
> —ALEXANDER HAMILTON, *REPORT ON THE ESTABLISHMENT OF A MINT* (1791)

Convention, where he contributed to the drafting of the U.S. Constitution and was the only representative from New York to sign the new document. In response to widespread criticism and opposition to the new constitution in New York in an effort to encourage its ratification, Hamilton, writing under the pseudonym Publius, collaborated with James Madison and Secretary of Foreign Affairs John Jay to write a series of political essays that would become known as the *Federalist Papers*. These commentaries and legal interpretations of the future U.S. Constitution have had a fundamental influence on national constitutional issues.

In 1789, Hamilton was appointed as Secretary of the Treasury in the administration of the first U.S. president George Washington. During 1790–1791, Hamilton presented the House of Representatives with a series of drafts for plans to address the need for managing the issue of public credit effectively; identifying sources of revenue for the federal government, primarily in the form of various import taxes and tariffs; establishing a mint and the financial apparatus through which it would be managed; and recognizing the importance of manufacturing industries to the national economy. Whereas some of his proposals were perceived as overly protectionist and nationalist, core features of his plans were adopted, such as the creation of the mint and the manner in which it operated, the establishment of a central bank that paved the way for the **Federal Reserve System**, and a public credit system that would provide the government with the fiscal stability essential to its growth and viability. In outlining his theories and rationales for the importance of a strong, central fiscal policy at the federal level, Hamilton relied to a significant degree on the writings of Scottish economists **David Hume** and **Adam Smith**. Despite his controversial legacy in the realm of historical perspectives on the Founding Fathers, in which Hamilton is perceived as something of a renegade, his economic policies are responsible for crafting the strong financial structure that encouraged and upheld the development of a modern democratic state and continues to support the emphasis on the importance of financial stability to the health of such a state.

In his "Report on the Establishment of a Mint," Hamilton outlined a system of national coinage that would be uniform and stable in the foreign exchange environment of international commerce. Although initially supporting a gold-based currency, out of concerns over the pitfalls of a **gold standard** relative to the contemporary market for gold, he outlined a bimetallic system that called for minting silver and gold coins. Above all, he stressed the importance of consistency in minting of the money unit in order to maintain credibility of U.S. markets in international commerce.

At the end of his tenure as treasury secretary, Hamilton's later political career was plagued by contention and scandal. Despite his continued influence on national politics, particularly as a leader of the Federalist faction in the era in which the first political parties began to take shape, a series of personal rivalries with his political opponents plagued his later years. Following a prolonged feud with Vice President Aaron Burr, the two engaged in a duel in which Hamilton was killed on July 11, 1804.

See also: Banking and Credit; Federal Reserve System; Gold Standard; Hume, David; Smith, Adam.

Further Reading

Chernow, Barbara, Jacob E. Cooke, and Harold C. Syrett, eds. 1961–1987. *The Papers of Alexander Hamilton.* 27 vol. New York: Columbia Univ. Press.
Chernow, Ron. 2004. *Alexander Hamilton.* New York: Penguin Books.
Gordon, John Steele. 1997. *Hamilton's Blessing: The Extraordinary Life and Times of Our National Debt.* New York: Walker.
Lind, Michael. 1994. "Hamilton's Legacy." *Wilson Quarterly* 18, no. 3 (Summer):40–52.
White, Richard D. 2000. "Political Economy and Statesmanship: Smith, Hamilton, and the Foundation of the Commercial Republic." *Public Administration Review* 60.
Wright, Robert E. 2002. *Hamilton Unbound: Finance and the Creation of the American Republic.* Westport, CT: Greenwood.

HARGRAVES, EDWARD HAMMOND

Edward Hammond Hargraves (1816–1891) was a British gold prospector from Australia credited with the discovery of gold that launched the Australian Gold Rush on February 12, 1851.

Hargraves was born on October 7, 1816, in Gosport, Hampshire, England, to John Edward Hargraves, a local militia lieutenant, and Elizabeth Whitcombe. After attending school in Brighton and later Lewes in East Sussex, he left England in 1832 to seek his fortune in New South Wales, where he worked on a cattle ranch on a sizeable plot he acquired near Wollongong after marrying Eliza Mackie in 1836. In 1839 Hargrave sold his ranch and moved to East Gosford to work for the General Steam Navigation Company. Amid the economic contraction of the 1840s, he left Australia to seek his fortune in the California Gold Rush shortly after gold was discovered there in 1848.

Hargraves's efforts to pan for gold for over two years in California were a disappointing failure, yet the terrain reminded him of the geological position around the region of Bathurst, where he had worked running cattle for a period of months in 1835. He returned to Sydney broke in January 1851. Full of hope, however, he brought his placer **mining** pans with him and set out immediately by horseback with local guide John Lister along a tributary of the Macquarie River, an area he perceived as geographically suited for alluvial gold deposits. On February 12, the initial pans they dipped in the river all produced flecks of gold. After exploring further inland with the assistance of guide James Tom, Hargraves returned to Sydney to stake his claim to the identified goldfield on April 3.

When the discovery was confirmed in May, news of gold spread rapidly and hopeful prospectors rushed in seek of riches. Within six months, 50,000 miners were prospecting for gold in the area around Bathurst, an influx that marked the beginnings of the Australian Gold Rush, which would result in significant social and economic changes to New South Wales in a short period of time due to the development and wealth produced in the wake of Hargraves's discovery.

Hargraves was awarded an initial reward for his claim of £10,000 by the government of New South Wales and received a salaried appointment as commissioner of crown lands. Additionally, he was offered an award of £5,000 by the government in Victoria, of which he only received £2,381 due to a dispute over his claim with guides John Lister and James Tom. The claims were later resolved in his favor.

Famous for having made this historic discovery, Hargraves returned in 1854 to England, where he was introduced to Queen Victoria as the man who had discovered gold in Australia. He subsequently toured on the lecture circuit and published *Australia and Its Gold Fields* in 1855. After he returned to Australia, Hargraves explored other areas of Australia in search of additional gold deposits on behalf of the government but had little success. He died at home near Sydney on October 29, 1891, and was survived by three daughters and two sons.

See also: California Gold Rush; Mining.

Further Reading

Goodman, David. 1994. *Gold Seeking: Victoria and California in the 1850s*. Stanford, CA: Stanford Univ. Press.

Hargraves, Edward Hammond. 1855. *Australia and Its Goldfields*. London: H. Ingram.

Hodge, Brian. 2004. "Hargraves, Edward Hammond (1816–1891)." *Oxford Dictionary of National Biography*. Oxford: Oxford University Press.

HATSHEPSUT

Hatshepsut (1508–1458 BC) was an Egyptian queen of the 18th dynasty who ruled as pharaoh in her own right contrary to Egyptian custom during 1479–1458 and thus became known as "the Female King" in a period later considered ancient Egypt's Golden Age. Daughter of Thutmose I and Queen Ahmose Nefertari, Hatshepsut's ancestry was considered to be directly descended from the god Amon by virtue of her birth and therefore she and her father presented her lineage as divine royalty in the pharaonic tradition. This would prove especially important for her right to the throne as she was the only heir of full royal blood produced by her parents.

Hatshepsut married her half brother, Thutmose II, a common strategy for the consolidation of royal lineages in Egypt. The only heir produced by their union was a daughter; however, Thutmose II conceived a son, Thutmose III, with one of

Head of the queen Hatshepsut, Hatshepsut Temple, Egypt. (Dreamstime.com)

his concubines. Upon the relatively early death of her husband in 1479, Hatshepsut initially ruled as co-regent with her young stepson, yet shortly thereafter usurped his authority and asserted her blood right to rule as pharaoh in the face of strict social and religious conventions against females assuming the throne. To circumvent this, Hatshepsut waged a strategic propaganda campaign reinforcing the fact that in contrast to her stepson, who was not a full-blood descendant of the pharaonic line, she was from birth acknowledged as a direct divine descendant of the gods. She also continually portrayed her strong ties with her father, Thutmose I, and reminded her subjects in inscriptions and monuments of his pronounced support of her as heir to the throne in her own right. Hatshepsut took the additional measure of presenting or depicting herself as a male king, wearing the traditional shendyt kilt rather than the female's ankle-length robe, along with the pharaoh headdress and false beard, a symbol of royal authority.

Propagandistic portrayals of her reign were strongly reinforced by her significant achievements in her own right as ruler and her resulting popularity among the priesthood and the common people. Hapshepsut pursued the reestablishment of essential trade routes and a series of wars in **Nubia**, Syria, and the Sinai Peninsula that would yield great wealth for Egyptian royal coffers. Her most successful expedition is considered to be her quest for precious spices and access to abundant gold mines in the "Land of Punt," the precise location of which remains a mystery to scholars; however, Punt is presumed to be somewhere in East Africa based on the trade items brought back from the Egyptians and details given in scarce written sources.

The vast riches of these campaigns financed a series of building projects acknowledged over the centuries as a watershed in the development of the classical architecture of ancient Egypt's Golden Age. The construction of her mortuary temple complex on the west bank of the Nile at Deir al-Bahri near the entrance to what would later become known as the Valley of the Kings is considered one of the architectural wonders of the world due to the sophistication of symmetry, position, and scale, unprecedented for this time period. The most famous of these buildings is Hatshepsut's own mortuary temple, the Djeser Djeseru, which translates as "holy of holies," a multilevel, colonnaded, terraced temple built into the cliffs that was originally adorned with lush gardens, polished granite obelisks, and elaborate statuary. According to contemporary inscriptions and records, Hatshepsut desired that much of the complex be adorned with elements of gold to enhance its splendor. Around 20 years after her death in 1458 BC, her successor Thutmose III attempted to erase all memory of her reign by defacing engravings and written records chronicling her rule as king, likely for political reasons and to ensure the succession of Thutmose's heir Amenhotep II as king.

Although Hatshepsut built the Deir al-Bahri complex to immortalize herself in a place of worship tied to her divine connections to Amon, she made arrangements for her burial elsewhere, in a tomb identified by archeologists as KV20, likely among the first ever constructed in the Valley of the Kings, where it was built to house her father's remains. Apparently desiring to join her father in the afterlife, or perhaps continue on their crucial alliance in the eyes of the people, an additional chamber was dug within that of her father so that they could be interred together. Scholars continue to debate exactly when and why, but in the following centuries, the mummies of both Hatshepsut and her father were moved to other tombs in the Valley of the Kings. In 1903, renowned archeologist Howard Carter excavated tomb KV20 finding only two empty sarcophagi with inscriptions indicating they were intended for Thutmose I and Hatshepsut, along with scarce funerary items, but no mummies. In 1989, an American Egyptologist was excavating another burial chamber in the Valley of the Kings, KV60, that of Hatshepsut's nursemaid, where he discovered the remains of two female mummies lying on the bare ground in an unsealed tomb that had been pillaged over many centuries. Despite this unassuming final resting place outside of a sarcophagus, one of the bodies was mummified in a royal pose and with the utmost perfection. The bodies were stored for future research until 2005, when Egyptologist Zahi Hawass, director of the Egyptian Mummy Project, undertook to examine all of the known female mummies from the 18th dynasty in an effort to finally identify that of Hatshepsut. With scarce forensic evidence to go on, the regal pose struck by the mummy from KV60 led researchers to compare these remains with a tooth found in a cartouche confirmed to contain Hatshepsut's mummified liver. Based on a match with these dental remains, the mummy has been preliminarily identified as that of the female pharaoh Hatshepsut and is now displayed among other pharaohs at the Egyptian Museum in Cairo.

See also: Mining; Nubia.

Further Reading

Brown, Chip. 2009. "Hatshepsut: The King Herself." *National Geographic*, April. http://ngm.nationalgeographic.com/2009/04/hatshepsut/brown-text/1.

Davis, Theodore M. 2004. *The Tomb of Hatshopsitu*. London: Duckworth. (Orig. pub. 1906.)

Tyldesley, Joyce. 1998. *Hatchepsut: The Female Pharaoh*. New York: Penguin.

HENRY VIII

King Henry VIII (1491–1547) reigned as the King of England from his coronation on April 21, 1509, to his death on January 28, 1547, as the second king in the new Tudor dynasty. Upon his ascendance to the throne, Henry inherited a wealthy treasury from the prudent economic policies of his father, King Henry VII. Yet Henry VIII's engagement in a series of costly wars with France and Spain on the Continent, along with his lavish expenditures on an indulgent court life, quickly eroded the treasury's surplus, resulting in an economic crisis for the crown. Henry's reign was also marked by his six marriages, all of which failed to produce a surviving male heir, and his divorce from Catherine of Aragon, which provoked the English Reformation.

> "Gold? Yellow, glittering, precious gold? . . . / This yellow slave / Will knit and break religions, bless th' accursed."
>
> —William Shakespeare, *The Life of Timon of Athens*

Henry was born on June 28, 1491, at the English royal palace in Greenwich, along the Thames River just south of London, the second son of the first monarch in the Tudor lineage, King Henry VII, and Elizabeth, daughter of King Edward IV of the York Dynasty. Upon the death of his older brother, Prince Arthur, in 1502, Henry became heir to the throne. King Henry VII subsequently sought to betroth Arthur's widow and daughter of Spanish monarchs Ferdinand and Isabella, Catherine of Aragon, to his younger son, which required a papal dispensation. After several years of negotiations, Henry VIII married Catherine on June 11, 1509, shortly after his coronation as King of England on April 21, 1509, upon his father's death.

During the first parliament convened by Henry VIII as king, in January 1510, the Tudor monarch contributed to the passage of a reformed act containing a wide variety of **sumptuary legislation** titled "An Act against wearing of costly Apparrell." Sumptuary laws were a common means of social control in late medieval and Renaissance Europe intended to limit displays of wealth and power not worthy of one's social status or noble title. Thus, Henry's laws prohibited any man under the rank of lord to wear cloth of gold or silver, along with other restrictions concerning banquets, dress, and adornment. A series of revisions to the laws continued throughout Henry's reign, including a 1533 revision that restricted the use of gold chains and other gold ornaments, with a penalty of up to three months imprisonment and a fine of £10 a day.

Henry's marriage to Catherine of Aragon was controversial from the start, a situation that worsened over the years as the queen failed to produce a male heir, bearing Princess Mary as the only healthy, surviving heir. After engaging in an affair with the sister of one of the queen's ladies-in-waiting, Anne Boleyn, Henry had Lord Chancellor Cardinal Thomas Wolsey attempt to obtain dispensation from the Pope for an annulment of his marriage to Catherine on the grounds of the Levitican prohibition against marrying one's brother's widow. Yet because the papacy had already provided a special dispensation authorizing the marriage along with the fact that Catherine was Pope Clement VII's aunt and such a dispensation would further undermine Papal relations with the Holy Roman Emperor Charles V, Wolsey's requests on behalf of the king were repeatedly denied. The controversy that evolved out of Henry's petition for a divorce was ironic in a sense as Henry had always been considered to be a loyal Roman Catholic amid the turmoil of the Protestant Reformation to the degree that he was awarded the distinction of "Defender of the Faith" for his defense of the Church against Lutheranism.

Wolsey's failure to obtain the Pope's permission to nullify the king's marriage undermined the king's faith in him and with Rome's final refusal, he was replaced as Lord Chancellor by Sir Thomas More, who was a devout Catholic and did not agree with the king on the issue. More resigned in 1532 amid increasing conflict with the king over the issue of Papal supremacy. That same year, Thomas Cromwell, a member of the Royal Council and one of the king's most trusted advisors, played a key role in implementing the separation of the English church from Rome, the liquidation of monastic assets into the royal treasury, and establishment of Henry as Supreme Head of the Church. Henry divorced Catherine, provoking the Pope to excommunicate him, and married Anne Boleyn in January 1533. He would subsequently marry four more times after Anne was beheaded for treasonous adultery, most likely because she also failed to produce a male heir. His subsequent marriages were to Jane Seymour, Anne of Cleves, Catherine Howard, and Catherine Parr.

Early in his reign, Henry attempted to assert England as a great power amid the political struggles and war on the continent between France, the Holy Roman Empire, and the Papal States, engaging in a series of very costly wars with little tangible political result. Lavish displays of royal power threatened to drain the royal treasury and yet produced little tangible diplomatic results. A prime example of such display of power was the 1520 encounter and treaty negotiations with France, at what has come to be known as the "**Field of the Cloth of Gold**," which involved several weeks of sumptuous entertainment and tournament amid encampments of silken tents and royal pavilions woven with gold and precious gems in northern France.

In response to the financial pressures of decades of costly wars and vulnerability to monetary trends in Europe that threatened to undermine the value of the British coinage, Henry was forced to pursue a series of measures to raise capital for additional military campaigns and replenish his treasury, particularly as he began to lose favor among his subjects in the late 1530s and 1540s, following the execution of Sir Thomas More and many other subjects for treason, events that many came to be seen as resulting from the king's increasingly irrational paranoia. After attempts to raise money by selling off demesne lands and monastic assets, increasing taxes, and selling royal privileges did not raise sufficient revenue, Henry undertook the desperate measure of establishing a form of bimetallic standard for the British currency in July 1544. He ordered the Lord Chancellor to adjust the ratio in production of the coinage at the Royal **Tower Mint** between gold and silver coinage by decreasing the weight of the silver coinage to prevent what he perceived to be the threat of an outflow of gold coinage and **bullion** to the European continent due to the relative prices of silver. What subsequently became known as the "great debasement" over the course of the following decade is cited by historian Christopher Challis as "one of the most extraordinary, unexpected, and unjustifiable intrusions by Tudor monarchy into the commercial and financial affairs" (1967, 465) of England.

In the final years of his life, Henry suffered from a variety of severe ailments and grew increasingly erratic and agitated. He died on January 28, 1457, at Whitehall Palace, leaving his nine-year-old son, who would succeed him as Edward VI with his uncle, Edward Seymour, serving as regent, and also his daughters Mary and Elizabeth, each of whom would assume the throne following Edward's early death.

See also: Bullion; Field of the Cloth of Gold; Sumptuary Legislation; Tower Mint

Further Reading

Challis, Christopher. 1967. "The Debasement of the Coinage, 1542–51." *Economic History Review* New Series 20, no. 3 (December):441–466.

Challis, Christopher. 1973. "A Contemporary Estimate of the Production of Silver and Gold Coinage in England, 1542–56." *English Historical Review* 88, no. 346 (October):821–835.

Challis, Christopher. 1979. *The Tudor Coinage*. Manchester, UK: Manchester Univ. Press.

Gould, J. D. 1970. *The Great Debasement: Currency and the Economy in Mid-Tudor England.* Oxford: Clarendon Press.
Hacket, Francis. 1929. *Henry VIII.* New York: Horace Liveright.
Hooper, Wilfred. 1915. "The Tudor Sumptuary Laws." *English Historical Review* 30, no. 119 (July):433–445.
Letters and Papers, Foreign and Domestic, of the Reign of Henry VIII preserved in the Record Office, the British Museum, and Elsewhere in England. 1862–1932. Arranged and catalogued by James Gairdner and R. H. Brodie, 23 volumes. London: Longman, Green, Longman & Roberts.
Milnes, Nora. 1917. "Mint Records in the Reign of Henry VIII." *English Historical Review* 32, no. 126 (April):270–273.
Richardson, W. C. 1954. "Some Financial Expedients of Henry VIII." *Economic History Review* New Series 7, no. 1:33–48.

HOLY GRAIL

The Holy Grail is a legendary sacred chalice, often depicted as crafted in gold and considered to have healing, life-giving properties. The legend of the Holy Grail originated as a motif in a genre of medieval romance literature strongly influenced by the patronage of knights returning from the Crusades and contemporary themes of chivalry. As the legend evolved from the late 12th to the 13th century, the holy vessel became increasingly identified with the Eastern Christian Eucharistic liturgy of Easter Week celebrating the Passion of Christ and was purported to be the cup used by Jesus Christ at the Last Supper.

The source of the Grail myth remains a disputed subject among scholars. A wide range of theories exist, including suggestions that the tale derives from pre-Christian Celtic folklore; apocryphal or Eastern rites of the Christian liturgy celebrating the Last Supper and the Passion of Christ; heretical sects such as that of the Cathars, which are thought to be connected to the affiliation with the Grail and the Knights Templar; the persistence of rites of ancient pagan mystery cults; the legend of a union of Jesus Christ and Mary Magdalene that produced descendants; and several actual reliquaries alleged at times to be the vessel used at the Last Supper. Although the most common depictions of the Holy Grail in popular culture represent it as gold chalice in the high medieval style, such a cup was unlikely the type used by Jesus and the disciples. Archeological evidence for early Christian banquets suggests that drinking vessels were typically glass, wood, or ivory. In the literary tradition, the Holy Grail is not always represented as a gold cup and has alternately been described as a stone, a flat dish, or even a symbolic reference to a sacred element or knowledge affiliated with mystic occultism and **alchemy**.

The first known reference to the Holy Grail is in an unfinished romance, *Perceval, ou Le Conte du Graal*, written by Chrétien de Troyes of the Champagne region of France ca. 1190 for his patron, Count Philip of Flanders, a crusader knight. In this chivalric tale, the young Arthurian knight Perceval sees a procession in which a beautiful girl holds the Holy Grail, a shining chalice of the purest gold adorned with the finest of precious jewels, in the Grail Castle, which is presided over by the Fisher King. The vessel contains a single wafer, possibly intended to be a Holy Communion wafer, which magically nourishes the Fisher King's ailing

Galaad, Perceval, and Bohort bring the Holy Grail to cure the beggar in Palestine. Illumination by Evrard d'Espinques from *Queste del Saint Graal*, ca. 1470. (Getty Images)

father. Although Chrétien de Troyes died before finishing the story, it was taken up by a series of subsequent authors who added to the adventures of the knights Perceval and Gawain in a series of romances known as *The Continuations* written over the course of the next four decades.

The literary tradition of the Holy Grail evolved from this chivalric adventure with an emphasis on the quest to a more spiritual emphasis in Burgundian poet Robert de Boron's trilogy of the Grail story written for his patron, also a crusader knight, the Lord of Montfaçon, in the 1190s. It is unclear whether this work emerged independently of that of Chrétien de Troyes or whether they influenced each other. Robert de Boron is the first to suggest that the Holy Grail was the vessel used by Jesus Christ at the Last Supper and that it later came into the possession of Joseph of Arimathea, who was imprisoned for collecting the blood of Jesus in a cup when Jesus was removed from the cross. In de Boron's tale, Jesus appears to Joseph with the cup, which magically nourishes Joseph for 40 years. When he is released, Joseph's heirs take the Holy Grail to Britain, where it is protected by the Knights of the Round Table with the aid of the wizard Merlin's magic.

Between 1200 and 1210, yet another variation was composed in verse by German poet Wolfram von Eschenbach whose epic poem *Parzival*, although clearly influenced by the original version of Chrétien de Troyes, introduces many mystical elements interwoven throughout the core of the story. Perhaps most significantly, he depicts the Holy Grail as a precious stone that is guarded by specially ordained guardians referred to as the "templars."

In addition to these stories, other tales and recast versions proliferated during the early 12th century, giving rise to a distinct literary genre that was likely influenced by the culture of the Crusades and the legends and reports of the miraculous properties of early Christian relics that the crusader knights brought back with them from the Holy Land. Despite rumors that the Holy Grail existed as one such relic (among the rumors is the legend that Charlemagne was an heir of Christ and Mary Magdalene through the Merovingian bloodline and that the cup was in the possession of his heirs for centuries), there is no factual evidence for the existence of a golden chalice used at the Last Supper. The Catholic Church has remained silent on the issue throughout history.

Among the wide-reaching origins and influences of the Holy Grail legend, ultimately derived from folk and literary sources, as well as from certain apocryphal texts such as the Gnostic Gospels, certain commonalities prevail among the variety of versions. Chief among them is the conception of the Grail as a vessel that has magical, life-restoring properties and is thus able to offer sustenance and healing to those who accept it with a pure heart, or the right intentions. This theme later converged with mystical elements of **alchemy** and the affiliation between the secrets of the philosopher's stone and the practice of **alchemy** and those with knowledge of the Holy Grail in occult sects and secretive chivalric orders and brotherhoods from the Middle Ages to the modern era, including the Cathars, heretics who practiced a dualistic type of Christianity tied to Gnosticism; the Knights Templar; the Kabbalists; the Rosicrucians; and the Freemasons. Similar to **alchemy**, these sects and organizations claimed to preserve a secret, esoteric knowledge passed down to their members since antiquity and often associated with certain rites of initiation, or a quest, that resulted in the inner realization of the perfection of man and the essence of creation in which the quest represented the pursuit of a mystical union with god and the promise of eternal life. Affiliation with modern chapters of many of these cults has continued to the present day and saw a particular popularity among followers of German reformer Rudolph Steiner's theosophy movement.

Following a period of relative quiet in the Renaissance, the Holy Grail theme experienced a resurgence in the Romantic literature of the 19th century and continued as a popular literary and cultural motif in modern versions of the story such as British poet Alfred Lord Tennyson's *Idylls of the King* (1869) and German composer Richard Wagner's musical drama *Parsifal*. The tale has been a popular theme in the history of cinema, perhaps most famously in the 1975 film *Monty Python and the Holy Grail*, a comedic spoof on the Arthurian legend. Symbolic elements of the Holy Grail myth have also been explored in great detail by modern psychologists affiliated with Jungian depth psychology, notably in studies by Carl Jung's wife Emma Jung and American mythologist Joseph Campbell.

Despite the fact that in medieval literary traditions, Holy Grail quests always eventually produced the desired item, the Grail subsequently remains hidden after its discovery in certain versions, and in modern-day colloquial usage, the term "holy grail" is often used to refer to a quest or desire that is highly sought after, yet likely unattainable.

See also: Alchemy.

Further Reading

Gaster, M. 1891. "The Legend of the Holy Grail." *Folklore* 2, no. 1 (March):50–64.
Godwin, Malcolm. 1994. *The Holy Grail: Its Origins, Secrets, and Meaning Revealed.* London: Labyrinth Books.
Klenke, M. Amelia. 1955. "Chrétien's Symbolism and Cathedral Art." *PMLA* 70, no. 1 (March): 223–243.
Loomis, Roger Sherman. 1959. *The Grail: From Celtic Myth to Christian Symbol.* Cardiff: University of Wales Press.
Weston, Jesse Laidlay. 2001. *Quest of the Holy Grail.* Mineola, NY: Courier Dover.
Wood, Juliette. 2000. "The Holy Grail: From Romance Motif to Modern Genre." *Folklore* 111, no. 2 (October):169–190.

HUME, DAVID

David Hume (1711–1776) was a Scottish philosopher and historian and one of the foremost proponents of the empirical viewpoints of the Scottish Enlightenment. Although many of Hume's works met with controversy due to his skepticism and atheistic approach to natural philosophy, his writings ultimately had a profoundly significant influence on the thought of many contemporary European and British scholars and philosophers, most notably economist **Adam Smith**, who was a close friend. Although Hume's legacy largely stands on his contributions to political philosophy and ethics, his antimercantilist views and theories of trade balance and price-specie flow contributed to the foundations of modern economic theory and still form the basis of economics to a large degree.

Hume was born May 7, 1711, in Edinburgh, Scotland, son of Joseph Hume (also spelled Home), lord of Ninewells, a modest estate near Berwick-upon-Tweed in the village of Chirnside, and Katherine Falconer, daughter of Scottish politician Sir David Falconer. David was the youngest of three children, with an elder brother and sister. Their father died when David was only three years old, leaving his mother to raise them on her own.

Hume entered Edinburgh University at the precocious age of 12. Whereas the common age at that time to attend university was 14, Hume had demonstrated a level of intellect that led his mother to enroll him early. Following a classical liberal arts education, his family hoped he would pursue a degree in law, yet Hume was much more taken with the study of letters and devoted himself to becoming a scholar and philosopher.

The passion of his intellectual pursuits led to an unfortunate nervous breakdown in 1729 that took several years to recover from. During the mid-1730s, Hume sought to broaden his experiences through a variety of vocations, working initially as a merchant in Bristol, then living in seclusion from 1734 to 1737 in the small village of La Flèche in the Anjou region of France, during which he wrote *A Treatise of Human Nature*, a work that would ultimately be his most renowned, yet one that the philosopher himself would soon denounce and repudiate as not representative of his mature philosophical viewpoints.

Hume returned to England in 1737 to prepare the manuscript for publication in three parts over the course of the next several years. The work met with very modest success, an outcome that disappointed Hume, who returned to Ninewells, where he began to write *Essays, Moral and Political*, which were published in several volumes during 1740–1741. With the increasing success of these works, Hume sought an academic chair in the department of philosophy at Edinburgh in 1744 and of logic at Glasgow, yet controversial accusations surrounding his theories that cast him as a heretic or atheist prevented him from ever obtaining such a position.

Disappointed, Hume proceeded to accept a curious post near St. Albans during 1745–1746 as tutor to the Marquess of Annandale, who was reputed to suffer from mental illness. Hume subsequently worked briefly as a secretary under General James St. Clair in 1746 and 1748, when he accompanied St. Clair on diplomatic missions to the courts of Vienna and Turin. During this period, a revised edition of his first treatise was published as *Philosophical Essays concerning Human Understanding*, which was eventually reprinted as a compendium with other essays as *An Enquiry Concerning Human Understanding*. In 1751, Hume published *An Enquiry Concerning the Principles of Morals*, and throughout the 1750s, he published a series of political and religious commentaries as well as his *History of England* while also working as a librarian at Edinburgh University, a post from which he resigned in 1757 amid heated controversy over his philosophical works.

From 1763 to 1765 he worked in Paris under Ambassador to France Lord Hertford and engaged in lively exchange with French Enlightenment philosophes such as Denis Diderot and Jean-Jacques Rousseau. He continued on in his political vocation as undersecretary of state in London during 1767–1768, and then retired to his home near Edinburgh in 1769, preparing and revising his works for additional publication. Hume died on August 25, 1776, of intestinal cancer, never having been married and leaving no heirs.

Hume's economic theories of the balance of trade and price-specie flow as elaborated in his 1752 essay "On the Balance of Trade," published as part of the *Essays, Moral, Political, and Literary*, challenged contemporary British mercantilist economic theories that focused on accumulating gold for the national treasury and populace by maintaining a favorable trade balance through protectionist trade restrictions as a means of consolidating and furthering national prosperity. According to Hume, such policies produced a false and fleeting effect because the money supply and its effect on domestic prices were directly correlated to the flow of trade. Thus, as exports increased and gold accumulated, prices in the nation with a favorable trade position would increase. The influx of gold also inevitably fueled spending, which in turn naturally led to more imports because of the influence of consumer behavior on market conditions, and this in turn could potentially lead to the reverse situation, namely, an outflow of gold and a higher percentage of imports than exports. At the foundations of this theory was Hume's overall philosophy that there was a natural balance among nations due to the currency mechanism cited in his theory of the automatic flow of currency that governments did not have the capacity to interfere with in a productive manner due to the cyclical nature of the balance of trade and flow of currency and its connection to the commodity (gold) to which it was tied. Hume cited as an example the disastrous effects the extreme influx of gold from the discovery and exploration of the New World had had on the Spanish and Portuguese treasuries within a very short period of time during the 17th century.

Hume's economic theory of money also contributed to the development of monetary policy in the modern era. Specifically, he argued that wealth is not tied to money, per se. He viewed gold as it related to money simply as a commodity and acknowledged the wider contributions of labor and the "art and industry of each nation" (*Essays* II.V.11), along with factors such as climate and raw materials, needed to obtain accurate measurements of wealth. With respect to a monetary policy that sought to manipulate the money supply, Hume observed:

> I scarcely know any method of sinking money below its level, but those institutions of banks, funds, and paper-credit, which are so much practised in this kingdom. These render paper equivalent to money, circulate it throughout the whole state, make it supply the place of gold and silver, raise proportionably the price of labour and commodities, and by that means either banish a great part of those precious metals, or prevent their farther encrease. What can be more shortsighted than our reasonings on this head? (*Essays* II.V.20)

See also: Inflation; Smith, Adam; Spanish Conquest.

Further Reading

"David Hume." *Stanford Encyclopedia of Philosophy.* http://plato.stanford.edu/entries/hume/.

Heilbroner, Robert L. 1953. *The Worldly Philosophers.* New York: Simon and Schuster.

Hume, David. *Essays, Moral, Political, and Literary.* http://www.econlib.org/library/LFBooks/Hume/hmMPL28.html.

Hume, David, and Penlhum, Terence. 1992. *David Hume: An Introduction to His Philosophical System.* West Lafayette, IN: Purdue Univ. Pr.

ICONOCLASM

Iconoclasm is a theological doctrine associated with the iconoclast controversies that erupted in the Eastern Christian Church in Byzantium during the eighth and ninth centuries when various emperors and Eastern ecclesiastics opposed the worship of sacred religious images, or icons, in any form. The controversy that ensued from the iconoclastic movement, deemed a heresy by the popes in Rome, contributed greatly to the eventual schism between the Eastern Roman Empire and the West. Although classic iconoclasm is associated with these Byzantine controversies, movements against the veneration of idolatrous images had existed in organized religions since biblical times, for example, in the Old Testament story of the **Golden Calf** and this and other proscriptions against image worship found in the Koran.

In fact, such ancient suspicions about image worship may have contributed to the emergence of the Byzantine iconoclasm episodes, as there is evidence that Byzantine emperor Leo III the Isaurian (r. 717–741) was concerned that an expansion of practices of venerating sacred religious images in the empire may have been a deterrent to the successful conversion of Jews and Muslims to Christianity due to cultural objections to idolatry. During the previous two centuries, Byzantine culture had developed more elaborate artistic and politico-cultural conventions for portraying sacred religious images, typified by the mosaics and other décor in the construction of the **Hagia Sophia** by Emperor **Justinian I** in the 530s and the blossoming of Christian art in such places as Ravenna. Iconoclasts also opposed the first coin minted featuring the bust of Jesus Christ, the gold solidus, or **bezant**, issued by Justinian II (r. 685–695). Such fears were part of the emperor's larger efforts aimed at consolidating imperial power by centralizing the administration of the church in coordination with that of the state. On the advice of certain theological councilors, particularly iconoclast Constantine, Bishop of Nacolia, Leo III issued an edict in 726 that defined all manner of image worship as idolatry, which he claimed was forbidden by passages in Exodus 20: 4–5:

> You shall not make to yourself a graven thing, nor the likeness of anything that is in heaven above, or in the earth beneath, nor of those things that are in the waters under the earth. You shall not adore them, nor serve them: I am the Lord your God, mighty, jealous, visiting the iniquity of the fathers upon the children, unto the third and fourth generation of them that hate me.

Leo III thus launched a widespread campaign to destroy all religious images in churches and monasteries and persecute those who did not adhere to the edict.

Leo also attempted to have the pope enforce his edict, a demand that was met with a recriminatory reply from Pope Gregory II, who clarified the theological basis for the legality of Christian reverential traditions involving images and warned Leo to stop disrupting the Church with his violent persecutions. Leo responded by issuing a second edict that expanded his prohibitions even further. During this period, St. John of Damascus remained a vigorous defender of the Christian tradition of venerating sacred religious images, and his writings on the subject served to rally opponents of iconoclasm in the East. Nevertheless, the controversy served as a catalyst for anti-imperialist sentiment among Christians in the West who rose in protest against the edict. Protests and riots also occurred in Constantinople against the military's destruction of mosaics, frescoes, monuments, and sacred Christian relics; and the persecution of clergy and monastics, many of whom appealed to the pope in Rome for help, among them the Patriarch Germanus of the Eastern Church, whose betrayal led Leo III to promptly divest him of his position.

After the death of Pope Gregory II in 731, his successor, Pope Gregory III, continued to oppose the iconoclastic movement, convening a synod that issued an edict excommunicating all those who participated in the destruction of sacred religious images or objects and in the violent persecution of those who venerated them. Yet Leo persisted to his death in 741, making even greater attempts to enforce his policies by encroaching on territories that were the jurisdiction of the Holy See in Sicily and southern Italy. Leo's successor and son Constantine V (741–775) expanded the movement even further and escalated the level of persecutions in the empire to such a degree that widespread protests led to violent civil conflict and a campaign to topple him from the throne. Undeterred, Constantine defeated his enemies and called for a synod of bishops to meet in Constantinople, which was beholden under the repressive environment of the persecution to concede to his demands to pronounce the theological basis for the iconoclastic edicts. Under the rule of Constantine, monks were particularly violated as the emperor attempted to forbid many of the conventions of monasticism, such as prohibiting monks from wearing a habit and closing down monasteries, out of contempt for a firm adherence to traditional religious traditions in monastic communities. The battle waged in the name of iconoclasm during the reigns of both Leo III and Constantine V was also connected to the larger military battles the emperors fought against the Bulgars to the north and the encroachment of Muslim rulers to the south. Constantine, known for his military prowess and success in battle, also wielded this power and authority in his rigid domestic campaigns against the veneration of sacred images.

With the death of Constantine in 775, his son succeeded to the throne as Leo IV (r. 775–780) and the scale of the persecutions diminished. Although Leo did not reverse the policies of his father, his wife Empress Irene may have influenced his more conciliatory attitude as she remained a staunch opponent of iconoclasm. The fate of the movement was also greatly dictated by Leo's premature death at the age of 30 from a fever while leading a military campaign. Irene subsequently ruled as regent for their son Constantine VI, who was only nine years old. Irene swiftly filled the military ranks and church administration with those loyal to her and to her efforts to restore icons and abolish iconoclasm in the empire once and for all, and she tactfully transferred imperial troops or clerics affiliated with the movement to more remote territories of the empire. In 787, she convened an ecumenical church council in Nicaea, known as the "Second Council of Nicaea," with the hope of finally healing the rift caused by the controversy. The council reversed the proclamations issued at the council in Constantinople under Constantine V and set the wheels in motion to restore relations with Rome. It also provided the means by which former iconoclasts might repent and in so doing reconcile themselves with the Church.

Despite the accomplishments of the Second Council of Nicaea, significant divisions and antipathy remained between Rome and Constantinople and Christians in the East and the West over the iconoclastic issue with respect to the question of political and religious allegiances and, especially, the restitution of the precious icons and art that had been destroyed and property that had been confiscated. Only a few decades later, supporters of iconoclasm experienced a resurgence in the context of the military failures of Emperor Michael I (r. 811–813), leading to the usurpation of his throne by a military general who was crowned Leo V (r. 813–820) on the basis of his military might as well as, to a degree, his support for iconoclastic doctrines. Leo once again launched a persecution against the worship of religious images on the same scale as previous iconoclastic emperors. Adherence to the edicts pronounced at the iconoclastic synod in 754 were also enforced by the subsequent emperor, Michael II (r. 820–829). After the death of Michael's successor and son Theophilus, the Byzantine Empire was once again ruled by a woman as regent under the Empress Theodora, who pronounced in 843 that the empire would return to the provisions set forth in support of icons at the Second Council of Nicaea and that iconoclasts were to be excommunicated. On the first Sunday of the following Lent, she orchestrated a ceremony in which the icons were returned to the church for mass, a ritual reenacted to this day in the Eastern Orthodox Church's "Feast of Orthodoxy," celebrated every year on the first Sunday of Lent.

Amid the political alliances and skirmishes that contributed to the dynamics of the controversy, through the alliance with the Frankish kingdom in the West, yet another foundation was forged in the division of church and empire by East and West as the Holy See in Rome anointed and crowned Frankish ruler Charlemagne as "Emperor of Rome" in the Basilica of St. Peter in 800. Despite this new alliance in the West, however, Charlemagne and his advisors remained somewhat ambivalent on their position with respect to iconoclastic doctrines and did not immediately accept the provisions of the Second Nicaean Council.

Versions of iconoclasm emerged in the early modern and contemporary periods in such movements as the Protestant Reformation, the leaders of which referred to the same passages of the Ten Commandments cited by Byzantine iconoclasts as prohibiting idolatry, and in modern Muslim extremists sects, for example, among the Taliban, who notoriously dynamited the ancient 6th-century statues of Bamyan, carved into cliff faces in Afghanistan's Hazarajat region. Since the early 20th-century, the term "iconoclast" has also entered into common parlance to refer generally to persons or movements who radically rebuke the accepted orthodoxy, for better or for worse.

See also: Bezant; Golden Calf; Hagia Sophia; Justinian I; Mosaic.

Further Reading

Bryer, Anthony, and Judith Herrin, eds. 1977. *Iconoclasm*. Birmingham, England: Univ. of Birmingham Centre for Byzantine Studies.
Gregory, Timothy E. 2010. *A History of Byzantium*. New York: John Wiley & Sons.
Grierson, Philip. 1982. *Byzantine Coins*. Berkeley: Univ. of California Press.
Herrin, Judith. 2009. *Byzantium: The Surprising Life of a Medieval Empire*. Princeton, NJ: Princeton Univ. Press.
James, Edward. 1930. *A History of the Iconoclast Controversy*. New York: Macmillan.
Oddy, Andrew, and Susan La Niece. 1986. "Byzantine Gold Coins and Jewellery: A Study of Gold Contents." *Gold Bulletin* 19, no. 1:19–27.

ILLUMINATED MANUSCRIPTS

Illuminated manuscripts were a product of decorative arts techniques in bookmaking associated primarily with the period 300–1600 in Judeo-Christian and Islamic cultures, and especially with monastic book production in the scriptoria of western Europe throughout the Middle Ages, from the 8th to the 15th century. During this period, advancements in the methods and materials used in book production along with an increasing regard for the sanctity and authority of the written word gave rise to a craft in which books were decorated with illustrations and elaborate calligraphic letters using colored inks made of precious metals like gold or other scarce elements such as lapis lazuli. This practice is known as "**chrysography**," the literal meaning of which is "writing in gold." In the High Middle Ages, **gilding** with gold leaf directly onto the page was another popular technique. Gold was thus used in manuscripts as a reflection of the book owner's wealth and status and was employed by artists in two ways, by grinding the metal into a powder to be used as an ink or by hammering gold leaf onto a prepared adhesive base and then embellishing or tooling the gold to create the desired effect.

March: Peasants at Work on a Feudal Estate, from the *Tres Riches Heures du Duc de Berry*. (Victoria & Albert Museum, London, UK /The Bridgeman Art Library International)

The earliest known examples of illuminated manuscripts have been identified in various archeological and written sources, but little exists before the Christian era. Although scarce sources indicate the existence of such practices in early antiquity, for example, a reference in *Pseudo-Aristeas* to an edition of the Hebrew Law Scrolls produced entirely in gold ink as a gift for Ptolemy I in the second century BC, the tradition did not emerge fully until late antiquity for two primary reasons: First, books were primarily produced as scrolls on papyrus, a more lightweight and fibrous surface that was not ideal for the application of heavier inks or decoration. Second, the culture of the written word was largely secular, residing in the realm of the intelligentsia or bureaucracy and did not merit the displays of wealth inherent in **gilding** and illumination but rather were deliberately modest and functional.

An ascendancy of the culture of the written word ran parallel to the institutionalization of the Catholic Church and the essential place of biblical scripture in that process. In addition to the increasing use of ornate books on public display for parishioners during the liturgy and for personal spiritual devotion or reflection, a shift in book production from papyrus scrolls to bound codices with pages of vellum, a more durable surface produced from the skins of sheep or calves, contributed to an environment in which the art of illumination would flourish. As Holy Scripture and the theology surrounding it assumed greater importance in the development of the Catholic Church during late antiquity and the early Middle Ages, the use of gold as embellishment for sacred texts was employed to symbolically reinforce the supremacy of the Church as well as the value, or "wealth," and eternal and unmalleable force of its teachings. Not every early Christian theologian agreed with the use of ostentatious display in asserting the power of the Church, a position also reflected in the **iconoclasm** controversy. In his *Epistles*, St. Jerome questions the intent behind the use of such decorative arts in manuscript production:

> Parchments are dyed purple, gold is melted into lettering, manuscripts are decked with jewels, while Christ lies at the door naked and dying.
>
> Let her treasures be not gems or silks, but manuscripts of the holy Scriptures; and in these let her think less of gilding and Babylonian parchment and arabesque patterns, than of correctness and accurate punctuation. (Epistle cvii.12)

Prior to the 13th century, illuminated manuscripts were produced largely by scribes working in monasteries to produce decorative liturgical books that were employed by the clergy during the mass as powerful symbols that contributed to ceremonial display and a sense of the mystery of the written word transmitted through the authority of the clergy. With the rise of universities in certain urban centers of scholasticism in Europe, however, there is evidence for an increase in literacy among lay scholars accompanied by more secular decorative book production among lay scribes and artists plying their craft in the towns and cities and forming guilds of skilled craftsmen. Two important manuals for illuminators exist from this period, an anonymous 14th-century manuscript of unknown origins, the *De Arte Illuminandi*, apparently produced in Italy, which provides a basic description of the techniques and materials employed in the production of manuscript **gilding** and chrysography, and 15th-century Florentine artist Cennino d'Andrea Cennini's *Libro dell'Arte*, a detailed treatise on a variety of decorative arts that provides essential information about the processes employed in the ancient practice of illumination.

Such manuals reveal the recipes and techniques for applying gold to manuscript pages. Whereas gold ink was created from specific amalgams of substances to create the desired technique directly on the prepared page, **gilding** involved preparing the parchment with an outline of the design and an undercoating of gesso along with a base pigment. A form of adhesive would then be applied to the vellum wherever the **gilding** would be placed, then left to dry. Immediately prior to the application of gold leaf, the adhesive surface would be very lightly moistened to activate it. Once the sheet of gold leaf is adhered to the page, it is burnished to produce a sheen. The scribe would then complete the desired decorative finish using a variety of methods such as tooling or punchmarking.

Yet again, the fate of illuminated manuscripts paralleled the history of book production: the introduction of printing with movable type forever changed the course of the transmission and use of the written word. The effects of the printing press on manuscript production in Europe were not immediate, however, and, ironically, this period saw an escalation in the demand and quality of illuminated manuscripts, perhaps because the cultural emphasis on this form of book, so familiar and treasured among its contemporaries, superseded the convenience of new printing technologies. By the early 16th century, the production of print books took off, leaving the art of illuminated manuscript to the realm of the decorative artists who have continued to produce fine examples of gilded books using timeless techniques to the present day.

Carefully preserved in monasteries, university libraries, and private collections of the nobility and aristocracy, exquisite examples of illuminated manuscripts are now conserved in national museums and archives. Among the most significant are the Godescalc Gospels at the Bibliothèque Nationale in Paris, a book of Gospels produced for Charlemagne ca. 781–783 by Frankish scribe Godescalc in the scriptorium at Aachen, birthplace of the Carolingian Renaissance and the miniscule style of calligraphic script. The **Gospel Books of Saint Médard de Soissons**, an early-9th-century illuminated manuscript conserved at the Bibliothéque Nationale in Paris, is considered to be one of the best examples of the Ada School illuminated books produced by the Ada School scribes and illuminators at Aachen. The Lindisfarne Gospels, conserved at the British Library, London, thought to have been created for Eadfrith, Bishop of Lindisfarne during 698–721, are an example of the Hiberno-Saxon style, a combination of Celtic, Byzantine, and classical elements. The *Très Riches Heures*, a sumptuous illuminated book of hours produced in France for the Duc de Berry in the early 15th century, considered the most famous example of high medieval illuminated books of hours, is conserved at the Musée Condé in Chantilly, France.

See also: Chrysography; Gilding; Gospel Books of Saint Médard de Soissons; Iconoclasm.

Further Reading

Alexander, Jonathan J. G. 1992. *Medieval Illuminators and Their Methods of Work*. New Haven, CT: Yale Univ. Press.

Calkins, Robert G. 1983. *Illuminated Books of the Middle Ages.* Ithaca, NY: Cornell Univ. Press.
De Hamel, Christopher. 1992. *Scribes and Illuminators.* London: British Museum Press.
Dodwell, C. R. 1993. *The Pictorial Arts of the West, 800–1200.* Pelican History of Art. New Haven, CT, and London: Yale Univ. Press.
Pächt, Otto. 1986. *Book Illumination in the Middle Ages.* Oxford: Oxford Univ. Press.
Whitley, Kathleen P. 2000. *The History and Technique of Manuscript Gilding.* New Castle, DE: Oak Knoll Press.

INCA

The Inca were an ethnic South American Indian peoples originating from an ancient tribal group in the Andean highlands who, by the 12th century, developed into a more organized civilization with a capital established by the legendary founder of the Inca dynasty, Maco Capac. He settled his tribe in the area around Cusco, from whence the development of an elaborate dynasty and highly stratified aristocracy, along with advanced technologies, architecture, infrastructure, and agriculture, facilitated the expansion and consolidation of the empire over the course of the next 400 years to stretch from the northern point of present-day Ecuador to southern Chile, flanked by the Andean mountain range and the Pacific Ocean. Archeological evidence indicates that the ancient Inca civilization developed advanced metalworking techniques as early as 3000 BC to produce precious objects in gold and silver for use in religious and royal ceremonies.

> "Treasure, gold in the earth, may easily overcome any man, hide it who will!"
> —*Beowulf*

Maco Capac was the first ruler known as Sapa Inca, a dynastic title designating the supreme ruler who was considered to be the son of the primary Inca deity, the sun god Inti. The title of Sapa Inca was passed father to son and as the empire expanded through the conquest of neighboring territories, this supreme ruler, acting as a type of religious, political, and military leader, commanded absolute allegiance from his subjects. An elaborate hierarchical social structure developed to support the power and authority of the emperor, with the royal family enjoying the privileges of an aristocratic class and levels of regional governors comprised of tribal heads from conquered ethnic groups in the empire and family clan heads leading groups within each tribe. The majority of the Inca population consisted of common farmers who were organized into communal groups, also with a designated leader, who helped to ensure that the needs of the group were met and also oversaw the submission of the *mita*, a form of tribute to the empire rendered as labor for military service, public works projects, or government farms.

The Inca developed a sophisticated network of roads, remarkable for their level of engineering, which helped centralize distribution of goods and internal migrations throughout the empire. They also demonstrated advanced engineering in the clearing and cultivation of farmlands on terraced plots in the Andes, as well in the

Inca gold headdress from the Museum of Gold in Lima, Peru. (Getty Images)

construction of elaborate temples and royal complexes with remarkable architectural elements and durable construction techniques, often in difficult terrain, such as the position of the Machu Picchu complex at 7,970 feet above sea level perched between two mountain peaks. At the time of contact, Spanish conquistadors marveled at the advanced metallurgical techniques among the Inca in their silver and gold craft. Gold- and silversmithing largely served a religious purpose, and the Inca considered precious metals as sacred gifts representing the sweat of the sun and the tears of the moon, as tangible elements of the divine that were also endowed with certain mystical, life-giving properties. Only the ruling class was permitted to adorn themselves with gold or own precious gold or silver objects, and elaborate rituals and restrictions governed the extraction of gold and silver from the region's abundant sources.

Gold artifacts served the elaborate state religion and were often used as funerary objects, symbolic totem pieces in religious ritual, or offered in sacrifice or tribute to the gods. The Inca religion combined worship of one powerful god, the sun god, who was directly linked to the royal family, with the worship of other nature gods and included rituals of human and animal sacrifice. History and religious myths were largely passed on through oral tradition, as the Inca civilization does not appear to have had a formal writing system.

The Inca Empire came to an abrupt end in the early decades of the **Spanish Conquest** when conquistador **Francisco Pizarro** ambushed the last reigning emperor **Atahualpa** in 1532 while he was stationed with his army in Cajamarca, in the midst of an Incan civil war over the royal succession with Atahualpa's brother Huáscar. A variety of elements in the structure and geography of the empire are considered to have contributed to its swift fall to the Spanish, including the Inca

Empire's advance road system, which facilitated the Spanish conquistadors' advances inland, and regional administrative districts, which governed a variety of ethnic groups subjected over time to Incan conquests and forms of tribute, a situation that provided opportunities for the Spanish to ally with certain native populations. The civilization of the Inca persists today in the culture of their descendants, the Quechua-speaking peasants who live on rural farms in the Andes, an ethnic group that represents around 45 percent of the total population of present-day Peru.

See also: Atahualpa; Pizarro, Francisco; Pre-Columbian Gold; Spanish Conquest.

Further Reading

Emmerich, Andre. 1965. *Sweat of the Sun and Tears of the Moon: Gold and Silver in Pre-Columbian Art*. Seattle: Univ. of Washington Press.
Hemming, John. 1970. *The Conquest of the Incas*. San Diego, CA: Harcourt and Brace.
Jones, Julie, and Heidi King. 2002. "Gold of the Americas." *Metropolitan Museum of Art Bulletin* 59, no. 4 (Spring).
Mackenzie, Donald Alexander. 1978. *Myths of Pre-Columbian America*. Boston, MA: Longwood Press.
Malpass, Michael A. 1996. *Daily Life in the Inca Empire*. Westport, CT: Greenwood Press.
Prescott, William H. 1850. *The Conquest of Peru with a Preliminary View of the Civilization of the Incas*. London: Oxford Univ. Press.
Winthrop, Samuel K., et al., eds. 1961. *Essays in Pre-Columbian Art and Archaeology*. Cambridge, MA: Harvard Univ. Press.

INFLATION

Inflation is a term used in the field of economics to refer to a short- or long-term sustained rise in overall price levels, a phenomenon that is generally considered to be the result of an increase in the money supply in conjunction with other macroeconomic or at times political factors.

There are presently four general economic theories used to analyze and predict price levels with a view to controlling monetary policy to maintain a desirable level of inflation targeted typically within the range of 1–3 percent.

1. The earliest theory of inflation originated with the work of **David Hume**, whose quantity theory suggests that price levels are affected by the money supply. This approach was perceived as problematic in the 1920s–1930s, however, as it was only accurate in the context of a full productive capacity. During the 1950s–1960s, Milton Friedman and other Chicago School economists revised Hume's theory to account for the velocity of circulation of money in short- and long-term cycles in a manner in which analysis of long-term price levels according to quantity theory was considered to be a useful tool for monetary policy.
2. The Keynesian theory of inflation developed by British economic advisor **John Maynard Keynes** calibrates inflation based on changes in consumer behavior relative to income levels, which accounts for the influence of the velocity of circulation of the money supply relative to interest rates and identifies data relative to a resulting inflationary gap. Keynes's theory helped policymakers to gauge how government spending as a category contributed to inflation dynamics yet was problematic in certain

contexts in which an inflationary gap did not exist, for example, in the case of persistent inflation during the 1950s and 1960s.
3. What is referred to as the "cost-push theory" evaluates inflationary dynamics in terms of the relationship between price levels and the cost of goods and the relative levels of money supply based on demand. This theory includes the concept of a "price-wage spiral," a phenomenon that is considered to be a result of inflation as wages are pressured to adjust to meet rising price levels, and vice versa, in a potentially harmful cycle that contributes to persistent inflation.
4. Structural theory of inflation combines various elements of all of these theories to analyze inflation rates relative to several elements of potential structural shifts in the overall economy, such as the balance of imports versus exports, money supply, or political phenomena, with a particular emphasis on the effects of increases in the rate of wages.

Central bankers and political economic advisors and analysts use such theories to avoid the potentially devastating effects of persistent inflation through monetary policy that controls the money supply by adjusting interest rates through the **Federal Reserve System** as well as through targeted government spending. Annual inflation rates are monitored and reported by government offices such as the United States Department of Labor Bureau of Statistics Consumer Price Index, which uses a wide range of price indexes to compute annualized growth percentiles.

The problem of inflation is considered a relatively recent one due to the fact that fixed exchange rate currencies tied to gold or silver were not as subject to inflationary patterns because of the limited parameters of currency supply and demand in such a system. During the 16th and early 17th centuries, however, Europe experienced persistent inflation on such a vast scale that the episode came to be referred to as the "**Price Revolution.**" Monetarist economists originally linked the inflationary tendencies of this period to the influx of gold and silver from the **Spanish Conquest** in Latin America, yet this theory is somewhat problematic because inflation began during the 1520s, before significant quantities of these precious metals had begun to arrive in Spain. Alternative explanations have included demographic factors such as population increase and urbanization or sociocultural trends, including the emergence of capitalist enterprises.

Widespread inflation also occurred in the 1950s and 1960s to such a degree that President Richard Nixon effectively ended the **Bretton Woods System** in 1971 by nullifying the convertibility of the dollar to gold in an effort to stem the tide of inflation caused by a general devaluation of the dollar and national deficits. The Oil Shock crisis of the remainder of the 1970s and other political factors contributed to a severe prolonged period of inflation through the 1980s. In the latter half of the 20th century, certain developing nations have been particularly vulnerable to sustained inflation, with around 18 countries with lesser-developed economies in regions such as Africa and Latin America experiencing the phenomenon of hyperinflation in which inflation rates escalate to levels as high as 50 percent per month. Since the 1990s, the inflation rate in the United States has remained relatively stable at levels between 1.5 and 4 percent, with the exception of certain brief reactions up or down to political or economic crises, such as Hurricane Katrina in 2005 and the onset of the subprime mortgage crisis in 2007.

See also: Banking and Credit; Bretton Woods System; Federal Reserve System; Gold Standard; Hume, David; Keynes, John Maynard; Price Revolution; Spanish Conquest.

Further Reading

Barsky, Robert, and Bradford DeLong. 1991. "Forecast Pre-World War I Inflation: The Fisher Effect and the Gold Standard." *Quarterly Journal of Economics* 106:815–836.
Bernholz, Peter. 2003. *Monetary Regimes and Inflation*. Cheltenham, UK: Edward Elgar.
Collard, Fabrice, and Harris Dellas. 2007. "The Great Inflation of the 1970's." *Journal of Money, Credit, and Banking* 39, no. 2/3 (March–April):713–731.
Friedman, Milton. 1992. *Money Mischief: Episodes in Monetary History*. New York: Harcourt Brace Jovanovich.
McCulloch, J. Huston. 1982. *Money and Inflation: A Monetarist Approach*, 2nd ed. New York: Academic Press.
Schwartz, Anna. 1973. "Secular Price Change in Historical Perspective" *Journal of Money, Credit, and Banking* 5:243–279.
United States Department of Labor, Bureau of Labor Statistics. "Consumer Price Index." http://www.bls.gov/cpi/.

INTERNATIONAL MONETARY FUND

The International Monetary Fund (IMF) is a regulating financing agency created at the Bretton Woods Conference in 1944 as an office of the United Nations to function as a multilateral financial institution overseeing the international monetary system with the purpose of ensuring global economic stability in the absence of an international **gold standard** through securing monetary cooperation among member nations, monitoring exchange rates, and promoting liquidity in lesser-developed economies.

The economic crisis of the Great Depression and the military conflicts of World War I and World War II put a wrench in the international monetary cooperation that had developed during the era of the classical **gold standard** in 1880–1914 as nations abandoned the **gold standard** and resorted to protectionist and deflationary trade and monetary policies that only exacerbated the contraction of global economies. Economists such as **John Maynard Keynes** argued for the establishment of a global financial entity that would manage monetary cooperation and regulate foreign currency exchange in a manner that contributed to global economic growth and stability.

During the Bretton Woods Conference in Bretton Woods, New Hampshire, in July 1944, representatives from 45 nations participated in the economic summit that created the foundations for the establishment of the IMF at the end of the war in a draft of the Articles of Agreement for the structure and operation of the multinational organization. These Articles of Agreement stipulated par values to be applied to all participating members based on the gold value of the U.S. dollar in July 1944 of 0.88867 grams of fine gold. Article I stated the organization's purposes as:

(i) To promote international monetary cooperation through a permanent institution which provides the machinery for consultation and collaboration on international monetary problems.
(ii) To facilitate the expansion and balanced growth of international trade, and to contribute thereby to the promotion and maintenance of high levels of employment and real income and to the development of the productive resources of all members as primary objectives of economic policy.
(iii) To promote exchange stability, to maintain orderly exchange arrangements among members, and to avoid competitive exchange depreciation.
(iv) To assist in the establishment of a multilateral system of payments in respect of current transactions between members and in the elimination of foreign exchange restrictions which hamper the growth of world trade.
(v) To give confidence to members by making the general resources of the Fund temporarily available to them under adequate safeguards, thus providing them with opportunity to correct maladjustments in their balance of payments without resorting to measures destructive of national or international prosperity.
(vi) In accordance with the above, to shorten the duration and lessen the degree of disequilibrium in the international balances of payments of members. (http://www.imf.org/external/pubs/ft/aa/aa01.htm)

The articles were signed and ratified by representatives from the 29 initial member nations on December 27, 1945, and a committee convened in Savannah, Georgia, in 1946 to appoint a board of directors, elect executive directors, and create bylaws. It was decided that the organization would establish its headquarters in Washington, D.C., and officially begin to operate in March 1947. **Exchange rates** were set according to the par value system, in which currencies were pegged to the U.S. dollar, which in turn was valued at a gold equivalent, an arrangement known as the **Bretton Woods System**. In 1971, President Richard Nixon suspended the convertibility of the dollar to gold amid international concerns that the dollar was overvalued. Member nations subsequently converted to various forms of floating exchange rates largely pegged to the U.S. dollar and monitored by the IMF.

With the expansion of IMF membership to include many African and other lesser-developed nations in the 1970s and 1980s, the organization established protocols and specific funds with concessionary lending standards intended to provide liquidity to the poorest nations with the purpose of stimulating global economic growth and promoting financial and political stability through monetary policy. The fund has since undergone a series of additional reforms affecting both governance and mandate that intended to address changing economic conditions due to geopolitical events, such as the oil shocks and attendant debt crisis of the late 1970s; and the breakup of the Soviet Union. The fund has sought to facilitate the international flow of capital in the era of globalization since 2006 and has adjusted its lending policies to assist with the recovery efforts following the devastating consequences of the global economic crisis resulting from collapse of the U.S. mortgage market in 2007.

Membership in the IMF expanded throughout the course of the latter 20th century and has grown to include a total of 186 member nations by the present day. Its board of governors is comprised of national finance ministers or central bank executives and 24 executive directors from representing member nations who oversee the organization's daily operations. Each member is required to submit a quota on a sliding scale relative to the size of its economy comprised of a percentage of both domestic currency and gold. The level of a member nation's quota corresponds to an equivalent level of voting power. The United States currently contributes the largest quota to the IMF and maintains the rights to 18 percent of the total votes, with other industrialized nations holding at least 50 percent of the total votes. In 2006, the IMF initiated a reform program of the allocation of voting rights according to quota intended to disperse the votes more widely among its members and support nations identified as having "dynamic emerging markets" and to provide low-income nations with a greater stake in the organization.

The policies and interventions of the IMF in global markets since the 1990s have generated a high level of criticism among some economists who see its lending practices, quotas, and conditions as potentially disruptive to free market factors that realign global inequities. Some economists argue that providing safety nets and bailouts to member nations may exacerbate cyclical and circumstantial economic crises that, if the IMF does not intervene to stimulate growth through monetary policies, may be avoided through domestic reform of banking and finance designed to minimize risk and encourage productive economic growth.

See also: Banking and Credit; Bretton Woods System; Exchange Rates; Gold Standard; Inflation.

Golden Opportunities for Developing Countries

According to a May 25, 2005, report by the World Gold Council titled "A Touch of Gold: Gold Mining's Importance to Lower-Income Countries," gold has become a crucial export product among highly indebted developing countries, who saw an average of an 84 percent increase in total gold production from 1999 to 2004. By 2004, developing countries were responsible for 72 percent of total global output of gold. Gold has become the primary export by significant percentages in such nations as Mali, Tanzania, Ghana, Guyana, and Guinea, where the gold mining industry has boosted not only export revenue but also improvements in the financial, social, and physical infrastructure of these developing nations. As a result of the study, the International Monetary Fund (IMF) retracted its plans in 2005 to liquidate a portion of its gold reserves as a means of offering more debt relief to such nations because such a measure would have had the contrary result of potentially reducing the market price of gold and therefore reducing the export revenue of the very nations the IMF sought to assist. According to chief executive officer of the World Gold Council James Burton, "The gold industry is of tremendous and growing importance to highly indebted countries and in many cases is directly assisting with the growth of both the industrial infrastructure and a skilled workforce. As a result, the future wealth and development of the world's poorest nations is more dependant than ever on a stable international market for gold."

Source: World Gold Council, http://www.gold.org/media/publications/start/80.

Further Reading

Bordo, M. D. and B. Eichengreen, eds. 1993. *A Retrospective on the Bretton Woods System: Lessons for International Monetary Reform.* Chicago: Univ. of Chicago Press.

Danaher, Kevin, ed. 1994. *50 Years Is Enough: The Case Against the World Bank and the International Monetary Fund.* Cambridge, MA: South End Press.

Driscoll, David D. 1997. *What Is the International Monetary Fund?* Washington, DC: International Monetary Fund.

Eichengreen, B. 1996. *Globalizing Capital: A History of the International Monetary System.* Princeton, NJ: Princeton Univ. Press.

International Monetary Fund. http://www.imf.org/external/.

McQuillan, Lawrence J., and Peter C. Montgomery, eds. 1999. *The International Monetary Fund—Financial Medic to the World?: A Primer on Mission, Operations, and Public Policy Issues.* Chicago: Hoover Institution Press.

ISABELLA I OF CASTILE

Isabella I (1474–1504) was a Spanish queen who ruled as Isabella I of Castile from 1474 to 1504 and of Aragon during 1479–1504, a period when both kingdoms were ruled jointly by Isabella and her husband Ferdinand II of Aragon during a marriage that reunited Spain and established the foundations for its ascendancy as one of Europe's greatest empires of the 16th century following the monarchs' support of **Christopher Columbus**'s expedition to the New World and the significant wealth in gold and other valuable commodities obtained through the **Spanish Conquest** of Central and South America.

Isabella was born on April 22, 1451, in Madrigal de las Altas Torres, Castile, daughter of King John II and Isabella of Portugal, the king's second wife. When Isabella was only three years old, her half brother Henry IV assumed the throne amid escalating political strife among the Castilian nobility. With the death of Isabella's younger brother Alfonso in 1568, civil war broke out among the Spanish magnates over the question of who would succeed to the throne. To placate the dissenting party, Henry designated Isabella as his heir on

Queen Isabella I, co-ruler of Castile and Aragon. (BiographicalImages.com)

September 19, 1468, in the agreement known as the Accord of Toros de Guisando. The question of Isabella's marriage thus became a very political one, and Henry favored a union with King Alfonso V of Portugal. Against Henry's wishes, however, Isabella married **Ferdinand II of Aragon** in October 1469. In the aftermath of Henry's death in 1474, the War of the Castilian Succession broke out among disputing factions over the competing claims of Isabella and Henry's daughter Juana la Beltraneja. In 1479, Isabella emerged victorious from years of civil war. That same year, Ferdinand ascended to the throne of Aragon upon the death of his father, and the kingdoms of Castile and Aragon were thenceforth united under the rule of both monarchs.

Together Ferdinand and Isabella undertook ambitious reforms and military campaigns to strengthen the Spanish Crown and territories: finalizing the Reconquest with the fall of Granada in 1492; utilizing the Spanish Inquisition as an instrument of social control and reinforcement of royal authority relative to the papacy; and supporting Christopher Columbus's historic voyage in search of a westward route to Asia in 1492, when the explorer discovered the New World and claimed new territories rich with gold, silver, and other precious resources in the name of the Spanish Crown. Their marriage also produced five surviving children who were strategically married to forge essential alliances for the monarchy: Isabella was married to Alfonso of Portugal and then remarried to Manuel I of Portugal. Juan was married to Princess Margaret of Austria. Joan married Philip, son of Habsburg Holy Roman Emperor Maximilian I. Maria married her sister Isabella's widower Manuel I, and Catherine married Arthur, Prince of Wales, and later his brother English king **Henry VIII**.

Throughout her reign Isabella was known for her piety and alliance with the papacy, for which she and Ferdinand were commonly referred to as the "Catholic monarchs." She also held a reputation as a shrewd and calculating ruler who wielded her authority with the utmost grace. The empire she had built seemed at risk, however, when she chose her daughter Joan, known as "Joan the Mad" due to mental illness, as her successor. Yet ultimately, the marriage between Joan and Philip I produced the future Holy Roman Emperor Charles V who would come to inherit the full legacy of Isabella's reign as the head of one of the world's largest and most powerful empires.

See also: Columbus, Christopher; Ferdinand II of Aragon; Henry VIII; Spanish Conquest.

Further Reading

Edwards, John. 2004. *Ferdinand and Isabella.* New York: Longman.

Liss, Peggy K. 1992. *Isabel the Queen: Life and Times.* Oxford and New York: Oxford Univ. Press.

Prescott, William H. 2003. *History of the Reign of Ferdinand and Isabella the Catholic.* Honolulu, HI: Univ. Press of the Pacific.

U.S. Library of Congress. "Spain, the Golden Years: Ferdinand and Isabella." *Country Studies Series.* http://countrystudies.us/spain/7.htm.

JEWELRY

Jewelry is a type of ornamentation for the body or clothing worn by men, women, and children that has existed since prehistory in most known human societies in a wide variety of forms, including hair ornaments such as hairpins, headdresses, and crowns; fasteners for clothing such as pins, belt buckles, and brooches; or for the body, such as rings, necklaces, bracelets, and earrings. Jewelry is made of many different materials depending on function, supply, and fashion; however, a gold or precious metal base with an inlay or setting of precious jewels such as pearls or gemstones has endured as among the most coveted form of the craft. Jewelry fulfills a range of functions in human society: as an indication of social status; as amulets to ward off evil spirits or invoke good fortune; as a demonstration of affiliation with a certain social, ethnic, or religious group or organization; for practical purposes such as fastening clothing or as a container for cosmetics or a portrait of a loved one; as a stamp for a wax seal; and as a sign of marital status or royalty in the case of wedding rings and **royal crowns** and crown jewels.

Archeological excavations have uncovered evidence of jewelry dating to as early as the Paleolithic period in Africa and the Middle East, consisting largely of wooden or stone beads, yet the earliest known origins of the development of gold jewelry date to the ancient Near Eastern civilizations of Egypt and Sumer as well as among pre-Columbian civilizations, during the fifth–third millennia BC. Jewelry uncovered from the tomb of Queen Puabi at the **Royal Tombs of Ur** from ca. 2900–2300 BC demonstrates a high level of craftsmanship and abundance of gold ornaments such as ornate beads, an early form of **filigree**, and cloisonné onto a gold base. Jewelry uncovered from the tombs of the Egyptian pharaohs sheds light on trade patterns, social and political conditions, and religious practices from the earliest dynasties ca. 3000 BC to the legendary riches and opulent jewels and amulets of New Kingdom pharaohs such as **Hatshepsut** and **Tutankhamun**, most famously perhaps the scarab, an Egyptian amulet signifying creation, rebirth, and power. In addition to remaining artifacts from the ancient Near East, written sources such as biblical texts, dowries, household and temple inventories, records of gifts and tribute, and art offer a range of details about the function, manufacture, and exchange of items of jewelry in early antiquity.

Gold jewelry also developed as a more sophisticated craft in the second millennium BC in China and India, where many ancient jewelry-making traditions still thrive and adornment of women and girls with gold jewelry remains a widespread custom. Throughout Indian history it was common for people of a variety of social levels to acquire jewelry as a means of consolidating and passing on wealth.

Throughout the ages, elaborate metalworking techniques developed to suit the tastes of royal courts, such as that of the Mughal Empire.

> "It feels almost soft, like something to be caressed. Only gold feels that way."
> —WILLIAM FAULKNER, *Land of the Pharaohs*

Despite a relative scarcity of evidence due to the fact that such a large percentage of goldwork was melted down by the conquering Spanish, the sophisticated and advanced techniques of the extant goldwork of pre-Columbian cultures of South America and select archeological evidence for early **mining** and **goldsmithing** have given scholars enough information to suggest that indigenous cultures such as the Inca, Mayans, and Aztecs had developed as early as 3000 BC among the Inca and then spread north and to the coast with increased contact over the following millennia, achieving technological advancements such as soldering, lost-wax **casting**, and types of **filigree** as early as 200 BC–AD 100.

During the eighth century BC, the ancient Etruscan civilization of present-day Italy achieved a highly advanced level of technical expertise and artistic genius. Etruscan tombs and archeological sites have provided scholars with a wealth of examples of fine jewelry that sheds light on a civilization whose origins still remain largely unknown. What is known as the Etruscan style demonstrates elements of motifs and methods that indicate these early inhabitants of the Italian peninsula had trade contact with cultures in Syria, Phoenicia, and the Greek world. Certain extant pieces of Etruscan jewelry were executed with such a high level of **goldsmithing** that modern craftsmen struggle to duplicate them. One of the most signature techniques employed in Etruscan jewelry was the method of **granulation**, which evolved into a more advanced process known as **filigree**.

The history of jewelry in ancient Greece similarly parallels the progression of cultural contact between Greece and the Near East during the reigns of Philip II of Macedonia and his son Alexander the Great in the mid-fourth century BC, as they expanded toward Persia and experienced both a greater supply of gold and advanced levels of technique and artistry. While there is significant evidence for the existence of gold jewelry dating back to the Mycenaean and earliest Greek civilizations, particularly references in Homer to the existence of artifacts from these earlier eras, the sixth–fourth centuries BC saw a significant development in gold prospecting and production, trade, and jewelry techniques. With Alexander's conquests of areas of Egypt, Syria, Asia Minor, and Persia, new motifs and styles were incorporated into Greek jewelry that would later influence the Roman style.

Roman traditions for personal adornment evolved from a more conservative approach to sumptuous display in a society in which Roman **sumptuary legislation** prohibited and restricted men and women of certain social levels from wearing lavish gold jewelry. With the expansion of the Roman Empire and increasing contact with Near Eastern cultures, Roman jewelry developed into a sophisticated art with strong influences from Etruscan, Hellenistic, and Eastern techniques and styles, one that can be found widely in excavations throughout the greater Roman sphere of influence in Europe and the Mediterranean Basin. Depictions of Romans wearing jewelry exist in frescoes recovered from ancient sites such as Pompeii and Herculaneum, and Romans also practiced the ancient tradition of burying the dead with their jewels, which is the source of most present-day examples of ancient Roman jewelry.

As the Roman era entered a period of decline, the craftsmanship in the eastern imperial territories of Byzantium evolved into one of the most famous and recognized styles of gold jewelry in late antiquity, reflecting the convergence of Roman, Greek, and Eastern influences along with the development of new techniques and styles that included elements such as repoussé, cloisonné, and enamel.

With the collapse of the Roman Empire, jewelry making saw yet another convergence of East and West during the medieval era as conquering nomadic chieftains from Central Asia swept in to conquer areas of Europe. As the pattern and migration of the barbarian tribes settled across former Roman regions, distinct styles emerged from the respective cultures of the Franks, Langobards, Celts, and Visigoths, with strong influences from Byzantine styles and elements from Eastern traditions as far as India due to the expansion of trade contacts in this period. During the Middle Ages, jewelry was largely worn by nobility and royalty for adornment, and members of the Church hierarchy for identification or ritualistic purposes; or, if worn among lesser classes, for a practical or amuletic function such as protection by a saint or a pendant containing herbs to ward off evil or prevent or curtail health conditions.

With an influx of gold on the Continent from voyages of discovery and exploration in Africa and the Americas, **goldsmithing** and the craft of jewelry making experienced a dramatic resurgence beginning in the late 15th century and continued with the emergence of the Renaissance period, in which a renewed interest in the styles and forms of classical antiquity and examples of goldwork from the Far East and South America elevated the creations produced in master metalworkers workshops to new levels of artistry. With the support of wealthy patrons from both the merchant class and among the nobility, famous Renaissance artists such as Benvenuto Cellini, Rosso Fiorentino, and Albrecht Dürer created masterful pieces of jewelry to fulfill the rising demand emanating from the increasingly lavish courts of Europe. The prevalence of ornate gold jewelry in Renaissance society grew to the degree that many Italian city-states and King **Henry VIII** of England sought to limit such forms of display by enacting **sumptuary legislation** prohibiting men and women of certain social classes from wearing ostentatious gold jewelry or fine fabrics.

The consolidation of courtly life around the figure of the king in France, England, and the Habsburg Empire during the 17th and 18th centuries was accompanied by the rise of the baroque and rococo styles, which incorporated a higher level of intricacy of sculptural elements such as cameos, carved or inlaid gemstones, and enamel work. The late 18th century and 19th centuries were marked by the neoclassical style popular during the age of Romanticism and technological developments that made precious jewelry available to a wider population who acquired jewelry for fashion as well as with a view to collecting and passing on family heirlooms. The supply of gold from the **California Gold Rush** of 1849 and other gold rushes during the mid-19th century greatly contributed to the expansion of the **goldsmithing** industry in the United States and abroad. The history of jewelry in 19th-century America can be viewed as a reflection of changing practices in social customs and relations between different classes, trends in fashion and art, and patterns of international trade. Some of the most famous modern jewelry makers founded their brand during this period, such as Pierre Cartier in Paris during the 1840s, Bulgari in Rome in the 1880s, and the establishment of Tiffany & Company in New York City in 1837 by Charles Lewis Tiffany. In the course of the late 19th and early 20th centuries, production by these jewelers and others often paralleled current art and cultural phenomena such as the Arts and Crafts Movement in the United States and the Art Nouveau style in Europe and, later, Art Deco.

In the contemporary era, the mechanization of **goldsmithing** processes such as **casting** has made gold jewelry less expensive to produce and more widely available in a range of alloys such as white gold, rose gold, and yellow gold with varying degrees of purity. Jewelry continues to function throughout the globe as a mark of status, group affiliation, commemoration, and superstition. Precious items of cultural significance among groups ranging from the hip-hop artists sporting bling in the late 1990s as a proud symbol of their identification with the black urban culture in the United States to the brides in India adorning themselves with gold jewelry to ensure a fortunate and happy marriage and to mark a measure of family social status continue to fulfill the timeless functions of jewelry in human society.

See also: Bling; California Gold Rush; Casting; Cellini, Benvenuto; Crowns, Royal; Filigree, Gold; Goldsmithing; Granulation; Hatshepsut; Henry VIII; Pre-Columbian Gold; Sumptuary Legislation; Tutankhamun; Ur, Royal Tombs of.

Further Reading

Campbell, Marion. 2009. *Medieval Jewellry in Europe, 1100–1500*. London: V & A Publications.
Clark, Charlotte. 1935. "Ptolomeic Jewelry." *Metropolitan Museum of Art Bulletin* 30, no. 8 (August):161–164.
Deppert-Lippitz, Barbara. 1996. "Late Roman Splendor: Jewelry from the Age of Constantine." *Cleveland Studies in the History of Art* 1:30–71.
Drost, E. and J. H. Hanbelt. 1992. "Uses of Gold Jewellry." *Interdisciplinary Science Review* 17:271–280.
Emmerich, Andre. 1965. *Sweat of the Sun and Tears of the Moon: Gold and Silver in Pre-Columbian Art*. Seattle: Univ. of Washington Press.
Evans, J. A. 1989. *A History of Jewellery 1100–1870*. London: British Museum Publications.
Fales, Martha Gandy. 1995. *Jewelry in America, 1600–1900*. Woodbridge, Suffolk, UK: Antique Collectors' Club.
Hemingway, Colette, and Seán Hemingway. 2007, April. "Hellenistic Jewelry." In *Heilbrunn Timeline of Art History*. New York: The Metropolitan Museum of Art. http://www.metmuseum.org/toah/hd/hjew/hd_hjew.htm.
Jennings, Anne M. 1988. "Women's Gold Jewelry from Egyptian Nubia." *African Arts* 22, no. 1 (November):68–71.
King, Heidi. 2002. "Gold in Ancient America." *Metropolitan Museum of Art Bulletin* New Series 59, no. 4 (Spring):5–55.

M. L. F. 1927. "Indian Jewelry." *Bulletin of the Museum of Fine Arts* 25, no. 149 (June): 39–43.
Oliver, Andrew, Jr. 1996. "Greek, Roman, and Etruscan Jewelry." *Metropolitan Museum of Art Bulletin* 24, no. 9 (May).
Scarisbrick, Diana. 1993. *Rings: Symbols of Wealth, Power, and Affection.* New York: Abrams.
Seidman, Gertrude. 1979. "Jewelry, Ancient to Modern." *Burlington Magazine* 121, no. 921 (December):828, 830.
Tait, H. 1986. *Seven Thousand Years of Jewellery.* London: British Museum Publications.
Wees, Beth Carver. 2004, October. "Nineteenth-Century American Jewelry." In *Heilbrunn Timeline of Art History.* New York: The Metropolitan Museum of Art. http://www.metmuseum.org/toah/hd/ajew/hd_ajew.htm.

JUSTINIAN I

Justinian I (482–565) reigned as Roman emperor of the Byzantine Empire based in Constantinople from 527 to 565. Justinian was born Petrus Sabbatius in the Latin province of Illyria, near modern-day Skopje, Macedonia. As a child, he was adopted by his uncle, the emperor Justin (r. 518–527), and brought to Constantinople for his education. In 525, Justinian married Theodora, who was a controversial choice as she was not from an aristocratic class and was reputed to be a courtesan. Despite this, Theodora proved to be an influential and savvy partner to Justinian during his rule. Throughout the majority of his uncle's reign, Justinian wielded significant power as first consul to the emperor, then later as co-emperor with Justin I beginning in April 527. Upon his uncle's death in August of that same year, Justinian assumed the title of emperor and embarked on a series of administrative and financial reforms, as well as ambitious building projects, in which strategic acquisition and utilization of gold in various factors was a deliberate and significant aspect of his expansion and consolidations of the Byzantine Empire's power and influence.

Medieval mosaic from the Hagia Sofia in Istanbul in which Emperor Justinian is seen offering a model of the church to Mary and Jesus Christ. (Pavle Marjanovic/Dreamstime.com)

During the first half of his reign, Justinian made his mark in the annals of history with his efforts to reconquer strategic portions of the western Roman Empire, including Rome itself, along with Persia and North Africa, and for his fiscal and administrative reforms of the Roman Empire, which culminated in an updated codification of Roman law, the *Corpus Juris Civiliis*, known as the "Codex Justinianus." He also undertook ambitious building projects famously chronicled by contemporary Procopius in his *De Aedificiis* (On Buildings), most notably the **Hagia Sophia** in Constantinople, the heart of the Eastern Orthodox Church until the Ottoman Conquest and an iconic symbol of Byzantine accomplishments to this day.

A crucial aspect of Justinian's strategy in both of these hallmarks of his rule, the reconquest and administrative reforms, was his careful attention to the establishment of the gold **bezant**, or solidus, as the premier currency exchanged in foreign markets, to be used as an instrument of establishing and solidifying the empire's power and influence. Originally established using a significant surplus of gold in the treasury that Justinian inherited from his predecessors, then minted from gold mines in the Balkans and Anatolia, as well as in regions beyond the empire's boundaries in **Nubia** and in the Caucasus and Ural mountain ranges, this influential new gold coinage fueled Justinian's building projects, reinforced diplomacy, promoted an expansion of Byzantine trade, and paid for the salaries of public officials and soldiers.

A series of unfortunate events in the 540s, including outbreaks of the bubonic plague and setbacks in war, precipitated an economic crisis that steadily eroded the empire's strength. In a desperate effort to respond to fiscal crisis and a marked decrease in the empire's supply of gold, mint master Peter Barsymes attempted to stem the ravaging effects of these circumstances on the treasury by **minting** a lightweight solidus and debasing the value of bronze coinage in order to maintain the value of debased gold coins. The empire could not recover and entered into a period of decline that reversed many of the advances of the previous two decades. Justinian died in Constantinople in November 565, leaving no heirs.

See also: Bezant; Hagia Sophia.

Further Reading

Barker, John W. 1966. *Justinian and the Later Roman Empire*. Madison: Univ. of Wisconsin Press.

Evans, J.A.S. 2005. *The Emperor Justinian and the Byzantine Empire*. Westport, CT: Greenwood.

Herrin, Judith. 2009. *Byzantium: The Surprising Life of a Medieval Empire*. Princeton, NJ: Princeton Univ. Press.

Sarris, Peter. 2006. *Economy & Society in the Age of Justinian*. Cambridge: Cambridge Univ. Press.

KARAT

The term "karat" is used to express the content of gold alloys relative to the fineness, or purity levels, of gold items. One karat equals 1/24th of the whole, therefore, 24-karat gold is the highest achievable fineness at 99.9 percent pure, 18-karat gold consists of 75 percent gold and 25 percent an alloy metal, and 12-karat gold is a fineness of 50 percent gold with 50 percent alloy.

The origin of the word karat can be traced to the Greek word *kerátion* in reference to carob seeds, which were commonly used as weights on scales in antiquity due to their relatively uniform and small size. The term also assumed the general meaning of "a grain," used colloquially to refer to a small fractional portion. In the 4th century, it took on the more specific and uniform meaning of a proportion of 1 out of 24 as it began to refer to a general measurement of 1/24th of the weight of a gold solidus coin, a denomination known as the *siliqua*, Latin for the carob seed pods. This system of measurement was also adopted in the Arabic world, where the word translated as *qirrat*, the source of its adaptation into the Italian word *carato* and the French *carat* in the environment of Mediterranean trade. By the mid-16th century, the term was used as the standard measure for indicating the fineness of gold.

Today the karat remains the standard unit of measurement for expressing the purity of gold in **minting**, manufacturing, and jewelry making, and gold items are typically stamped with a hallmark certifying their levels of fineness. In many cases these hallmarks are legally required and monitored by trade officials. Additionally, karat standards are regulated in many countries to ensure their legitimacy, and some governments allow only certain standardized percentages and disallow alloys under 12 karats to be marketed as gold. In the United States, the Federal Trade Commission regulates karat hallmarks on gold products and criminally prosecutes misrepresentations of purity levels. In 1994, the European Court ruling in the Houtwipper case established a legal requirement of standardized hallmarking in European Union countries to protect consumers and facilitate regulated international trade of gold products.

See also: Jewelry.

Further Reading

Bachman, H. G., and Robert B. Cook. 2003. *Gold: The Noble Metal*. Ann Arbor: Univ. of Michigan Press.

Evans, J. A. 1989. *A History of Jewellery 1100–1870*. London: British Museum Publications.

World Gold Council. "The Caratage (Karatage) System for Gold Jewelry." http://www.utilisegold.com/jewellery_technology/caratage/.

KEYNES, JOHN MAYNARD

John Maynard Keynes (1883–1946) was a British economist, financier, political advisor, journalist, and patron of the arts whose economic theories on the causes of and solutions for persistent unemployment and economic crisis, particularly his argument for government intervention by means of spending, even at the expense of a fiscal deficit, as the most efficient means of stimulating recovery in the face of economic recession, remain a central part of modern economic policy. Still considered one of the most significant economic theorists of the 20th century, Keynes's work is also often hailed as fundamental to the development of contemporary macroeconomics.

Keynes was born on June 5, 1883, in Cambridge, England, to father John Neville Keynes, professor of economics and later academic administrator at King's College, Cambridge, and mother Florence Ada Brown, who had the distinction of being one of the first females to graduate from Cambridge. Following a period of years as a day student at Saint Faith's School, the young Keynes obtained a scholarship to preparatory school at Eton and later enrolled at King's College in 1902 amid a period of reform at Cambridge in which academic departments and majors were expanded and refined. Keynes was thus encouraged by economist Alfred Marshall to pursue his studies in the field of political economy rather than mathematics and the classics, the subjects which he had initially emphasized. After earning his bachelor's degree in 1905, he took the civil service entrance exams and worked in the India Office in Whitehall, London. Based on these experiences, he wrote his first major work of economics, *Indian Currency and Finance*, published in 1913.

In 1909 Keynes returned to Cambridge to accept a position as lecturer in economics, teaching until 1915. Through his network of colleagues at Cambridge, Keynes became involved with the Bloomsbury Group, an elite society of bohemian intelligentsia associated with such figures as writers Virginia Woolf and Lytton Strachey and various other artists and intellectuals. Following the outbreak of World War I, Keynes remained a fellow of the economics department at Cambridge but returned to a government position as advisor to the treasury on the problem of Britain's limited foreign currency reserves. In this capacity, he joined British Prime Minister David Lloyd George at the deliberations for the Versailles treaty at the conclusion of the war, yet resigned his post due to his disagreement over Allied leaders' imposition of harsh reparations against Germany, which he asserted would likely never be repaid yet nevertheless have potentially harmful effects on the global economy. Keynes outlined his criticism of events in *The Economic Consequences of the Peace*, published in 1919, which begins with the following prophetic warning:

> The power to become habituated to his surroundings is a marked characteristic of mankind. Very few of us realise with conviction the intensely unusual, unstable, complicated, unreliable, temporary nature of the economic organisation by which Western Europe has lived for the last half century. We assume some of the most peculiar and temporary of our late advantages as natural, permanent, and to be depended on, and we lay our plans accordingly. On this sandy and false foundation

we scheme for social improvement and dress our political platforms, pursue our animosities and particular ambitions, and feel ourselves with enough margin in hand to foster, not assuage, civil conflict in the European family. Moved by insane delusion and reckless self-regard, the German people overturned the foundations on which we all lived and built. But the spokesmen of the French and British peoples have run the risk of completing the ruin which Germany began, by a peace which, if it is carried into effect, must impair yet further, when it might have restored, the delicate, complicated organisation, already shaken and broken by war, through which alone the European peoples can employ themselves and live. (1919, 1)

In the face of rebukes from government leaders for his book's harsh portrayals of the public figures engaged in negotiations at Versailles, Keynes gave up his Cambridge lecture seat but remained affiliated with the university as a fellow. Despite the disfavor the book provoked among government circles, Keynes's views were significant in contributing to public perceptions that the treaty's treatment of Germany was too harsh.

During 1920–1921, Keynes divided his time between his teaching and administrative commitments in Cambridge and his work in London as a financier and private investor, writing for a series of publications, and also acting as a political and financial advisor. In the spring of 1921, while visiting North Africa he was stoned by the locals for refusing to pay a price for a shoe shine that he deemed to be above market price, and he is reported to have remarked while leaving the scene that he stood his ground so as not "be a party to debasing the currency" (Heilbroner 1999, 284). In 1921, he published his *Treatise on Probability*, in which he argued for the human element in mathematical understandings of probability. In the subsequent years, he continued to publish widely on the subject of monetary reform and the condition of prolonged unemployment in Great Britain as he began to formulate the foundations of his stimulus through spending recommendations, the cornerstone of what would become "Keynesianism." In his 1923 work *A Tract on Monetary Reform*, Keynes argued that Great Britain should go off the **gold standard**, the constraints of which were forcing the treasury and the Bank of England to pursue deflationary policies during a period when economic expansion offered the best solution to Britain's economic ills. Over the course of 1924, he increasingly argued for forms of government intervention in the economic crisis that ran contrary to contemporary emphasis among economists that laissez-faire policies would lead the market to self correct in time.

In April 1925, Winston Churchill returned the British currency to the **gold standard** at parity with its pre–World War I value, a dollar-to-pound ratio of $4.86, through the **Gold Standard Act of 1925**, a decision that Keynes voiced vehement public opposition to, calling gold a "barbarous relic" in modern financial systems. Churchill would later concede in hindsight that this legislation was misguided and had largely been undertaken out of pride. In August of that year, Keynes would marry Russian ballerina Lydia Lopokova.

With the onset of the Great Depression in the 1930s, Keynes, having just published his two-volume work *A Treatise on Money* in 1930, contributed to a series of economic advisory councils between the United States and Great Britain dedicated to addressing the financial crisis. Among the more immediate actions taken was the removal of British currency from the **gold standard**. During 1934, Keynes advised U.S. President Franklin D. Roosevelt directly as the American leader crafted what was to become the New Deal, arguing that the situation required a plan focused on recovery and reform using aggressive government spending, even at the expense of a growing fiscal deficit.

Keynes's theory of the importance of spending to a healthy economic recovery were further systematized with the 1936 publication of *The General Theory of Employment, Interest, and Money*, which laid out the structure of his argument that government spending programs offered the most viable solution to the problem of prolonged, increasing unemployment and monetary crisis. The model put forth by Keynes was widely accepted among contemporary economists, especially in times of crisis, and *The General Theory* would become one of the most influential works of economic theory in the modern era.

After suffering a serious heart attack in 1937, Keynes greatly limited his activities as he recovered in Wales. Two years later, however, he returned to a teaching position at Cambridge and once again advised the British treasury on monetary policy and war finance while also fostering crucial economic relations with the United States during World War II, obtaining a director position on the board of the Bank of England. In 1942, he was awarded a peerage title for his service to the crown, now assuming the title of Lord Keynes, Baron of Tilton. In June 1944, Keynes attended the summit at which the Bretton Woods Agreement was articulated at the Mount Washington Hotel in Bretton Woods, New Hampshire. Keynes's contributions to the debate included his suggestion that the problem of the international **gold standard** be solved by creating an international banking institution that would oversee the settlement of foreign currency exchange not by means of **gold reserves**, but through a type of bank-issued money guarantee extended to member nations. Along with the director of monetary research at the U.S. Treasury, Dexter White, Keynes also proposed that the U.S. dollar be maintained as a benchmark currency convertible into gold at a fixed rate, with member nations' currencies then tied to the dollar within a margin of 1 percent of the fixed rate. Keynes hoped that such policies would also promote the expansion of trade and discourage the protectionist policies that had developed during the 1930s. These proposals were crucial to the formation of the **International Monetary Fund** and the World Bank.

After suffering another heart attack at the Bretton Woods Conference, Keynes continued to work as a liaison with the United States to ensure the maintenance of British financial interests, brokering a multibillion dollar loan with highly advantageous repayment terms from the United States to Great Britain intended to help rebuild the British economy in the face of the wartime devastation. Less than a year later, Keynes died of a heart attack on April 21, 1946, at his home in East Sussex.

Throughout his life, Keynes had been an avid collector of 20th-century art and historical artifacts and documents. Of particular significance was his fascination with the esoteric and alchemical works of **Sir Isaac Newton**. Following the sale of Newton's Portsmouth Papers by Viscount Lymington at a Sotheby's auction in the summer of 1936, Keynes methodically sought to acquire as many of the documents as he could from dealers and private collectors in order to reassemble them into a coherent archive, with a particular view to obtaining documents and correspondence related to **alchemy**. Upon his death, Keynes generously donated this private archive to King's College, Cambridge, where it is conserved today as a part of the John Maynard Keynes Papers. Historians interpret Keynes's interest in Newton's work as indicative of his own moral philosophy and the overarching ethical premises that underscored his economic theories, suggesting that Keynes proposed solutions to fiscal problems not to perpetuate wealth, per se, but to achieve a deeper sense of balance as the key to living a fulfilled life.

Although widely acknowledged for its contributions to the systematization of modern macroeconomic theory, Keynesianism fell dramatically out of favor in the late 20th and early 21st centuries, when it became an almost derogatory term in the world of free-market economics and finance. Not surprisingly, however, U.S. Treasury advisors turned to Keynesianism once again following the economic collapse of the national financial systems in the United States and subsequent global economic recession in 2008. After monetary policy efforts by the Federal Reserve Bank failed to prevent the deepening of the crisis, under the stewardship of President Barack Obama, the U.S. Congress passed an economic stimulus bill designed to increase spending and reduce unemployment that was decidedly Keynesian in its approach, risking the accumulation of trillions of dollars in national deficit out of the hope that spending would indeed chart the path to economic recovery.

See also: Banking and Credit; Bretton Woods System; Gold Standard; Gold Standard Act of 1925; Inflation; International Monetary Fund; Newton, Sir Isaac.

Further Reading

Buchan, James. 2008. "When Keynes Went to America." *New Statesman*, November 10: 27–28.
Clarke, Peter. 2009. *Keynes: The Rise, Fall, and Return of the Twentieth Century's Most Influential Economist.* London: Bloomsbury Publishing.
Davidson, Paul. 2007. *John Maynard Keynes.* New York: Palgrave Macmillan.
Dillard, Dudley D. 1948. *The Economics of John Maynard Keynes: The Theory of Monetary Economy.* Upper Saddle River, NJ: Prentice-Hall.
Heilbroner, Robert L. 1999. *The Worldly Philosophers: The Lives, Times, and Ideas of the Great Economic Thinkers.* New York: Simon and Schuster.
Keynes, John Maynard. 1919. *The Economic Consequences of the Peace.* http://socserv2.soc sci.mcmaster.ca/~econ/ugcm/3ll3/keynes/peace.htm.
Keynes, John Maynard. 1936. *The General Theory of Employment, Interest, and Money.* http://homepage.newschool.edu/het//texts/keynes/gtcont.htm.
Keynes, John Maynard. 1946. "Newton, the Man." Lecture written for Royal Society of London but never presented. http://www-history.mcs.st-and.ac.uk/Extras/Keynes_Newton.html.
Keynes, John Maynard. 1971–1989. *The Collected Writings of John Maynard Keynes*, ed. Austin Robinson and Donald Moggridge. 30 vol. London: Macmillan.
Kindleberger, Charles. 1985. *Keynesianism versus Monetarism and Other Essays in Financial History.* London: George Allen and Unwin.
Lawler, Joseph. 2009. "We're All Keynesians Again." *American Spectator* (April):52–53.
Moggridge, D. E. 1992. *Maynard Keyes: An Economist's Biography.* London: Routledge.

Parsons, Wayne. 1997. *Keynes and the Quest for Moral Science: A Study of Economics and Alchemy.* Cheltenham, UK: Edward Elgar.

Patinkin, Don. 1976. *Keynes' Monetary Thought: A Study of Its Development.* Durham, NC: Duke Univ. Press.

KLONDIKE GOLD RUSH

The discovery of gold in a tributary of the Klondike River in Canada's Yukon Territory in the summer of 1896 sparked a gold rush that brought thousands of prospectors to the region over the following three years. In August 1896, Skookum Jim Mason and his cousin Dawson Charlie of the Tagish First Nation and Californian George Washington Carmack were fishing for salmon near the mouth of the Klondike River when, based on rumors of gold in the region, they discovered a gold nugget in the silt of Rabbit Creek, later to be known as "Bonanza Creek." The next day, Carmack hiked 50 miles downriver to the town of Forty Mile to register his claim. News that the friends had struck gold spread quickly among prospectors in the Yukon, and within days, the creek was packed with goldseekers and an even richer deposit of alluvial gold had been discovered at another tributary of the Klondike later known as "Eldorado Creek."

Due to the challenges of traveling in the winter months, fast approaching in the immediate aftermath of the discovery, the official gold rush to the Klondike River

"For without a doubt there is in these lands a very great amount of gold."
—CHRISTOPHER COLUMBUS

Miners and their pack burros, laden with supplies, are shown at Dyea Point near Skagway, Alaska, during the Yukon gold rush, ca. 1897. (Library of Congress)

area did not occur until the summer of 1897, when thousands of prospectors set sail from Seattle and San Francisco in pursuit of gold after witnessing the earliest miners return by steamship to Seattle in mid-July with suitcases full of gold. The fever for gold was fueled even more urgently by the economic recessions of the 1890s, when financial hardship motivated people to drop everything and depart for the rugged northern territory in the hopes of a better life. The years 1897 and 1898 witnessed what is now called the "Klondike stampede" as prospecting towns were populated so quickly along the route that Canadian mounted police had to regulate strictly the flow of human traffic to the area, requiring that parties of prospectors had adequate food and supplies.

The Alaskan boomtowns of Skagway and Dyea grew rapidly as the point of arrival by sea for most of the stampede and the point of departure for either the inland route to the Yukon by way of the Chilkoot Pass or White Pass on trails that were difficult and treacherous for both man and pack animals. *Harper's Weekly* writer Tappan Adney observed of the Chilkoot Pass, "There is nothing but the grey wall of rock and earth. But stop! Look more closely . . . The mountain is alive. There is a continual moving train; they are perceptible only by their movement, just as ants are . . . They are human beings, but never did men look so small" (Adney, 114). Detachments of the Canadian Mounties were established at the head of each trail to collect customs duties on the items prospectors brought into Canada. Goldseekers would then continue on by water in crudely constructed boats from Lake Lindemann through a series of lakes to the Yukon River, disembarking at Dawson City. Those who could afford the expensive ship passage could travel the route inland by water. The small Yukon settlement of Dawson City, the closest town to the gold fields, grew from a small First Nations settlement to a population of 40,000 in 1898, becoming the largest Canadian city west of Winnipeg at that time. The population then plummeted to 8,000 at the end of the gold rush a year later.

A small percentage of prospectors profited from the Klondike Gold Rush as much of the gold was mined in the immediate aftermath of its discovery. Among some of the more famous prospectors were author Jack London and future U.S. President William Howard Taft. By all estimates, the total value of gold mined in the Klondike amounted to $50 million, which is roughly equivalent to the total sum spent in travel and development to facilitate the gold rush. The stampede came to an end in 1899 with the onset of the Spanish-American War and the discovery of gold in Nome, Alaska. Despite its swift decline, the Klondike Gold Rush was a watershed event for the development of Western Canada, Alaska, and the transportation networks of the Pacific Northwest, which contributed to the settlement of towns and a flow of goods and migration and the population of a territory that had thus far been viewed as too rugged to develop on this level.

During the early 2000s, several dozen **mining** companies pursued mineral exploration in search of gold in the Klondike region, with around 7,900 claims staked from 2006 to 2009 and average annual spending by **mining** companies in the area at $60 million for 2009. Assisted by modern digital mapping technologies and computer geographic analysis, contemporary prospectors have identified deposits with significant yields averaging one gram of gold for every metric ton of ore.

See also: Gold Nuggets; Mining.

Further Reading

Adney, Tappan. 1899. *The Klondike Stampede*. New York: Harpers.
Berton, Pierre. 2003. *The Life and Death of the Last Great Gold Rush*. Carroll and Graf Publishers.
Blackwood, Gary L. 1997. "Klondike Gold Rush." *Wild West Magazine*, August. http://www.historynet.com/klondike-gold-rush.htm.
Weber, Bob. 2009. "Once Again, Gold Miners Rush in to the Klondike." *Anchorage Daily News*, August 17.

KUBLAI KHAN

Kublai Khan (1215–1294) was emperor of the Mongol empire during 1260–1294 and founder and ruler of the Yüan Dynasty in China after the Mongol conquest of the remainder of China in 1279, after which he ruled an empire that spanned from the Black Sea, as far north as the Golden Horde, to the south from the Mediterranean to Central Asia, and finally encompassing the extent of China. Kublai Khan's encounters with Italian merchant explorer **Marco Polo** earned him fame among the courts of Western Europe, where Polo's stories of the Mongolian emperor's lavish lifestyle and the flow of commerce based on the issue of paper money contributed to Western perceptions of the gold and riches of the East.

Kublai Khan was born in Mongolia in 1215, son of Tolui, the fourth son of the empire's founder Genghis Khan. During the 1250s, he was entrusted by the emperor, at the time his brother Möngke, with the civil and military responsibilities for organizing the completion of the Mongol conquest of the Chinese Song Dynasty. In this capacity, Kublai Khan became immersed in Chinese culture and philosophy, which is considered a critical factor in the success of his later administration as emperor of such a vast empire. Following the death of his brother, Kublai Khan was unanimously elected to succeed his brother as khan by the Mongolian *kuriltai*, or great assembly. In opposition to the succession of Kublai Khan, his brother Arigböge held his own great assembly in Karakorum naming himself as khan, and civil war ensued, lasting into the mid-1260s. Arigböge's supporters viewed Kublai Khan's assimilation to Chinese culture as a betrayal of traditional Mongolian traditions.

In 1271, Kublai Khan completed the consolidation of his vast empire by conquering the remaining Chinese kingdom of the Southern Song Dynasty and establishing himself as head of the Yüan Dynasty, founding a capital in what is now Beijing. With the Chinese territory fully subdued by 1279, Kublai Khan set his sites even further with military advances in neighboring kingdoms such as Burma, Java, and even over to Japan. According to **Marco Polo**, the khan's expedition to Japan was based largely on rumors of the island's great wealth in gold. These campaigns brought little success and significant loss, yet the khan persisted in efforts to expand his empire's sphere of influence.

During a merchant expedition to the East in the late 1260s, Polo's father Niccolò and uncle Maffeo had been invited by a Mongolian imperial envoy to meet the khan to satisfy his curiosity about Western cultures. Kublai Khan entrusted the Polo brothers to deliver a letter to the pope in Rome, in which he requested that 100 learned men be sent back to the Mongolian court to instruct the emperor in their ways. In 1269, Niccolò and Maffeo met 17-year-old Marco for the first time when they returned to Venice. Upon receiving a letter from the newly elected pope, Gregory X, the brothers were joined by Marco in 1271 as they set out once again for the Khan's court, failing to convince any scholars to join them on their difficult journey. Marco Polo stayed on at the Mongolian court for nearly 17 years and recounted his experiences in a travel narrative that would serve as the primary source of European knowledge of the Far East for many centuries due perhaps in part to the intrigue of Polo's depictions of the Khan's sumptuous lifestyle and vast palaces of gold. He even described the people of the Mongol city of Vochan, where men made casts of their teeth in pure gold to wear as adornment covering their upper and lower teeth.

Kublai Khan consolidated his power through strategic social and administrative reforms, such as an ethnic system of social class that favored Mongolians, patronage of Buddhist temples, extensive development of infrastructure and imperial buildings, and fiscal reforms that established the circulation of paper money made from mulberry trees as the primary method of currency. Polo was particularly impressed with the authority of paper money in the empire, commenting in his travel account:

> The emperor's mint then is in this same city of Cambaluc, and the way it [money] is wrought is such that you might say he has the secret of **alchemy** in perfection, and you would be right. For he makes his money after this fashion. He makes them take of the bark of a certain tree, in fact of the mulberry tree, the leaves of which are the food of the silkworms, these trees being so numerous that the whole districts are full of them. What they take is a certain fine white bast or skin which lies between the wood of the tree and the thick outer bark, and this they make into something resembling sheets of paper, but black. When these sheets have been prepared they are cut up into pieces of different sizes.
>
> All these pieces of paper are issued with as much solemnity and authority as if they were of pure gold or silver; and on every piece a variety of officials, whose duty it is, have to write their names, and to put their seals. And when all is prepared duly, the chief officer deputed by the Khan smears the seal entrusted to him with vermilion, and impresses it on the paper, so that the form of the seal remains imprinted upon it in red; the money is then authentic. Anyone forging it would be punished with death. And the Khan causes every year to be made such a vast quantity of this money, which costs him nothing, that it must equal in amount all the treasure of the world.
>
> Furthermore all merchants arriving from India or other countries, and bringing with them gold or silver or gems and pearls, are prohibited from selling to any one but the emperor. He has twelve experts chosen for this business, men of shrewdness and experience in such affairs; these appraise the articles, and the emperor then pays a liberal price for them in those pieces of paper. The merchants accept his price readily, for in the first place they would not get so good a one from anybody else, and

secondly they are paid without any delay. And with this paper money they can buy what they like anywhere over the empire. (Polo 1903, 587)

Yet many of these reforms were superficial and would not ultimately survive the reign of Kublai Khan himself. Despite the security and consistency of the initial issues of paper money, the currency became devalued rapidly during the 1280s due to poor fiscal management and overissue of paper money.

Toward the end of his reign, Kublai Khan suffered prolonged illness due to gout. He died on February 18, 1294, in Tatu (Beijing). His body was subsequently returned to Mongolia for ceremonial burial. He had produced an estimated 15 children with his many wives and women of the harem. One of his sons, born in 1243 and designated as crown prince, had been given the name Zhenjin, meaning "True Gold" in Chinese, by Chinese Buddhist monk Haiyun.

See also: Alchemy; Polo, Marco.

Further Reading

Polo, Marco, and Rustichello of Pisa. 1903. *The Travels of Marco Polo*, trans. Henry Yule. London: John Murray. http://www.gutenberg.org/ebooks/10636.

Rossabi, Morris. 1990. *Kubilai Khan: His Life and Times*. Berkeley: Univ. of California Press.

Wardwell, Anne E. 1992. "Two Silk and Gold Textiles of the Early Mongol Period." *Bulletin of the Cleveland Museum of Art* 79, no. 10 (December):354–378.

LACE, GOLD

The traditions and techniques used to create the variations of linen lace commonly produced today have their origins in early types of lace work with thread or very fine wire made of precious metals such as gold and silver. The word "lace" derives from the Latin root *laquem*, which refers to forms of ornamental netting. While sometimes used exclusively to describe a specific modern technique of lace making dating only to the 15th century, the term lace is broadly used in reference to the intertwining of threads of various materials to form ornamental patterns using a needle or a bobbin or other methods such as darning, crocheting, or drawing the threads of a mesh net into a pattern.

The earliest archeological evidence for gold lacework exists among ancient Egyptian and Assyrian tombs from ca. 1500–1000 BC in which ornamental gold netting adorned heads and clothing of the deceased. In the *Odyssey*, Homer makes reference to "veils of golden netting." During later antiquity, gold lace was commonly produced by nuns in convents for the vestments of the high clergy, as well as for the garments of the nobility. One of the most significant examples of gold lace vestments from the early Middle Ages was discovered in the coffin of England's St. Cuthbert (ca. 634–687), Bishop of Lindisfarne, at Durham Cathedral. Techniques in the production and patterns of gold lace making likely spread during the Byzantine period from the eastern Mediterranean to the West through trading centers such as the Venetian Republic and Spain.

During the High Middle Ages, several Italian city-states became centers of lace production in intricate patterns and designs, with perhaps the most famous among them the Venetian "Point de Venice," an elite pattern coveted by the French nobility. The wealthy states of Genoa and Lucca were also significant centers of lace manufacture during the 14th century. In the mid-15th century, however, the increasing appearance of sumptuous gold lace ornamentation on the clothing of wealthy common people of the merchant class led to a series of **sumptuary legislation** throughout Europe restricting the use of gold lace. For example, in 1476, Venice enacted a law forbidding the use of gold or silver embroidery on fabric and the production of the Point de Venice with gold or silver threads. At the same time, heavy export taxes imposed on the gold lace trade during the 15th century crippled the industry. To avoid suffering such fees and work restrictions, many skilled artisans either substituted lesser threads such as flax or linen to produce traditional patterns of lace or emigrated to other countries, particularly France.

During the early modern era, several regions of France such as Aurrillac and Arras became renowned centers of gold lace production due to the popularity of this luxury commodity among the French monarchy. Yet, whereas gold and silver threads continued to be interwoven with linen threads for ornamentation, lace making largely converted to linen materials after the 15th century.

The present-day demand for gold lace continues to be most prevalent in ceremonial uniforms or Church vestments, theatrical costumes, or decorative household textiles. The proportion of gold used today is relatively small, with an average of 3 parts gold used in an alloy also comprised of 90 parts silver and 7 parts other metals such as copper. Decorative lace of gold wire and thread is manufactured on a large scale in contemporary India, where a type of embroidery in gold or silver threads has existed since antiquity.

See also: Alloying; Sumptuary Legislation.

Further Reading

Clifford, R. 1913. *The Lace Dictionary*. New York: Clifford & Lawton.
Jackson, Mrs. F. Nevill. 1910. "Gold and Silver Lace." *Every Woman's Encyclopedia*. London.
Levey, Santina M. 1983. *Lace: A History*. London: Victoria and Albert Museum.
Naik, Shailaja. 1996. *Traditional Embroideries of India*. New Delhi: A.P.H Publishing Corp.
Pollen, Maria. 1908. *Seven Centuries of Lace*. New York: Macmillan.
Pollen, Maria. 1910. Lace. In *The Catholic Encyclopedia*. New York: Robert Appleton Company. http://www.newadvent.org/cathen/08729b.htm.
Watt, Melinda. 2000. "Textile Production in Europe: Lace, 1600–1800." In *Heilbrunn Timeline of Art History*. New York: Metropolitan Museum of Art. http://www.metmuseum.org/toah/hd/txt_l/hd_txt_l.htm.

LEPRECHAUNS

Leprechauns are a mythical fairy-type creature from Irish folklore, perhaps deriving from the legend of the Tuatha Dé Danann, a tribal people said to have inhabited Ireland before the arrival of the Celts. The word "leprechaun" may derive from either the Irish word for shoemaker (*leath bhrogan*) or the Irish word for a type of pygmy or sprite (*luacharma'n*). Leprechauns are said to have buried crocks filled with gold

Spider Web Tightrope, a mixed media children's book illustration by contemporary artist Wayne Anderson, 1982. (The Bridgeman Art Library International)

coins, **bullion**, and other treasure and are considered to possess certain magical powers, which they use to protect their gold from potential thieves. Some say that the myth of the leprechauns' treasure originated in local folklore about little people entrusted with guarding the cache of gold hidden in Ireland by the Vikings during their conquest of Ireland. Jovial in nature, leprechauns were also associated with good luck, and yet another legend associates the name with the Celtic god Lugh (pronounced "luck").

These creatures are typically depicted in legend as tiny old men wearing colored suits with waist coats, often green, hats, and buckled shoes, often bearded and sometimes carrying or working on a shoe, hence their association with shoemakers. Various tales attribute their constant labor on shoes as the source of their reputed wealth. Legend also has it that leprechauns enjoy drinking and dancing late into the evening so much that they are constantly wearing out their shoes and must repair them. Known as clever troublemakers who love to make mischief and play practical jokes on humans, leprechauns also take care to avoid humans, fearing that if they are caught, the location of their gold might be revealed. If caught, leprechauns would promise to grant their captor three wishes in exchange for their freedom but, once freed, unravel the promise through a riddle or a ruse. The leprechauns' precious pot of gold was said to be located at the end of the rainbow, an impossibility as rainbows have no end.

> "When it's raining gold, get a bucket."
> —WARREN BUFFETT

Irish American depictions of leprechauns diverge vastly from the mystical spirit-like creatures of ancient Celtic lore that have endured in Ireland. In the United States, leprechauns are associated with the St. Patrick's Day holiday and appear as jovial, comical creatures whose appearance mirrors derogatory cartoon caricatures and stereotypes of Irish immigrants from the waves of immigration in the late 19th and early 20th centuries. Leprechauns thus appeared wearing old shabby clothes and silly caps and displaying reckless, drunken behavior. Such stereotypes are seen in the 1959 Disney film *Darby O'Gill and the Little People*, in which Sean Connery plays the role of Irish storyteller Darby O'Gill, who takes on leprechaun King Brian in a contest of wits to outsmart the tiny rogue and steal the leprechauns' gold. In Ireland, on the other hand, the legend of the leprechauns was transmitted for centuries as an offshoot of the pre-Christian folk religion practiced by the Celts in which spirits or mythical creatures such as leprechauns served to maintain social norms and behaviors by warning of the threats that lurked about if one transgressed normal behavior.

See also: Bullion.

Further Reading

Croker, T. C. 1862. *Fairy Legends and Traditions of the South of Ireland*. London: William Tegg.
Koch, John T. 2006. *Celtic Culture: A Historical Encyclopedia*. Santa Barbara, CA: ABC-CLIO.
Yeats, W. B. 1888. *Fairy and Folk Tales of the Irish Peasantry*. London: W. Scott.

MIDAS, KING

King Midas ruled the Anatolian kingdom of Phrygia from the capital city of Gordion, about 80 kilometers southwest of modern-day Ankara, Turkey, during approximately 738 to 696 BC. There is substantial historical evidence for Midas's rule over a wide expanse of central and southwestern Anatolia in inscriptions on monuments uncovered by archeologists in the region, contemporary Assyrian texts during the reign of King Sargon, and in Greek and Roman art and literary sources dating from as early as the sixth century BC. Although Midas has also been referred to in sources as Mita, ruler of the Mushki peoples, this may have been due to a spelling variant of the name of the same Phrygian king who also ruled over the neighboring Mushki. Hailed by his contemporaries for his vast wealth and famous for his lavish lifestyle, Midas and his riches became immortalized centuries later in Greek and Roman myths relating a legend in which on account of his greed, Midas was cursed by the gods with the golden touch and everything he touched became physically transformed into gold.

According to the earliest sources, Midas earned a reputation throughout the eastern Mediterranean region as an influential and powerful ruler of vast wealth, an image that led to his swift elevation to a figure of legend by the mid-seventh century BC, when there is evidence that his name was already used proverbially to imply great wealth. Following a period of prosperity and political dominance, Midas is thought to have poisoned himself during an attack by the Kimmerians that destroyed Gordion. Greek sources from the sixth to fifth centuries BC tend to depict Midas admirably as a man of wealth and seeker of wisdom. According to Greek historian Herodotus of the fifth century BC, Midas was the first "barbarian" ruler to present an offering to the oracle at Delphi in the form of an ornate throne. Greek sympathy and admiration for Midas during this period may have been based on their shared common enemy in the Kimmerians as well as a position of cultural respect in the context of good diplomatic relations. Midas's prominent reputation as a ruler of magnificent wealth and influence among Greek chroniclers is credited with launching him to the status of legend in centuries of Greek and Roman texts.

By the fourth century BC, however, portrayals of Midas's wealth began to assume an element of irony, as legends and literary references cast early tales into stories of folly and greed. Prior to the appearance of the "golden touch" legend, sources as early as the sixth century BC describe an encounter between Dionysus's companion, the satyr Silenos, and Midas in Midas's rose garden, where Midas questioned the mythical creature about the meaning of life and the path to happiness, to which the satyr replied with a lengthy fantastical tale illustrating the moral that true happiness is impossible to find on earth. Whereas this story casts Midas in a role of seeker of wisdom, it also highlights the ironic point that one who has unimaginable worldly riches is not content and therefore searches for a more profound meaning or satisfaction, a story that parallels later Anatolian king **Croesus** of Lydia's encounter with the philosopher Solon. King Midas was also the center of another important tale that would later become an element of the golden touch legend. Paintings on early Greek vases and subsequent literary sources narrate the involvement of Midas, said to have been instructed in music by Orpheus himself, in a music contest between the satyr Marsyas, later referred to as Pan, and the god Apollo. When Midas judged Marsyas to be the better musician, in anger Apollo had Midas sprout ass ears. The depiction of Midas as a donkey is often associated with the transition of his legend to a man of foolish greed, yet there is also evidence that ass ears were a remnant of a Bronze Age attribute of kingship in the Anatolian region.

Interpretations of Midas as vulnerable to the false allure of material wealth appear widely in Greek sources of this period, notably in Plato's *Laws* (ca. 360 BC) in the warning that even if a person is as wealthy as Midas, such material riches are worthless if the person is not just. In his excursus on the true meaning of wealth in *Politics* (written ca. 335–323 BC), Aristotle makes reference to Midas's folly when he warns that although a man may be wealthy in terms of material riches, he is still vulnerable to dying of starvation as Midas did, evidence that the tale of the golden touch was already in circulation to the degree that the legend was used in this proverbial manner.

The first textual source for the golden touch legend most commonly known throughout history was set forth in Ovid's *Metamorphoses* (ca. AD 8), which represents a compilation of a wide range of legends involving Midas into one single narrative. According to Ovid, Midas's servants encountered the inebriated satyr Marsyas (or Pan) in the rose garden and brought him to the king. Delighted with his notorious surprise guest, Midas feasted with the satyr for 10 days, and then returned him safely on day 11 to Dionysus, who out of gratitude offered to grant Midas his greatest wish. Midas fatefully wished that everything he touched would turn to gold. When the wish was granted, he set about gleefully turning things around him into gold, but eventually despaired when this gift also prevented him from drinking or eating. Wishing to reverse the gift, he prayed to Dionysus, who instructed him to purify himself in the waters of the Pactolus River, which assumed the power of the golden touch and thenceforth had sands rich with gold. Ovid then incorporates the tale of the music contest into the story, ending the legend with a further reminder of Midas's humiliation.

In 1957, American archeologists uncovered a tomb thought to be that of Midas at an excavation on the site of ancient Gordion. In the early 1980s, archeologist Elizabeth Simpson identified a curious combination of artifacts at the site as the remains of a funerary feast directly outside of the tomb, possibly in front of Midas's body as it lay in state on a bier. An incredible quantity of vessels and platters were uncovered with significant food residue on them. In 1997, chemists collaborated with this archeological record to reconstruct what may have been the menu at this feast: a stew of barbequed lamb, lentils seasoned with fennel or anise, and honey mead, wine, and beer. While scholarly debates still linger over whether this is indeed the tomb of King Midas, the wealth of funerary sites and excavated monuments at the site of ancient Gordon confirm the extent of the ancient Phrygian king's legendary wealth. The name "Midas" and the phrase "the Midas touch" continue to be used millennia later in the contemporary Western world as common proverbial references to a propensity for wealth.

See also: Croesus.

Further Reading

Gorman, Jessica. 2000. "King Midas' Modern Mourners." *Science News* 158, no. 19 (November 4):296–298.
Lloyd, Seton. 1999. *Ancient Turkey: A Traveller's History*. New York: Longitude.
Ramage, Andrew, and Paul Craddock. 2000. *King Croesus' Gold: Excavations at Sardis and the History of Gold Refining*. London: British Museum Publications.
Roller, Lynn E. 1983. "The Legend of King Midas." *Classical Antiquity* 22:299–313.

MINING

Gold mining is the process of extracting gold from lode ore and placer deposits from the earth through a variety of techniques, including placer mining, hydraulic mining, and collecting gold as a byproduct of extraction processes involving other metals such as copper. Since prehistory, the pursuit of gold sources throughout the world has had an effect on geopolitics as the predominance of kingdoms, empires, and nations often depend on control or access to precious natural resources such as gold. Developments in methods and techniques for identifying the characteristics of gold ore deposits and their placement and for extracting and processing gold contributed to advancements in the natural sciences and the origins of modern chemistry.

> **Gold Mine**
>
> In contemporary colloquial usage, the term "gold mine" is commonly used to refer to something that has the potential to be a source of significant wealth.

Geologists maintain a variety of theories for how gold deposits were formed in the earth. Ancient and medieval treatises in the natural sciences on how gold was formed and where it was located in the terrain were often tied to Aristotelian earth sciences and the literature of **alchemy**, derived to a significant degree from ancient Greek and Arabic sources, which posited various theories typically premised on the idea that gold derived from aqueous sources and was produced through transmutative interaction with the sun, to which was typically ascribed a certain spiritual or astrological aspect. Such ideas took shape in a variety of manners, for example, in works by the 11th-century Persian polymath Avicenna; in the scholastic writings on the origins of precious metals by the 13th-century Dominican scholar Albertus Magnus and his student **Thomas Aquinas**; and in the works of a

13th-century alchemist known as Geber. Similar theories existed among philosophers in India and China. These ideas developed even further during the Renaissance along with the emergence of more scientific inquiries into the specific origins of gold and the types and attributes of gold sources in works such as Italian metallurgist Vannoccio Biringuccio's *De la Pirotechnia* (1540) and Saxon scholar Georgius Agricola's *De Re Metallica* (1546). Systems of classification and evaluation of types of rocks and minerals and their geological positions developed in this period constituted the foundations upon which modern scientific theories would evolve during the 17th and 18th centuries, when new methods of scientific inquiry and technological/industrial advances led to improvements in identifying gold deposits and methods of gold extraction. During this period, theories of sulfuric vapors, their interaction with aqueous elements in the earth's crust, and the occurrence of cooling or oxidation constitute the foundations of modern theories of sedimentation processes, chemical reactions that form metals, and magmatic hydrothermal explanations for the geological origins of gold in which metallic minerals are thought to originate when volcanic magma is crystallized deep in the earth's crust to form veins of ore deposits.

Understandings of how gold formed contributed throughout history to the prospecting and extraction of gold. Archeologists have uncovered evidence of mining operations from as early as the second millennium BC in ancient Arabia and Mesopotamia. A wide variety of written sources attest to ancient mining practices: ancient Egyptian inscriptions and tablets contain references to gold mines in **Nubia** and along the Red Sea; the Old Testament contains numerous references to gold sources and gold extraction in the ancient world; the 11th-century manuscript known as the "Turin Papyrus" contains a map considered to be the oldest extant written geological survey, which outlines gold mining operations in eastern Egypt, a site where archeological excavations have uncovered evidence of an extensive quartz vein mining complex; the ancient Greek tale of the **Golden Fleece** was considered by fourth-century-BC scholar Palaephatus to have been inspired by placer mining methods in the Black Sea region; and the Roman naturalist Pliny documented gold sources and extraction technologies in Europe, India, Tibet, and the Far East. These artifacts and texts reveal the extent of trade in gold and the importance of mining from antiquity to the present in patterns of settlement, cultural exchange, and political power. Gold sources and mining operations in West Africa and Eurasia during antiquity and the Middle Ages led to the development of established trade routes and transportation networks, while the pursuit of legendary "cities of gold" in the New World revealed the ancient wealth in pre-Columbian gold ornaments and decorations. The discovery of gold in the gold rushes of the mid-19th century in California, Australia, the Klondike, and at Witwatersrand in South Africa likewise contributed to settlement patterns of these regions and technological developments in the modern mining industry.

Gold deposits are found in several forms. In alluvial sources, or placer deposits, gold from lode ore sources has eroded or decomposed and then is concentrated in streambeds by means of gravity. Gold is extracted from these types of sources by several methods for placer mining, such as panning, sluicing, hydraulics, and dredging. The largest source of gold ore exists in bedrock quartz vein deposits at a range of depths from the earth's surface. These are extracted through underground mining operations using a system of shafts and tunnels or open-pit mining to excavate the rock and then chemically extract the gold through methods such as the **cyanide** process. Gold is also mined as a byproduct of other precious metals, such as copper.

Since the **California Gold Rush** in the mid-19th century, the environmental effects of mining extraction and processing techniques such as dredging, surface mining, and potentially harmful chemical byproducts from the **cyanide** process and the use of other toxic substances has led to close regulation of the impact of industrial mining operations on surrounding environments. Gold mining also poses serious occupational hazards and has resulted in tragic incidents throughout its history.

In the present day, South Africa remains one of the world's largest gold producing region: starting with the discovery of gold near Johannesburg in the late 19th century, South Africa produced as much as 40 percent of the world's total output of gold. Advances in prospecting and mining technology in the late 20th century led to a more than 50 percent increase in total global output of gold, from 1,280 tonnes in 1980 to 2,540 tonnes in 2007. According to the U.S. Geological Survey, as of 2007, South Africa, China, the United States, Australia, and Peru were the top five gold-producing countries in the world and were responsible for approximately 60 percent of total world output. Nevertheless, the contributions of mining operations in countries with smaller production levels remain a significant factor in the world's supply. Since 2007, the annual output of gold produced from South Africa mines has steadily decreased while at the same time that of China has increased. According to the U.S. Geological Survey, China was the world's largest gold producer in 2009, having collected 300 tonnes of gold.

> **Fool's Gold**
>
> Pyrite is an iron sulfide mineral commonly referred to as "fool's gold" because of its yellow-toned lustrous appearance and its occurrence in quartz vein lodes and sedimentary or metamorphic rock that often also contain real gold. The term fool's gold is often used idiomatically to refer to all nature of things that are thought to constitute treasure or wealth but are in fact worthless.

See also: Alchemy; Aquinas, Thomas; Assaying; California Gold Rush; Cyanide; Golden Fleece; Inca; Klondike Gold Rush; Nubia; Pre-Colombian Gold; Witwatersrand Gold Rush.

Further Reading

Adams, E. D. 1938. *The Birth and Development of the Geological Sciences*. Baltimore: Williams & Wilkins.

Boyle, Robert W. 1987. *Gold: History and Genesis of Deposits*. New York: Van Nostrand Reinhold.

Crane, Walter Richard. 1908. *Gold and Silver: Comprising an Economic History of Mining in the United States.* New York: John Wiley and Sons.
Feathering, George. 1997. *The Gold Crusades: A Social History of Gold Rushes, 1849–1929.* Toronto: Univ. of Toronto Press.
Forbes, Robert James. 1964. *Studies in Ancient Technology.* Boston: Brill.
Green, Timothy. 1993. *The World of Gold: The Inside Story of Who Mines, Who Markets, Who Buys Gold.* London: Rosendale Press.
Guilbert, John M., and Charles F. Park. 2007. *The Geology of Ore Deposits.* Long Grove, IL: Waveland Press.
Hahn, Emily. 1980. *Love of Gold.* New York: Lippincott and Crowell.
Harrell, J. A., and V. M. Brown. 1992. "The World's Oldest Surviving Geological Map: The 1150 B.C. Turin Papyrus from Egypt." *Journal of Geology* 100:3–18.
Ingalls, W. R. "Chronology of the Gold and Silver Industry, 1442–1892." *Mineral Industry, Its Statistics, Technology, and Trade* Vol. I (1893), pp. 225–231.
Kirkemo, Harold, William L. Newman, and Roger P. Ashley. "Gold." *U.S. Geological Survey.* http://pubs.usgs.gov/gip/prospect1/goldgip.html.
Marsden, John, and Iain House. 2006. *The Chemistry of Gold Extraction.* Littleton, CO: Society for Mining Metallurgy and Exploration.
Meyer, C. 1995. "A Byzantine Gold-Mining Town in the Eastern Desert of Egypt: Bir Umm Fawakhir, 1992–1993." *Journal of Roman Archaeology* 8:192–224.
Mohide, Thomas Patrick. 1981. *Gold.* Toronto: Ontario Minister of Natural Resources.
Morteani, Gioiulio, and Jeremy Peter Northover. 1995. *Prehistoric Gold in Europe: Mines, Metallurgy, and Manufacture.* New York: Springer.
Rose, T. K. 2008. *The Metallurgy of Gold: Mining and Mineral Processing,* 6th ed. Palm Springs, CA: Wexford College Press.
Tylecote, R. E. 1976. *A History of Metallurgy.* London: The Metals Society.
World Gold Council. "Chinese Gold Production Rises." http://www.goldipedia.gold.org/news/2010/07/20/story/15224/chinese_gold_production_rises/.
Yannopoulos, John C. 1991. *The Extractive Metallurgy of Gold.* New York: Van Nostrand Reinhold.

MOSAIC

Mosaic is a decorative art form with ancient origins in which small, typically square-shaped pieces of stone, marble, colored glass, or ceramic embellished with gold leaf or bright pigments are assembled and inlaid onto a surface using a type of cement and in some cases grout to form a pattern or image as décor on floors, ceilings, or other architectural elements of sumptuous villas, religious buildings, or public buildings and spaces.

The earliest archeological evidence for primitive forms of mosaic in which stones are inlaid to create floor panels or adorn architectural columns or arches is found in certain sites in Mesopotamia dating from as early as the third millennium BC, as well as in certain Persian sites. In the ninth and eighth centuries BC, decorative pebble flooring in Asia Minor and Assyria demonstrated a greater use of color and some instances of simple patterning. The technical practice of mosaic as it is defined by archeologists and art historians, however, is evident from sites in Greece dating to around the fifth century BC. Excavations of sites containing the ruins of private homes of the wealthy elite from this period in various sites in Greece, such as at Corinth and Olynthos, contain examples of decorative stone mosaic floor coverings in geometric patterns that represent the emergence of the classical

In this mosaic from the Hagia Sophia in Istanbul, Turkey, Byzantine emperor Constantine IX (r. 1042–1055) is offering a purse filled with gold coins to Christ, who is seated to his left. (Pavle Marjanovic/Dreamstime.com)

mosaic style, which would develop by the next century to include specific floral or geometric motifs, a greater use of color, more precise use of cut stone, and even depiction of specific images or stories. Archeologists have also discovered extensive examples of pre-Columbian mosaic art pieces and floor and wall decoration among the Aztec people of Mexico, the Inca of coastal Peru, and the Pueblo cultures in the American Southwest.

The influence of Greek mosaic techniques on Roman decorative arts was profound, and the use of mosaic in luxurious private homes among the Roman elite became widespread. Evidence of floor mosaic and in rare cases wall mosaic from Roman villas exist in sites throughout the former Roman Empire, from the villas of Herculaneum and Pompeii, to Africa, across the Mediterranean, northern Europe, and the British Isles. These mosaics were used as interior decoration and depicted everyday life in the Roman Empire as well as popular decorative motifs such as birds, flowers, or geometric patterns in their borders and images of the legends and heroes of Roman mythology and history. In the absence of an abundance of written sources, these mosaics are of great significance to historians in providing details about the practices and customs of Roman antiquity. With the development of glassmaking in the later Roman period, mosaic artists began to use pieces of colored glass, known as "tesserae," in addition to stone.

With the advent of Christianity in the Roman Empire under the Emperor Constantine, the use of mosaic as a decorative art shifted in significant ways from its predominance in private homes to one of the primary means of decorating early Christian churches and monasteries. With the prohibition against the use of sacred images or representations on floor coverings in the Codex Theodosianus of 427, however, mosaics adorned walls, ceilings, and architectural features such as niches or columns, rather than floors. During this period mosaic craftsmen also switched to the use of colored glass or gold tesserae to assemble their elaborate patterns in a style intended to illuminate the holy figures depicted on the walls by filling the background of the images with gilded glass pieces. Examples of the growth of the technique in the interior decoration of early Christian architecture exist in many sacred churches and monasteries in Italy and the Levant, perhaps most notably in the Church of St. George (Rotunda of Galerius) in Thessaloniki, commissioned by Constantine himself in the fourth century; the nave of the Church of Santa Maria Maggiore in Rome (430–440); the Church of Santa Sabina (422–433) and the Lateran (446–462) in Rome; and in Ravenna, where some of the finest examples of early Christian mosaic dating from the fifth and sixth centuries abound throughout the city's churches.

The art of mosaic flourished during the Byzantine Empire from the 6th to 15th century, a period when the Byzantine mosaic style and craftsmen greatly influenced the development of the art throughout Europe and the Middle East. Some of the most iconic and elaborate gold-embellished Byzantine mosaics were those commissioned by the Emperor **Justinian I** to adorn the walls, ceilings, and niches of the **Hagia Sophia** in Constantinople during the 6th century. When Ravenna was absorbed under the Byzantine Empire in 539 and established as the seat of the Exarchate, these mosaic styles developed to an advanced level, for example, in the mosaic depiction of Justinian I and the Empress Theodora at the Basilica of San Vitale. The Byzantine style also had a significant effect on Islamic mosaic art in the mosques and palaces of the great caliphates during the 6th and 7th centuries, such as those found in Jerusalem's Dome of the Rock and the great mosque of Damascus, as well as in mosques and palaces throughout Muslim Spain. Due to Islamic cultural and religious prohibitions against depicting sacred images, mosaics in these buildings depicted elaborate decorative geometric patterns and motifs containing in many cases complex mathematical patterning elements. Decorative mosaics were also found in Jewish synagogues as floor decoration.

The **iconoclasm** controversy and subsequent persecution of the 8th and 9th centuries in the Byzantine Empire tragically destroyed many examples of early Byzantine mosaic art. Yet the practice came to thrive once again by the 10th century and would soon predominate in influence in the decoration of Christian churches of the High Middle Ages throughout Europe. The 12th and 13th centuries can be characterized as the pinnacles of mosaic art, with artists such as Cimabue and Giotto creating more elaborate, precise images illuminated by dramatic, extensive backgrounds and highlights of gold. Examples of these masterful creations are found in basilicas throughout Italy and Europe, from the detailed mosaic paneling inside St. Mark's Basilica in Venice to the ceiling of the Baptistery to the Florence Cathedral and to mosaics found in the Eastern Orthodox churches of medieval Russia and the Balkan region.

A general shift toward decorating church interiors with fresco during the late 14th and 15th centuries led to a decline in the popularity of mosaic during the later Middle Ages. Despite this, however, many Renaissance masters the likes of Raphael, Tintoretto, and Titian, among others, created designs for detailed, brilliant mosaics that were executed by skilled artisans trained in the ancient technique. Classical mosaic returned to popularity during the baroque period of the late 18th and 19th centuries amid a general appreciation of decorative arts techniques and would be inaugurated into the modern period at the turn of the 20th century as it was employed by art nouveau artists and architects such as

Antoni Gaudi in Spain and Italian glassmaker Antonio Salviati in Venice. Mosaic continues to thrive in the contemporary era in a wide variety of mediums and contexts, ranging from small mosaic jewelry to a continued tradition of decorating Christian churches; artisanal decorative arts for the home and garden; public art adorning architecture and public spaces or statuary; and fine art.

See also: Hagia Sophia; Iconoclasm; Justinian I.

Further Reading

Bowersock, G. W. 2006. *Mosaics as History: The Near East from Late Antiquity to Islam.* Cambridge, MA: Belknap.

Chavarria, Joaquin. 1999. *The Art of Mosaics: A Guide to the History, Material, Equipment, and Techniques.* New York: Watson-Guptill.

Dunbabin, Katherine. 2001. *Mosaics of the Greek and Roman World.* Cambridge: Cambridge Univ. Press.

Joyce, T. A. 1914. "Ancient American Mosaic." *Burlington Magazine for Connaisseurs* 25, no. 135 (June):134–137.

Müller, Valentin. 1939. "The Origin of Mosaic." *Journal of the American Oriental Society* 59, no. 2 (June):247–250.

NANOTECHNOLOGY

Nanotechnology is a relatively recent interdisciplinary scientific field in which researchers are exploring the uses of a variety of substances at the nanoscale, or in a range from 1 to 100 nanometers. A nanometer is measured at one-billionth of a meter, a level just slightly larger than that of an individual atom. This technology is currently being studied and applied in the fields of chemistry, biology, physics, and engineering and offers great potential benefits. Nanotechnology emerged in the late 20th century in the research of physicists and biologists who recognized the benefits of exploring various scientific questions from a bottom-up approach starting at the molecular level. It has since evolved as a major discipline that is viewed as having great potential for improving the human condition through scientific innovations. Since the 1980s, gold in its nanoparticulate form has been increasingly optimized for use in a variety of applications across many scientific and technological sectors.

Although gold is chemically inert in its typical scale, when reduced to a nanoparticle, it becomes soluble and can be used as an incredibly stable, efficient, and affordable **catalyst** and conductivity mechanism. In 2008, a research team under the direction of Professor Hannu Häkkinen of the Nanoscience Centre of the University of Jyväskylä in Finland contributed a significant advancement to understandings of the nanotechnology of gold when they identified the precise atomic structure of gold nanoparticles. Knowledge of how gold nanoparticles are structured provides scientists with more accurate information about features of gold in this form in terms of its stability, composition, and chemical and physical properties. Whereas the majority of gold nanoparticles are manufactured in labs by means of chemical processes, scientists recently discovered the existence of naturally occurring nanoparticles of gold in the environment in areas where saline water had interacted with gold deposits to create nanoparticles over time. This discovery has great potential for improving the effectiveness of industrial gold prospecting and **mining** operations.

Gold nanoparticles are currently being employed in nanotechnology research in a wide variety of industries, with the potential for their use predicted to increase significantly in the near future in the development of biomedical treatments and diagnostic technologies; environmental technologies that reduce toxic automobile emissions and purify ground or drinking water of harmful toxins; efficient **fuel cell** technology as an improved clean energy power source; for economical and improved conductivity in a wide range of computer technologies; and as a stabilizing element and pigment enhancer in industrial paints, dyes, and textiles.

See also: Biomedical Research; Catalysis; Fuel Cells.

Further Reading

CSIRO Australia. 2008. "Natural 'Invisible' Gold Found In Nanoparticles." *ScienceDaily*, June 24. http://www.sciencedaily.com/releases/2008/06/080623105020.htm.

Georgia Institute of Technology. 2010. "Using Gold Nanoparticles to Hit Cancer Where It Hurts." *ScienceDaily*, February 18. http://www.sciencedaily.com/releases/2010/02/100216140402.htm.

Keel, Trevor, Richard Holliday, and Tim Harper. 2010. "Gold for Good: Gold and Nanotechnology in the Age of Innovation." *World Gold Council Bulletin*, January.

Schmidbaur, Hubert, ed. 1999. *Gold: Progress in Chemistry, Biochemistry, and Technology*. Chichester, UK: Wiley and Sons.

University of Jyvaeskylae. 2008. "Structure Of Gold Nanoparticles Solved." *ScienceDaily*, August 21. http://www.sciencedaily.com/releases/2008/08/080820081154.htm.

NEWTON, SIR ISAAC

Born in Woolsthorpe, Lincolnshire, in 1642, Sir Isaac Newton (1642–1727) was a British mathematician, scientist, and natural philosopher credited with establishing the foundations of modern science through his discoveries and inventions in physics, mathematics, and optics. Newton enrolled in Trinity College at Cambridge University in 1661 and later served as professor of mathematics at Cambridge from 1669 to 1696. In 1696 he moved to London to assume an appointment as warden of the Royal Mint, where he investigated and prosecuted counterfeiters and implemented the procedures mandated by what would come to be known as the **Great Recoinage Act of 1696**. In 1700 he was promoted to Master of the Mint, a post he held until his death in 1727. During this period, he was instrumental in implementing modernized **minting** techniques that helped to prevent problems such as **coin clipping**, which debased the currency in circulation. Queen Anne knighted Newton in 1705 as the first scientist ever to receive this honor. He died in London in 1727 without having ever married or produced any heirs and was buried in Westminster Abbey.

Among Newton's works, the *Philosophiae Naturalis Principia Mathematica* (1687) and *Opticks* (1704) contain two of his most renowned discoveries: the laws of motion, which established the foundations of mechanical physics, and the invention of the reflecting telescope. Less well known is Newton's wide range of works in the ancient field of **alchemy**, a process in which alchemists sought to convert base metals into precious metals such as gold and silver through the use of an **alloying** agent perceived as having mystical properties. Many technical aspects of the science of **alchemy** as it was practiced during this period are now considered precursors to modern chemistry. Until recently, very little research has touched on Newton's alchemical interests. This may be due in part to the fact that many of Newton's notes on his alchemical experiments were destroyed in a lab fire, or Newton may have concealed his research in this area because of the crown's ban on **alchemy** and the stigma ascribed to alchemical arts by modern scientists. Among the evidence we do have for his interest in alchemical arts is the *Index Chemicus*, an annotated index of known alchemical texts compiled by Newton during the late 1660s, which contains references to the works of over 100 authors; and the notebooks contained in the Cambridge University archive known as "The Portsmouth

Papers," a collection of miscellaneous documents by Newton donated to the university by the 5th Earl of Portsmouth in 1872, a significant portion of which is focused on various aspects of **alchemy**. Prior to the discovery of these papers, Newton's interest in **alchemy** had remained largely unknown.

Whereas historians of science have largely dismissed Newton's fascination with alchemy as insignificant with respect to his other works, it is important to note that he pursued his studies of **alchemy** during the same period in which he was working on the *Philosophiae Naturalis Principia Mathematica* and *Opticks*. Admittedly, Newton's fascination with the occult aspects of **alchemy** seems contrary to his rationalistic philosophies and the principles of hard science and fixed laws of nature expounded upon in his most famous texts. Yet, at the same time, Newton's ruminations on material theory likely led him to search for a universal first cause responsible for setting all other laws into motion. According to historian of science Craig Chalquist, "for Newton, alchemy held out the possibility of uncovering what animated the laws themselves" (2009, 202).

Newton sought to reveal the key to interpreting the mysterious forces of the universe through an extensive series of alchemical experiments, drawing from the vast range of texts collected in his personal library. These experiments attempted to identify the essence of matter, what is often referred to in **alchemy** as the "philosopher's stone." Newton's engagement with these types of experiments suggests that he believed in the existence of an essential essence that animated all things, something inexplicable by linear, empirical theorems. Newton followed classical alchemical recipes to experiment with the metalloid antimony, a strong alloy used commonly in the process of purifying gold. Antimony was believed by alchemists to be the philosopher's stone, because in the process of transformation by means of an alloy, it produced a crystalline star pattern known as the "regulus," proof in alchemy that one metal could be transformed into another.

Contemporary scholars now tend to view Newton's 30 years of research in the field of alchemy not as a trivial detour into occultism or a quest to transmute lead into gold, but rather as experiments in the emergent field of chemistry and as the logical extension of the interest in one of the world's greatest scientific minds in uncovering the full breadth of what makes the world go round. In the words of Newton himself:

> It is well known, that Bodies act upon another by the Attractions of Gravity, Magnetism, and Electricity; and these Instances show the Tenor and Course of Nature, and make it not improbable but that there may be more attractive Powers than these. For Nature is very consonant and conformable to her self. (*Opticks*, Book 3, Part 1)

The scope of Newton's interest in **alchemy** remained largely unknown to modern scholars until 1936, when an additional portion of documents from the Portsmouth Papers was put up for private auction at Sotheby's in London by Viscount Lymington. Upon learning of the content of these notebooks and papers, economist **John Maynard Keynes** subsequently sought to acquire as many of Newton's alchemical works as possible with the purpose of assembling them into one archive. Keynes's estate contained a provision for the transfer of the archive to the collections at King's College, Cambridge, where it remains today as part of the John Maynard Keynes Papers. In a 1942 essay titled "Newton, the Man," originally intended to be read as a lecture at Cambridge, Keynes hailed Newton as a philosopher with an intellectual legacy rooted in the very origins of man:

> Newton was not the first of the age of reason, he was the last of the magicians. He was the last of the Babylonians and Sumerians, the last great mind which looked out on the visible and intellectual world with the same eyes as those who began to build our intellectual inheritance rather less than 10,000 years ago. (Keynes 1946)

See also: Alchemy; Coin Clipping; Great Recoinage Act of 1696; Keynes, John Maynard; Tower Mint.

Further Reading

Chalquist, Craig. 2009. "Sir Isaac Newton, Alchemist." *Psychological Perspectives* 52, no. 2:199–218.
Craig, John Herbert, Sir. 1946. *Newton at the Mint*. London: Univ. Press.
Dobbs, B. 1991. *The Foundations of Newton's Alchemy, or, "The Hunting of the Green Lyon."* Cambridge: Cambridge Univ. Press.
Dobbs, B. 1992. *The Janus Face of Genius: The Role of Alchemy in Newton's Thought*. Cambridge: Cambridge Univ. Press.
Keynes, John Maynard. 1946. "Newton, the Man." Lecture written for Royal Society of London but never presented. http://www-history.mcs.st-and.ac.uk/Extras/Keynes_Newton.html.
Li, Ming-Hsun. 1963. *The Great Recoinage of 1696 to 1699*. London: Weidenfeld and Nicolson.
Newton, Sir Isaac. 1704. *Opticks*. Oxford: Oxford Univ. Press.
The Newton Project. http://www.newtonproject.sussex.ac.uk/prism.php?id=21.
Principe, Lawrence M., ed. 2007. *Chymists and Chymistry: Studies in the History of Alchemy and Early Modern Chemistry*. Darby, PA: Chemical Heritage Foundation.
Shere, Jeremy. 2006. "Sir Isaac's Alchemy." *Humanities, Then and Now* 29, no. 1. http://www.indiana.edu/~rcapub/v29n1/alchemy.shtml.
White, Michael. 1977. *Isaac Newton: The Last Sorcerer*. Reading, MA: Addison-Wesley.

NOBEL PRIZE MEDALS

Nobel Prize medals are awarded as part of the prize given to recipients of the annual award administered by the Nobel Foundation established by the will of Swedish entrepreneur and inventor Alfred Bernhard Nobel in 1895. Now considered one of the highest international awards in intellectual, diplomatic, and scientific fields, the Nobel Prize is given to acknowledge excellence in areas of scholarship that have provided "the greatest benefit to mankind" in the fields of physics, chemistry, physiology, medicine, literature, and peace. In addition to a diploma and prize money paid in different sums each year by the Nobel Foundation, recipients are honored with a gold medal. The first prizes were awarded in December 1901.

The gold medals are now trademarked symbols of the foundation and are minted at the Mint of Norway and AB Myntverket, a private coin and medal manufacturer in Sweden. Medals produced before 1980 were crafted entirely of 23-karat gold. Since 1980, however, the medals are minted using 18-karat green gold and plated with a coating of 23-karat gold, measuring 66 millimeters in diameter with a depth of between 2.4 and 5.2 millimeters. Medals in the original categories of physics, chemistry, physiology or medicine, and literature are struck with an image of Nobel's profile and life dates (1833–1896) on the obverse and on the reverse different images according to the discipline.

The obverse of the medals contain the same inscription for all the disciplines—*Inventas vitam juvat excoluisse per artes*, a quote from Virgil's *Aeneid* that translates as "they who bettered life on earth by newfound mastery" (6.663). The reverse of the medals for physics, chemistry, physiology or medicine, and literature were designed by Erik Lindberg.

Photograph of the Nobel Peace Prize awarded to United Nations peacekeeping forces in 1988. (Corel)

The physics and chemistry medals, administered by the Royal Swedish Academy of Sciences, are marked on the reverse with an image of the goddess Isis, a symbol of nature who is seen coming forth from the clouds holding a cornucopia while the epitomization of the Genius of Science holds back the veil covering her face. The medal for the award in physiology or medicine, administered by the Nobel Assembly at the Karolinska Institute, depicts on the reverse an image of the Genius of Medicine sitting with an open book on her lap and collecting water from a rock for a sick girl. On the reverse of the medal for literature, awarded by the Swedish Academy, there is an engraving of a seated young man listening to the muse's song and writing it down. Each of these medals also includes a plate on the bottom of the image with an engraving of the recipient's name and the Latin abbreviations for the respective institutes. The medal for the Nobel Peace Prize given by the Norwegian Nobel Committee is minted on the reverse with an image designed by Gustaf Vegeland of three men joined in an embrace signifying the fraternal bond, with the Latin inscription *Pro pace et fraternitate gentium* "For the peace and brotherhood of men." The name of the recipient and the words *Parlamentum Norvegiae* ("Norwegian Parliament") are engraved along the edge. The Nobel Prize in Economics in Memory of Alfred Nobel, adopted later, contains an engraving designed by Gunvor Svensson-Lundqvist that includes the north star emblem of the Royal Swedish Academy of Sciences and the name of the academy in Swedish on the reverse, with a different profile of Nobel above an image of the Bank of Sweden's crossed horns of plenty and the words "The Bank of Sweden, in memory of Alfred Nobel, 1968" engraved along the upper edge.

See also: Karat; Olympic Gold Medal.

Further Reading

Feldman, Burton. 2001. *The Nobel Prize: A History of Genius, Controversy, and Prestige.* New York: Arcade Publishing.

Levinovitz, Ageneta Wallin, and Nils Ringertz, Nils, eds. 2001. *The Nobel Prize.* London: Imperial College Press.

NUBIA

Ancient Nubia was a region in the Nile River Valley that spanned at its peak an area of over 1,000 miles along the Nile from present-day Aswân in southern Egypt in the north to Khartoum in central Sudan to the south, the Red Sea to the east, and the Libyan Desert to the west. Nubia's geographic position was of great significance for the trans-Saharan trade routes that brought gold from Africa to Egypt and the West, as well as for having ownership over an alluvial plain and valley and mountain ore shafts with bountiful gold supplies. The region was continuously inhabited by the ethnic and cultural group known as the Nubians from the Paleolithic era ca. 8000 BC to the 1960s, when the Aswân High Dam flooded this portion of the Nile, forcing the Nubians to relocate.

As an important intersection between the cultures of north Africa, sub-Saharan Africa, and the Mediterranean region, Nubia developed some of the earliest sophisticated urban settlements in the ancient world. Due to the fact that this area of the Nile River Valley was relatively narrow with a limited territory for cultivating crops, urban centers on the scale of the cities of Egypt did not develop here. Yet an abundance of natural resources such as gold and copper fuelled the local economy and contributed to the development of Nubia as an essential trading outpost and hub of cultural exchange between Africa, Asia, and Europe. The name "Nubia," although used in the present day to refer to the ancient region, did not emerge until the Middle Ages, when it was used in reference to the Nuba (or Noba) tribal peoples of the area. Old Testament sources refer to the region as Kush, and the ancient Greeks called it Ethiopia. Some scholars have suggested that the etymology of the term Nubia derives from the Egyptian word for gold—*nub*—although this remains a debated theory. Following the Arab conquest of the region, it was referred to generally as the Sudan, from the Arabic *Bilad es-Sudan*, "Land of the Blacks."

The earliest written records indicate that the Egyptians looked toward ancient Nubia for opportunities to exploit its natural resources, particularly gold and electrum. Control of the Nubian kingdom passed alternately between the Egyptian pharaohs and local rulers, with the Nubians at one time overthrowing the Egyptians ca. 900 BC and conquering Egypt to establish the 25th dynasty, which ruled for over 100 years and extended its sphere of influence to Palestine. The Egyptians relied on Nubian trading centers to receive goods from the East and exploited the area to capture slaves to work on their building projects. The Nubians also faced threat of invasion during this period from the Assyrian and Persian empires.

The gold mines of ancient Nubia were scattered among the valleys and mountains surrounding the Nile with the most important mine based at Wadi Allaqi, in a valley directly east of the city of Elephantine, which was referred to by the Egyptians as the "gateway to the south" due to its strategic position for Egyptian trade and commerce. When the Romans replaced the Ptolemaic dynasty as effective rulers in Egypt in 30 BC, they actively sought control of Wadi Allaqi for access to its rich mines, which led to a series of military skirmishes between the Nubians and the Romans. During the next three centuries of Roman control over Egypt, however, friendly trade relations were established between the neighboring territories, and a significant Roman influence is evident in the archeological record for Sudan. After Roman rule, the former great Nubian kingdom fragmented into three smaller kingdoms. In the 6th century, these kingdoms converted to Christianity and persisted as Christian kingdoms that were strongly allied with the Byzantine court at Constantinople for 1,000 years until the 14th century, when they converted to Islam after coming under Arabic Muslim rule. During the 16th century, Nubia was conquered by the Ottoman Turks. In 1821, the region was invaded by Egyptian ruler Muhammad Ali and annexed to Egypt, which at that time was a Turkish viceregency. Under Egyptian-Turkish rule until 1883, this area of the Sudan was brutally exploited for its natural resources and slave labor, which resulted in political unrest among the locals. In 1898, the region was established as a British protectorate that lasted until independence in 1956.

In 1958, President Abdel Nasser of Egypt constructed the Aswân High Dam to generate electric power and control the flooding of the Nile River Valley. During the 1960s, the lake created by the dam tragically flooded all of lower Nubia, submerging its ancient sites and forcing the relocation of the Nubian peoples, who largely fled to Egypt. Yet the international community responded with a large-scale coordinated effort to rescue the archeological treasures that continues to the

---

Turin Papyrus

Around 1820, Italian diplomat and antiquarian Bernardino Drovetti discovered an ancient papyrus in Deir el-Medina, Egypt, in the area of ancient Thebes, containing a map of ancient Egyptian gold mining operations in Nubia at the site of present-day Wadi Hammamat in Egypt's central eastern desert dating from the reign of Rameses IV (1151–1145 BC). The map was drawn on a papyrus scroll measuring 41 centimeters wide by 2.82 meters long with descriptive hieratic script accompanying the images. The map illustrates the topography and geology of the mining area that served as one of the primary sources for Egyptian gold. The illustrations depict the distribution of the different types of rock in the mountains, indicating the igneous rock areas in pink and the metamorphic rock in black. The map also identifies the location of gold-bearing quartz veins; the location of the Bir Umm Fawakhir gold-processing settlement; and other elements of geologic interest for stone quarry operations. Translations of the hieratic text, a type of bureaucratic script used by ancient Egyptian functionaries, refers to the respective graphic details according to their function, for example, "mountains of gold," "houses of gold-working establishment," and "road to the sea." It is thought that the map was created by the Theban scribe Amenhakte, son of Ipuy. Today, the papyrus is conserved at the Egyptian Museum in Turin, Italy.

Source: Harrell, James A., and V. Max Brown. 1992. "The World's Oldest Surviving Geological Map: The 1150 B.C. Turin Papyrus from Egypt." *Journal of Geology* 100, no. 1 (January):3–18.

present day and ultimately has expanded to include extensive archeological excavations in areas of northern Sudan. Among the artifacts recovered were many tools used for gold **mining**, as well as precious gold objects from royal and noble burials and religious temples and churches. In June 2007, archeologists from the University of Chicago discovered the "first ever evidence of large-scale gold processing in the ancient Nubian kingdom of Kush, more than 55 3,500- to 4,000-year-old grinding stones along the Nile River" (Morrison, 2007).

In modern-day Sudan, political upheaval has disrupted **mining** activities for several decades. However, several state-run and international **mining** operations continue to explore and mine the region's rich natural resources of gold, copper, and other minerals. Simple placer mining also continues to this day along this area of the Nile, where ancient wealth in gold and even a vibrant Nubian culture still thrive despite centuries of conquest and subjugation driven largely by pursuit of gold.

See also: Mining; Trans-Saharan Gold Trade.

Further Reading

Edwards, David N. 2004. *The Nubian Past: An Archaeology of the Sudan*. New York: Routledge.
Forbes, Robert James. 1964. *Studies in Ancient Technology*. Boston: Brill.
Garrard, Timothy F. 1982. "Myth and Metrology: The Early Trans-Saharan Gold Trade." *Journal of African History* 234:443–461.
Harrell, J. A., and V. M. Brown. 1992. "The World's Oldest Surviving Geological Map: The 1150 B.C. Turin Papyrus from Egypt." *Journal of Geology* 100:3–18.
Morrison, Dan. 2007. "Gold Mining, Burial Objects Unearthed on Nile." *National Geographic*, June 19. http://news.nationalgeographic.com/news/2007/06/photogalleries/gold-nile/.
Taylor, John H. 1991. *Egypt and Nubia*. London: British Museum Press.
Welsby, D, and J. Anderson, eds., 2004. *Sudan Ancient Treasures*. London: British Museum Press.

NUESTRA SEÑORA DE ATOCHA

The *Nuestra Señora de Atocha* was an armed Spanish galleon of the *Tierra Firme* treasure fleet that transported gold, silver, and valuable raw materials from the ports of Spanish America and the West Indies back to Spain. The ship sunk in a hurricane off the coast of Florida on September 6, 1622, laden with an extraordinary bounty of treasure. After centuries of recovery efforts, the shipwreck was discovered by a treasure salvage team on July 20, 1985, and yielded an estimated $450 million in precious artifacts and gold and silver **bullion**.

During the colonial era of the late 16th to the early 18th centuries, the Spanish monarchy sent two treasure fleets once a year from the port of Cadiz to New Spain carrying supplies from Spain, and on the return trip, the fleets brought back rich cargoes of gold, silver, and other products such as tobacco and precious gems. These voyages faced many dangers and had to be planned carefully to avoid bad weather or navigational delays as well as the constant threat of pirates or hostile navy ships. The fleets would depart together from Spain early in the year with the aim of returning before the onset of hurricane season in late July, charting a route similar to that of **Christopher Columbus**. Once they reached the Caribbean, the *Nueva España* fleet would head north to the ports of Mexico, while the *Tierra Firme* fleet stopped at the ports of Portobelo on the Isthmus of Panama and Cartagena in Colombia. The two fleets would then meet again in Havana and reassemble as a convoy for the voyage back to Spain.

The *Nuestra Señora de Atocha* was built in Havana in 1620 by master shipwright Alonso Ferreira to serve as an amaranth, or rear guard, to the *Tierra Firme* treasure fleet. The ship stood at a height of 112 feet, weighed 550 tons, and was outfitted with 20 bronze cannons. On March 23, 1622, the ship departed from Cadiz as part of the annual treasure fleet convoy, stopping on the island of Dominica in the Caribbean, then Cartagena, and on to Portobelo, where treasure arriving by mule-train from Lima and Potosi was significantly delayed. The *Atocha* sat in the port for nearly two months as the shipments were loaded and recorded, then departed for Havana on July 22, making an additional stop in Cartagena to load more gold and silver. The fleet did not reassemble in Havana until August 22.

Out of fear of rumors of pirates assembling on the return route, a large majority of the treasure was loaded on the *Atocha* and its sister ship the *Santa Margarita*, as these ships were heavily armed. Despite the windfall to the Spanish treasury of decades of shipments of gold and silver collected from New Spain since its discovery and conquest, newly crowned King Philip IV faced financial strain due to the costs of ongoing war with the French and the Dutch. It was essential that the treasure fleet arrive safely to replenish the royal treasury and pay off debts.

On September 4, 1622, the convoy set sail for Spain, with the *Atocha* and the *Santa Margarita* bringing up the rear as the fleet pursued a northward course toward the Florida Keys to take advantage of favorable currents along the Gulf Stream. By the next day, however, winds increased dramatically from the northeast as a strong hurricane approached. The violence of the seas grew throughout the day, and by evening a change in the winds to the south proved advantageous for the advance ships in the fleet, steering them to the calmer waters of the Gulf of Mexico, while the *Atocha* and the *Santa Margarita* and three other smaller ships, now stripped of their sails and their mast, were thrust helplessly toward the treacherous reefs along the Florida Keys. According to testimonies in documents in the Archives of the Indies in Seville, a giant wave lifted up the *Atocha*, smashed the ship upon the reef, and then dragged the hull down. Because of the heavy weight of its cargo, the ship sank quickly, resting 55 feet below the water with only the tip of its mast remaining above water. Five survivors (three crewmen and two black slaves) out of the 265 estimated passengers and crew clung to the mast and were rescued the next day by another ship.

Upon hearing of the shipwrecks, Spanish authorities launched immediate recovery efforts, sending a small fleet to salvage the desperately needed treasure; however, a subsequent hurricane made such efforts even more difficult by scattering the remains of the wreckage. Over a period of several years, recovery efforts continued using divers and equipment to drag the sea floor. The loss of the cargo had serious financial consequences for Spain as the king was now forced to sell assets and borrow additional money to finance Spanish involvement in the Thirty Years War.

In the 1970s, treasure hunter Mel Fisher undertook what would become a 15-year search for the legendary contents resting in the hull of the *Atocha* shipwreck. After first searching an area off the Florida Keys to no avail, Fisher received new information about the whereabouts of the shipwreck from colonial historian Eugene Lyon, who was researching his doctoral dissertation in the Archives of the Indies in Seville. Based on the accounts of Francisco Nunez Melián, who had led a salvage expedition from Havana in 1626 that successfully recovered a significant portion of the cargo from the *Santa Margarita*, Lyon suggested to Fisher that the shipwrecks had occurred in the Marquesas Keys. After a period of finding sporadic artifacts that hinted at the wreck's location, Fisher's team of treasure hunters discovered what he called the "mother lode" on July 20, 1985, when the team's magnetometer indicated the exact parameters of the ship on the ocean floor beneath.

Despite their incredible good fortune, the team faced a series of challenges following their discovery, as both the state of Florida and the U.S. government asserted claims to significant portions of the treasure and Fisher dealt with honoring and administering the rights of hundreds of shareholders in the recovery venture under the aegis of his company Treasure Salvors. After an eight-year court battle, Fisher and his associates received a favorable judgment from the U.S. Supreme Court, and the booty was catalogued and administered accordingly. Among the vast treasures retrieved were 78 gold coins; 3,100 uncut green emeralds; 160,000 silver coins; 32 tons of silver **bullion** ingots; 115 gold bricks and disks; 78 gold chains; and incredibly rare gold **jewelry** and religious artifacts. Interestingly, a significant amount of contraband items not listed on the ship's manifest were also recovered, including several pounds of green emeralds smuggled by a low-level sailor and crudely crafted gold ingots.

See also: Bullion; Columbus, Christopher; Spanish Conquest.

Further Reading

Lyon, Eugene. 1979. *The Search for the* Atocha. New York: Harper & Row.
Matthewson, R. Duncan. 1986. *Treasure of the* Atocha. Hialeah, FL: Dutton.
Stall, Sam. 1986. "Treasures of the *Atocha*." *Saturday Evening Post* 258, no. 8 (November): 50–100.

OLYMPIC GOLD MEDAL

Olympic gold medals are conferred upon athletes who win first place in the quadrennial sporting events of the modern-day Summer and Winter Olympic Games. Second- and third-place winners receive silver and bronze medals. Whereas athletes competing in the ancient Olympic Games in Athens, which date back to 776 BC, received olive wreaths as their symbolic prize, from the beginning of the modern Olympic Games in 1896, held in Athens, winners received medals made originally out of silver and bronze for first and second place. In subsequent years first-place winners began to receive gold medals, with the silver and bronze medals being awarded to second- and third-place athletes. After 1912, however, the term "gold model" remained in acknowledgment of the status achieved by the athlete as the medals were produced largely of silver with a small percentage of gold alloy.

A gold medal from the Vancouver 2010 Winter Olympics. (Seregal/Dreamstime.com)

The Olympic Charter requires that gold medals consist of 92.5 percent silver and a minimum of 6 grams of 24-**karat** gold with a diameter of no less than 60 millimeters and a thickness of 3 millimeters. The host city is responsible for minting the medals. In 1928, Italian artist Giuseppe Cassioli created the design that was used on the obverse side accompanied by the name of the host city with a representative image of an Olympic champion depicted on the reverse. Beginning in 1972, host cities began to create original designs to be minted on the reverse, and when Athens hosted the Summer Games of 2004, a new image was created to replace Cassioli's design because it featured a Roman amphitheater and was considered by the Greeks to be an anachronistic representation of the ancient Greek origins of the Olympic Games. In the contemporary period, an estimated 3,000 medals are minted for the Olympic Games.

See also: Karat; Nobel Prize Medals.

Further Reading

Arends, Brett. 2010. "What's A Gold Medal Worth?" *Wall Street Journal*, February 18. http://online.wsj.com/article/SB10001424052748704804204575069743500426152.html.
Findling, John E., and Kimberly D. Pelle, eds. 2004. *Encyclopedia of the Modern Olympic Movement*. Westport, CT: Greenwood.
Suddath, Claire. 2008. "A Brief History of Olympic Gold Medals." *Time*, August 7. http://www.time.com/time/magazine/article/0,9171,1830391,00.html.
Young, D. 2004. *A Brief History of the Olympic Games*. Malden, MA: Blackwell.

OSCAR STATUETTE

The Oscar statuette is a gold trophy awarded to professionals in the filmmaking industry who are recognized with an Academy of Merit award. The awards are announced at an annual award ceremony hosted by the American Academy of Motion Picture, Arts, and Sciences. Known as the Academy Awards, this event grew from modest origins in its first year in 1929 to become one of the most famous and widely broadcast award ceremonies in the world.

In 1929 Metro Goldwyn Meyer (MGM) studio head Louis B. Mayer organized the first award banquet, held May 16, 1929, at the Roosevelt Hotel in Hollywood, California. Emil Jennings was the first person to receive an Academy Award when he was acknowledged at the ceremony in the Best Actor category for his performances in *The Last Command* and *Way of All Flesh*. The trophies for the Academy of Merit award were designed by MGM art director Cedric Gibbons, who created a stylized art deco image of a sleek knight holding a crusader's sword in both hands that rested on a base consisting of a film reel with five spokes to represent the

Illustration of an Oscar statuette awarded by the Academy of Motion Picture Arts and Sciences. (Connie Larsen/Dreamstime.com)

five original branches of the academy—actors, writers, director, producers, and technicians. The design was then sculpted in clay by sculptor George Stanley, cast in bronze at the C. W. Shumway and Sons Foundry in Batavia, Illinois, and gilded with gold. A few years later, the bronze cast was replaced with a metal alloy resembling pewter known as britannia, and plated with copper, nickel, and, finally, 24-**karat** gold to create a statuette measuring 13.5 inches tall and weighing 8.5 pounds. For three years during World War II, however, due to a metal shortage the members of the academy decided to substitute the traditional statuettes with plaster painted replicas and offer the recipients metal replacements at the end of the war.

Since 1983, the statuettes have been manufactured to these dimensions, with the exception of a slight change to the base, at the Chicago, Illinois, trophy manufacture R. S. Owens & Company, who presently produces 50 statuettes a year, a process that takes 3 to 4 weeks.

In response to the sale of several trophies at auction or in private transactions in which the statuettes fetched high prices, in 1950, the academy required recipients to sign an agreement restricting them and their heirs from selling Oscars. The exception in the agreement is that recipients or their heirs can sell them back to the academy for one dollar.

Although there is much dispute about the origins of the name "Oscar," it is thought to derive from a comment by academy librarian Margaret Herrick, who remarked that the figure featured on the trophy looked like her Uncle Oscar. In 1939, the academy officially named the statuette the Oscar.

See also: Gilding; Karat; Nobel Prize Medals; Olympic Gold Medal.

Further Reading

Academy of Motion Picture Arts and Sciences. "The Oscar Statuette." http://www.oscars.org/awards/academyawards/about/awards/oscar.html/?pn=statuette.

Holden, Anthony. 1993. *Behind the Oscar: The Secret History of the Academy Awards*. New York: Simon and Schuster.

Kinn, Gail, and Jim Piazza. 2002. *The Academy Awards, the Complete History of Oscar*. New York: Black Dog and Leventhal Publishing.

Osborne, Robert. 2003. *75 Years of the Oscar: The Official History of the Academy Awards*. New York: Abbeville Press.

PIZARRO, FRANCISCO

Francisco Pizarro (ca. 1475–1541) was a Spanish conquistador who conquered the **Inca** Empire of Peru and established the territory as a Spanish colony that would later become the Viceroyalty of Peru with its capital in the city of Lima, founded by Pizarro in 1535. Pizarro's conquest and settlement of the vast, wealthy **Inca** territories greatly enriched the treasury of the Spanish Crown in silver and gold.

> "Curst greed of gold, what crimes thy tyrant power has caused."
> —Virgil

Pizarro was born around 1475 in Trujillo, Extremadura, Castile, as the illegitimate son of a Spanish soldier, Colonel Gonzalo Pizarro, and Francesca Gonzalez, a woman of common origins. Pizarro fought in regional wars in Spain as a young man, and in 1502 embarked on his first voyage to the New World as part of an expedition of the new governor of the Spanish colony in Hispaniola. In 1510 he joined explorer Alonso de Ojeda on a voyage to Urabá, in present-day Colombia, and later joined conquistador **Vasco Núñez de Balboa** on a 1513 expedition across the Isthmus of Panama that is heralded as the first European encounter with the Pacific Ocean.

Pizarro became a close associate of Balboa's successor as governor of the Panamanian settlement of Castillia de Oro, Pedro Arias de Avila, and through Avila's patronage obtained his seat as mayor and magistrate of the newly founded Panama City during 1519–1523, a period when he began to consolidate his own fortune. During 1523–1524, now in his early forties, Pizarro formed a partnership with priest Hernando de Luque and fellow soldier Diego de Almagro with the aim of exploring further south along the west coast of South America based on rumors circulating among fellow explorers and native interpreters that there was a kingdom of vast riches headed by a powerful, wealthy emperor.

The trio set out on two expeditions during 1524–1525 and 1526–1528 that would prove challenging and cost a significant amount of lives due to bad weather, a shortage of food, and encounters with hostile natives. Despite these difficulties, in their encounters with the natives they observed items of gold, silver, and precious jewels that fueled their aspirations to explore further. In 1528, Almagro returned to Panama for reinforcements while Pizarro and the remainder of the crew waited on an island off the coast of present-day Ecuador. Having heard of the loss of life sustained by the expedition thus far, the Spanish governor ordered Almagro to abandon the voyage and return with the crew. In response to this command,

Pizarro is infamously reputed to have drawn a line in the sand, declaring, "There lies Peru with its riches. Here, Panama and its poverty. Choose, each man, what best becomes a brave Castilian." Only 13 men crossed the line to accompany him for the remainder of the exploration, later to become known as *"Los Trece de la Fama,"* or, the Famous Thirteen.

Pizarro and Almagro circumvented the governor's orders by sailing directly to Spain to seek the support of Holy Roman Emperor Charles V himself in the spring of 1528. In a charter signed by the emperor's wife, Queen Isabella, known as the *Capitulación de Toledo*, Pizarro was granted permission to proceed with the conquest of Peru in the name of Spain and was conferred with a noble coat of arms along with various rights and privileges, among them the title of governor and captain general of the lands discovered along the coast south of Panama, which would be designated as the Province of New Castile, in which capacity he had authority equivalent to a viceroy. Although the Famous Thirteen and his original partner Almagro would also receive significant rights and privileges, Pizarro's seniority over Almagro would remain an issue as the conquest progressed forward.

In January 1530, Pizarro thus sailed to Panama from Seville with a crew that included his three brothers, Hernando, Juan, and Gonzalo, along with his cousin Pedro. They departed the following January for Peru with one ship, a crew of 180 men, and around 30 horses. Unable to disembark in the area they had explored previously due to the hostile natives, Pizarro headed slightly north, where he founded the settlement of San Miguel de Piura in July 1532 as a base from which to launch an inland expedition. He was joined here by fellow conquistador Hernando de Soto, who joined Pizarro as he head inland toward the Andes. Along the way they learned through their interpreter that **Inca** emperor **Atahualpa** was embroiled in a civil war with his brother Huáscar over the succession to the throne in the **Inca** capital of Cusco. When Pizarro arrived with his men, **Atahualpa** was in the town of Cajamarca with nearly 30,000 soldiers as he ruled over a northern territory based in Quito.

Upon hearing of Pizarro's approach, **Atahualpa** dispatched an envoy with a small tribute, a ceremonial drink in a golden chalice, and an invitation to meet the **Inca** ruler. The wealth in gold displayed by the **Inca** envoy seemed to confirm the rumors of the region's riches. When the Spanish approached Cajamarca and saw the size of Atahualpa's army, Pizarro knew they would have to plan to ambush and capture the king, as their tiny band of men could not engage such a force in battle. With the permission of the emperor, he sheltered his men at a nearby complex and invited **Atahualpa** to a meeting at the Spanish camp the following day. At the appointed time, when the emperor entered the courtyard with ceremonial splendor, they were surprised to find it empty. Pizarro sent out Dominican friar Vicente de Valverde with an interpreter to attempt to explain the tenets of Christianity to the **Inca** ruler and express their mission to convert him. The emperor replied in anger, throwing the priest's breviary to the ground. Pizarro's surprise attack ensued, and amid the chaos of the horses and gunfire, both unfamiliar to the **Inca**, **Atahualpa** was captured.

During his encounters and imprisonment among the Spaniards, **Atahualpa** observed their lust for gold. In an effort to obtain his release by ransom, he offered to fill a large room once with gold to the ceiling and a smaller room twice with silver in exchange for his release. The bargain was struck and over a period of two months gold and silver was deposited in the rooms in a wide variety of forms, ranging from decorative objects to elaborate vessels and plates, instruments, and other treasures. Once **Atahualpa** had honored his promise, however, Pizarro feared that setting the emperor free would present too great a risk to the conquest, and he had the emperor arrested for sedition, idolatry, and murder and sentenced him to burn at the stake. In exchange for agreeing to convert to Christianity, Atahualpa's sentence was commuted to death by strangling.

One year later, in 1533, Pizarro continued to consolidate Spanish authority over Peru when he captured the **Inca** capital of Cusco. In January 1535, he founded the city of Lima as the regional capital, a city that would become a hub of shipments of gold from New Spain back to Europe.

In the wake of these achievements, however, a dispute broke out between Almagro and Pizarro over Almagro's claims to jurisdiction in Cusco. Following a violent skirmish in which Pizarro's brothers Hernando and Gonzalo were imprisoned, Almagro was arrested and executed in July 1538. Several years later, supporters of Almagro's son, Diego, who had been stripped of all possessions and privileges, attacked Pizarro in his palace in Lima and assassinated him on June 26, 1541. Pizarro had four illegitimate children with various mothers, three sons and one daughter, Francisca, whose birth was acknowledged as legitimate by royal decree in 1537. Pizarro's remains lay in the Lima Cathedral.

See also: Atahualpa; Inca; Spanish Conquest.

Further Reading

Berg, Richard. 1976. "Conquistador: Pizarro and the Conquest of Peru, 1524–1533." *Strategy & Tactics* 58:4.

Hemming, John. 1970. *The Conquest of the Incas*. London: Macmillan.

Pizarro, Francisco. 1986. *Testimonio: Documentos oficiales, cartas y escritos varios*, ed. Guillermo Lohmann Villena. Madrid: Consejo Superior de Investigaciones Científicas, Centro de Estudios Históricos, Departamento de Historia de América "Fernéndez de Oviedo."

Stern, Steve J. 1993. *Peru's Indian Peoples and the Challenge of Spanish Conquest: Huamanga to 1640*, 2nd ed. Madison: Univ. of Wisconsin Press.

Varon Gabai, Rafael. 1997. *Francisco Pizarro and His Brothers: The Illusion of Power in Sixteenth-Century Peru*. Norman: Univ. of Oklahoma Press.

Wood, Michael. 1968. *Conquistadors*. Berkeley: Univ. of California Press.

POLO, MARCO

Marco Polo (ca. 1254–1324) was a 13th-century Venetian merchant and explorer who traveled with his father and uncle from Venice to China, where he remained for many years at the court of the Mongol Empire. Polo's account of his journey and the wonders of the East, the product of a collaboration with writer Rustichello of Pisa, was an immediate success throughout Europe, earning Polo widespread fame for centuries despite certain doubts about the legitimacy of some of his tales. Yet for those who found his depictions of Asia credible, the geographic and cultural information Polo provided in his accounts inspired other adventurers, in particular **Christopher Columbus**, to set out on voyages of discovery and exploration in pursuit of the legendary riches of the lands of Asia recounted by Marco Polo.

Polo was born ca. 1254 in Venice while his merchant father, Niccolò, was in the Middle East with his brother Maffeo on a trading expedition. During their journey, as they traveled east from their base in Constantinople, they encountered an imperial envoy of the Mongolian court and traveled to meet the emperor **Kublai Khan** himself, who received Niccolò and Maffeo warmly and expressed a great curiosity about the customs of the Christian West. The great khan entrusted the merchant brothers with letters addressed to the pope and a request that they return with 100 learned Christian men to instruct him in the ways of the West and holy oil from the lamp at Jerusalem. They were accompanied on their journey home by an emissary of the Mongol court. Upon reaching Acre in 1269, however, they learned from the papal legate that Pope Clement IV had died. They thus continued on to Venice, where they would remain in anticipation of news of the newly elected pope.

Impatient after two years of a stalemate over the papal election, Niccolò and Maffeo departed once again for Acre in 1271, bringing along the 17-year-old Marco. In Acre, they requested letters of reply for the great khan from the papal legate Teobaldo of Piacenza and departed once again for the Mongol empire. Yet shortly after their departure, Teobaldo himself was elected as Pope Gregory X, so they returned to Acre to obtain an official papal reply to **Kublai Khan** as well as to connect with two Dominican monks who agreed to accompany them to the Mongol court. After a treacherous overland journey of two years, passing from Turkey through the lands of Persia and then through Central Asia along the Silk Road, during which the monks returned home out of fear, the Polos reached the court of **Kublai Khan** in China near present-day Beijing in 1275. They would remain as guests and foreign emissaries of the great khan for 24 years, during which they traveled throughout the empire freely, entrusted with a gold engraved medallion that served as a form of "passport" signifying they were to be given safe passage with the emperor's blessing throughout Mongol territory.

During their sojourn as guests of the emperor, the Polos engaged in very lucrative trade, traveling widely throughout Asia conducting business that greatly increased their personal wealth. Marco Polo also claims to have served as a type of foreign official in the province of Yangzhou, although scholars question this claim and the translation of the term "official" in this context. There is evidence, however, that **Kublai Khan** did engage a variety of foreign officials to assist him with establishing the Mongol administration of the conquered Chinese territories and that he embraced a cosmopolitan environment at the imperial court.

In 1292, the Polos received the emperor's permission to return to Venice. For their voyage home, they were entrusted with accompanying a Mongol princess to Persia, where she was to be wed to a prince, with a fleet of 14 ships traveling throughout Southeast Asia and along the western coast of India, reaching the

Persian Gulf through the Strait of Hormuz. Once they delivered the princess at the end of their extremely treacherous journey, during which a majority of the sailors perished and the Polos lost a significant portion of the wealth they had acquired during their stay in the Mongol Empire, they crossed over land to Trebizond, on the Black Sea, and then by ship to Venice, where they arrived in 1295 to the utter disbelief and amazement of their Venetian friends and family who had long since assumed they were dead.

One year after his return to Venice, Marco Polo was taken prisoner by the Genoese while commanding a Venetian galley during a naval battle in the war between Venice and Genoa. While imprisoned, Polo met the writer Rustichello of Pisa, to whom he recounted the tales of his travels throughout the East during a year of imprisonment. Written in the vernacular Franco-Italian and published under a variety of titles, including *Il Milione, A Description of the World*, and, most commonly, *The Travels of Marco Polo*, his story met with instant popularity throughout Europe. By the early 14th century, the manuscript had been translated into a variety of European languages. Upon his return to Venice, Polo soon married Donata Badoer, daughter of a wealthy merchant, with whom he had three daughters. He went on to develop his family's trading business into an increasingly lucrative enterprise. During his own lifetime, Polo faced doubts and criticism about the truths of his claims about his voyage. Asked to recant his stories and acknowledge the tale as fiction upon his deathbed, Polo instead is claimed to have answered that in fact he had only told of half of what he truly had seen. He died on January 8, 1324, in Venice, where he was buried in the church of San Lorenzo.

Whether fact or fiction or a combination of both, *The Travels of Marco Polo* proved to be of profound historical significance in large part due to the book's elaborate descriptions of the wealth of gold to be found in the East. As a merchant trader, Marco Polo described the points of interest of his travels from the viewpoint of economic possibility, political organization, geographic detail, and religious and cultural considerations, information of use to explorers intending to pursue foreign trade. Among the many wonders described by Polo, detailed descriptions of the availability of gold dust and ore in the geographic terrain and the many sumptuous golden textiles, treasures, adornments, and architectural features abound in the book. In fact, reference to gold can be found on almost every few pages. Of his travels from Persia through Central Asia to Cathay, Polo describes cities with towers of gold, rivers, lakes, and mountains containing "vast quantities of gold"; abundant textile production of luxurious cloth of gold; cultures where men adorned themselves with teeth cast in pure gold; and, finally, upon meeting the great khan, a court adorned with so much gold and silver that "none without seeing it could possibly believe it." As he passed through Persia, he encountered the alleged tomb of the **biblical Magi**, three kings who were said to have traveled to Jerusalem to present three offerings to a newborn prophet, of gold, frankincense, and myrrh. If the prophet accepted the gold, he would be acknowledged as a king, the incense, as a god, the myrrh, as a prophet—the young Jesus accepted all three gifts and was thus revered by the magi as at once a king, a god, and a prophet.

In his accounts of the period of his stay in the Mongol empire, Marco Polo provides insights about the khan's use of the wealth of gold politically and in connection to a form of paper currency, and he tells secondhand of the legends of the gold to be found in Cipangu, or Japan, where he reports there is a widespread abundance of gold, including palaces covered in two-inch-thick layers of gold. Of his sea voyage home through Southeast Asia and the archipelagos of the Indian Ocean, Polo observes that gold is to be found "in almost incredible quantities."

These tales of wealth in gold and the geographic position of the capital cities of the East relative to Europe inspired advances in cartography over the course of the next century and fueled the interests of navigators and explorers in the possibility of a shorter, safer ocean passage to Asia. Perhaps most notoriously, **Christopher Columbus** carried an annotated copy of *The Travels of Marco Polo* with him on his watershed voyage in search of the riches of the Indies. Columbus's manuscript is conserved today at the Bibliotheca Colombina in Seville.

See also: Columbus, Christopher; Biblical Magi; Kublai Khan.

Further Reading

Belliveau, Denis, and Francis O'Donnell. 2008. *In the Footsteps of Marco Polo*. Lanham, MD: Rowman & Littlefield Publishing Group.

Bergreen, Laurence. 2008. *Marco Polo: From Venice to Xanadu*. New York: Knopf.

Jackson, A. V. Williams. 1905. "The Magi in Marco Polo and the Cities in Persia from Which They Came to Worship the Infant Christ." *Journal of the American Oriental Society* 26:79–83.

Metropolitan Museum of Art. *In the Footsteps of Marco Polo: A Journey through the Met to the Land of the Great Khan*. http://www.metmuseum.org/explore/marco/get.html.

Polo, Marco. 1903. *The Travels of Marco Polo*, trans. Henry Yule. http://www.gutenberg.org/catalog/world/readfile?fk_files=44328.

Wardwell, Anne E. 1992. "Two Silk and Gold Textiles of the Early Mongol Period." *Bulletin of the Cleveland Museum of Art* 79, no. 10 (December):354–378.

PONTE VECCHIO

The Ponte Vecchio is a medieval bridge on the Arno River in Florence, Italy, in a location that featured a bridge as early as Roman antiquity where the Via Cassia crossed the river. The current structure is believed to have been built in 1345 by architect Taddo Gaddi, although conflicting sources credit Neri di Fioravante for the bridge's design, which consists of three segmental stone arches and a two-story gallery of shops lining either side. An opening in the shops at the midpoint forms a kind of piazza with views of the river on either side.

The earliest sources indicate that the bridge had always hosted shopkeepers who sold their wares from tables extending out from the storefronts, originally consisting of butchers and food merchants as was customary for such bridges in other Italian cities of the era. In 1565, after the Medici family had relocated across the river to Palazzo Pitti from Palazzo Vecchio, Duke Cosimo I de Medici commissioned Giorgio Vasari to design and construct a covered corridor extending from Palazzo Pitti to Palazzo Vecchio so that he and his family might reach the offices of the state more securely without navigating through the crowds and

The Ponte Vecchio is a medieval bridge straddling the Arno River in Florence, Italy, renowned for retaining the characteristic shops and apartments that were a common feature of European bridges during this period. Since the 16th century, the shops have largely housed goldsmiths and money changers. (Edward Phillips/Dreamstime.com)

stench of the bridge's market. Built atop the eastern side of the bridge, what is now known as the Vasari Corridor, this corridor created a sense of architectural uniformity to the bridge's previously jumbled array of colorful second-story buildings and rooftops.

In 1593, Duke Ferdinand I proclaimed that the bridge should be free of "unclean" vendors who contributed to the foul odors and noise. He also decreed that henceforth the shops were to be filled only with gold- and silversmiths or money changers with the aim of establishing the bridge as a popular destination for "gentlemen and foreigners" (Calabi 2004, 33) rather than common citizens doing their daily food shopping. Ferdinand I thus transformed the reputation of the bridge into one of the most esteemed locales for buying and selling gold. A bust of renowned Renaissance goldsmith **Benvenuto Cellini** (1500–1571) sits proudly as a symbol of the bridge's enduring association with this art.

During the retreat of the German army through Florence in August 1944, the Ponte Vecchio was the only bridge across the Arno spared from bombing by direct order of Hitler, although the city suffered a tragic loss of ancient buildings on either side of the bridge which were bombed to block the passage across the river.

The Ponte Vecchio has stood as an iconic symbol of this famous Renaissance city since its construction. In his *History of Florence* (1423), 15th-century chronicler Goro Dati observes that "there are four bridges in the city all of stone and gracefully composed. Among them is one on which there are beautiful shops everywhere, all made of stone, so that you would not think you were on a bridge were there not a piazza in the middle, from where you can see the river above and below."

Today the Vasari Corridor links the bridge with the Uffizi Gallery off Piazza Signoria. Having been restored in 1973 and reopened to the public, the corridor serves as an art gallery with over 1,000 paintings from the 17th and 18th centuries on display and one of the most significant collections of self-portrait paintings by European master artists from the 16th century to the modern era, an extension of the original collection cultivated by Cardinal Leopoldo de' Medici during the Grand Ducal period in the 17th century. The shops and buildings lining both sides of the bridge are still inhabited today by gold- and silversmiths waiting to greet the influx of tourists that cross the bridge every day. The statue of **Benvenuto Cellini** erected in the bridge's small piazza has also became something of a phenomenon as tourists began the practice of attaching padlocks to the fence surrounding the statue as a gesture of romantic love. Unfortunately, this tradition ultimately became a burden to the city and a potential threat to the statue, and the local government began to impose a significant fine against this nostalgic practice.

See also: Cellini, Benvenuto; Goldsmithing.

Further Reading

Calabi, Donatella. 2004. *The Market and the City: Square, Street, and Architecture in Early Modern Europe*. Burlington, VT: Ashgate Publishing.
Dati, Goro. 1904. *L'Istoria di Firenze dal 1380 al 1405*. Ed. L. Pratesi. Norcia.
Fletcher, Sir Banister. 1897. *A History of Architecture*. New York: C. Scribner's Sons.
Graf, Bernhard. 2005. *Bridges that Changed the World*. New York: Prestel Publishing.
Lewis, R.W.B. 1996. *The City of Florence: Historical Vistas and Personal Sightings*. New York: Macmillan.

PRE-COLUMBIAN GOLD

In 1492, **Christopher Columbus** set out on a voyage to discover a new, swift route to Asia in pursuit of the legendary riches recounted in Marco Polo's travel account in which the island of Cipangu, or Japan, was said to have an "endless" quantity of gold. Upon arriving in San Salvador, Columbus's earliest encounters with the natives of the island confirmed that his quest for gold was on the right track as he observed carefully the abundance of gold adornments worn by the islanders, inquiring further about the availability of gold among the local chieftains. Golden objects were produced as gifts or means of exchange in such abundance that Columbus reported excitedly back to King Ferdinand and Queen Isabella of Spain chronicling the vast gold sources of the region and announcing that he had claimed the island for Spain.

> "The desire of gold is not for gold. It is for the means of freedom and benefit."
> —Ralph Waldo Emerson

Figure pendant cast in a gold and copper alloy referred to by the Spanish as *tumbaga*, likely depicting a member of the chieftain class known as the *caciques* among the Tairona people of northern Colombia, ca. 1000–1400. (Werner Forman/Art Resource, NY)

Chronicles and correspondence from initial contact between Western explorers and conquistadors serve as the primary historical source for the marvels of pre-Columbian **goldsmithing**, along with artifacts recovered largely from tombs or temples in modern archeological excavations. The treasures dispatched by explorer **Hernán Cortés** in 1520 from the Aztec empire to the court of Charles V, king of Spain and emperor of the Holy Roman Empire, inspired awe among the artists and connoisseurs of Renaissance Europe. In a journal entry from August 27, 1520, master German artist Albrecht Dürer describes in wonder his impressions of the Aztec goldwork exhibited in Brussels:

> I saw the things which have been brought to the King from the new land of gold, a sun all of gold a whole fathom broad, and a moon all of silver of the same size, also two rooms full of armor of the people there, and all manner of wondrous weapons of theirs . . . These things were all so precious that they are valued at 100,000 florins. All the days of my life I have seen nothing that rejoiced my heart so much as these things, for I saw amongst them wonderful works of art, and I marvelled at the subtle Ingenia of men in foreign lands. Indeed I cannot express all that I thought there. (Dürer 1973, 53–54)

The riches of the New World were chronicled even further during the conquest of **Inca** Peru **by Francisco Pizarro** in 1531–1533 as the Spanish conquistador marched inland in pursuit of the Inca's reputed cities of gold. During their sack of Incan emperor **Atahualpa** and his retinue at Cajamarca, Spanish chroniclers described the mastery of Incan **goldsmithing** in detail, citing intricate works of design and craftsmanship in pieces used as self adornment among the monarchy and nobility and decorative pieces for temples, including intricate miniature gardens cast entirely in gold and the endless variety of gold treasures collected during Atahualpa's infamous ransom.

Tragically, none of the original pieces of pre-Columbian gold sent to Europe by Cortés or Pizarro is known to exist today. These treasures were likely melted down to create gold **bullion** for the Spanish Treasury, and the pre-Columbian capitals in ancient cities such as **Tenochtitlan** and Cusco were plundered during the conquest. However, from a combination of the detailed descriptions in Spanish chronicles and exquisite examples of gold decorative art and **jewelry** recovered from tombs and ancient building sites throughout Latin America, scholars have been able to identify the progression of the ancient craft across the region starting from as early as 3000 BC. These extant pieces allow archeologists and art historians to establish a chronology of the advancement of metalworking techniques and artistic style; however, with the exception of Spanish descriptions of various tools and methods used in pre-Columbian goldwork, mysteries remain about exactly how such magnificent levels of craftsmanship were achieved at such an early date using rudimentary tools of wood and stone and basic technologies.

Pre-Columbian gold production evolved in three identified periods associated with core regions of production, originating in the first phase in the Peruvian highlands perhaps as early as 3000 BC and developing in the period from ca. 500 to 200 BC to include technique such as hammering and annealing and decorative elements achieved by incising, chasing, and embossing. In the second phase, these techniques apparently spread north from Peru to Colombia and Venezuela and then on to the coasts of Panama and Costa Rica from ca. 200 BC to AD 100, a period that also saw significant technological advancements with evidence that furnaces were built that could sustain temperatures high enough to achieve enhanced methods such as open-back casting with the lost wax method and soldering. During the third phase from AD 100 to 400, the craft developed among the civilizations of Mesoamerica. This period saw advancements in soldering techniques and the development of master **goldsmithing** elements such as **filigree** and hollow **casting** that continued to progress up until Western contact, at which point pre-Columbian metalworking had advanced to a degree equal to if not exceeding the craftsmanship of Europe's most talented goldsmiths. Throughout these phases, placer **mining** was evidently the most common method for extracting gold from rich alluvial sources by panning for gold in the river or otherwise trapping the gold deposited by the river, along with limited underground **mining** in the period immediately preceding the conquest.

The cultural role of gold artifacts among the ancient inhabitants of Latin America was in large part a religious one. According to the **Inca**, gold represented the sweat of the sun and silver, the tears of the moon. Precious metals and gemstones were thought of in spiritual terms and considered to be imbued with aspects of the divine and to contain properties that could sustain or prolong human life in a manner consistent with ancient theories of **alchemy**. Columbus and other explorers noted that natives approached the task of extracting gold from abundant alluvial sources as a sacred process that required those who sought this spiritual treasure to fast and uphold religious rituals. Archeologists have discovered that gold offerings were concealed within the walls and foundations of sacred temples. In the ancient town of Atzcapotzalco, in modern-day Mexico, an elite caste of goldworkers were believed to be able to transmit the sacred properties of gold and create elixirs of gold to cure various diseases. In the ancient Aztec language Náhuatl, the word for gold, *teocuitlatl*, translates literally as "excrement (*cuitlatl*) of the gods (*teote*)."

Gold also served as an indication of power and status in pre-Columbian social hierarchies. In the **Inca** Empire, gold was worn exclusively by the ruling class of elites; common people were not permitted to mine or exchange gold. Likewise in the Aztec Empire, rulers held exclusive rights to mine gold and produce precious gold objects. In a similar way to that of many other ancient civilizations, gold decorative ornaments were produced as funerary objects to accompany the deceased in the afterlife as an indication of their status and as offerings, or idols, to the gods. Gold was never coined or used as a medium of exchange as its value was tied to more mystical elements. Gold figurines and pendants often depicted humans and animals as symbolic representations of their authority and importance, some of which had movable parts, a remarkable testament to the ingenuity and craftsmanship of ancient pre-Columbian goldsmiths.

See also: Alchemy; Atahualpa; Columbus, Christopher; Cortés, Hernán; Ferdinand II of Aragon; Isabella I of Castile; Pizarro, Francisco; Spanish Conquest.

Further Reading

Dürer, Albrecht. 1973. *Albrecht Dürer: Diary of His Journey to the Netherlands*, ed. J. A. Goris and G. Marlier. Greenwich, CT: New York Graphic Society.

Emmerich, Andre. 1965. *Sweat of the Sun and Tears of the Moon: Gold and Silver in Pre-Columbian Art*. Seattle: Univ. of Washington Press.

Jones, Julie, and Heidi King. 2002. "Gold of the Americas." *Metropolitan Museum of Art Bulletin* 59, no. 4 (Spring).

Lleras, Roberto, Clara Isabel Botero, and Santiago Lodono. 2007. *The Art of Gold: The Legacy of Pre-Hispanic Colombia*. New York: Skira.

Mackenzie, Donald Alexander. 1978. *Myths of Pre-Columbian America*. Boston: Longwood Press.

Winthrop, Samuel K., et al., eds. 1961. *Essays in Pre-Columbian Art and Archaeology*. Cambridge, MA: Harvard Univ. Press.

PRICE REVOLUTION

During the period from ca. 1520 to the early 17th century, Europe experienced a secular decline in the value of money and rapid, sustained **inflation** so dramatic it soon came to be known in the historiography of the period as the Price Revolution.

Early monetarist theories of the cause of the Price Revolution, a phenomenon of which contemporaries were very aware and which royal advisors and merchants sought to comprehend as it was occurring, tied the escalating price levels to the dramatic influx of gold and silver **bullion** from the New World territories to Spain, which spent its newfound wealth rampantly and thus passed this surplus of available specie along to other European economies. This situation was thought to have been further exacerbated by what the classical quantity theory of money labels as a balance-of-payments deficit due to the fact that the Spanish demand for foreign imports greatly exceeded its own exports, which in turn led to an expansion of the money supply in the foreign European markets that were provisioning the Spanish consumption of goods and thus driving up prices throughout Europe.

Although many noted economists subscribed to the theory that the influx of gold **bullion** from the **Spanish Conquest** of Latin America caused a precipitous rise in **inflation**, including such figures as Jean Bodin, **John Maynard Keynes**, Earl Hamilton, and **Adam Smith**, certain elements of this theory are problematic when applied to modern econometric models. Among the flaws is the fact that the trend toward increasing, sustained inflation began around 1520, several decades before gold **bullion** began to arrive at a substantial enough level to have the identified effects. Additionally, price increases were not consistent across contemporary consumer price indexes, and different commodities increased at different levels. Data for both the exact quantities of gold and silver imported during the period is not as accurate as it should be to support such an argument, and scholars have used contrasting data to argue the reverse—namely, that an increase in prices fueled the need to mine and prospect for specie on a greater level. Finally, there are significant challenges to the compiling of accurate consumer pricing data across Europe.

As an alternative theory, economists and historians have proposed that demographic factors such as population increase and urbanization, in conjunction with socioeconomic trends such as the expansion of capitalist enterprise and the development of modern banking and more sophisticated credit instruments, fueled the rise in prices. Other theories consider how increased silver mining operations and bimetallic currency exchanges may have affected European economies, and some economists point to the possible effects of Portuguese merchants' importation of increasing levels of gold mined in West Africa through African port cities rather than from the traditional overland routes of the **Trans-Saharan Gold Trade**.

Interestingly, the value of gold as a commodity increased at a slower pace than the price-level increases across other categories of commodities. Scholars have thus evaluated examples of debasement of various currencies during this period and considered the significance of their individual effects on domestic economies; for example, the Great Debasement resulting from King of England Henry VIII's desperate attempt to raise revenue by ordering an adjustment of the ratio of gold and silver coins produced at the **Tower Mint** during 1543–1551, a policy that led to crippling inflation and widespread hardship and civil unrest among his subjects. The Price Revolution also had a significant impact on economic conditions in the Ottoman Empire and is linked to a general increase in the flow of gold to Asia, where the gold was hoarded rather than re-circulated in global trade.

Scholars still widely debate the causes of the Price Revolution in 16th-century Europe as new understandings of cyclical economic patterns and monetary circumstances continually emerge. Significant research remains to be done to identify a more sufficient level of accurate data about price levels and the inflow and outflow of specie during this period as well as about the increasingly centralized governments' policies toward the regulation of commerce.

See also: Bullion; Henry VIII; Inflation; Keynes, John Maynard; Smith, Adam; Spanish Conquest; Trans-Saharan Gold Trade; Tower Mint.

Further Reading

Braudel, Fernand, and Franck Spooner. 1967. "Prices in Europe from 1450 to 1570." In *The Cambridge Economic History of Europe*, vol. 4, eds. E. E. Rich and C. H. Wilson, 338–442. Cambridge: Cambridge Univ. Press.

Fischer, David Hacket. 1996. *The Great Wave: Price Revolution and the Rhythm of History*. London: Oxford Univ. Press.

Fisher, Douglas. 1989. "The Price Revolution: A Monetary Interpretation." *Journal of Economic History* 49, no. 4 (December):883–902.

Gould, J. D. 1964. "The Price Revolution Reconsidered." *Economic History Review* New Series 17, no. 2:249–266.

Grice-Hutchinson, Marjorie. 1952. *The School of Salamanca: Readings in Spanish Monetary Theory*. London: Oxford Univ. Press.

Hamilton, Earl J. 1970. *American Treasure and the Price Revolution in Spain, 1501–1650*. New York: Octagon Books. (Orig. pub. 1934.)

Kindleberger, Charles. 1989. *Spenders and Hoarders: The World Distribution of Spanish American Silver, 1550–1750*. Pasir Pajang, Singapore: ASEAN Economic Research Unit.

Ramsey, Peter H. 1971. *The Price Revolution in Sixteenth-Century England*. London: Metheun.

Schwartz, Anna. 1973. "Secular Price Change in Historical Perspective" *Journal of Money, Credit, and Banking* vol. 5:243–279.

Tóth, Sándor László. 1988. "The 'Price Revolution' in the Ottoman Empire at the End of the Sixteenth Century." *Acta Universitatis Szegediensis de Attila Jozsef Nominatae: Acta Historica* 87 (July):35–47.

RICARDO, DAVID

David Ricardo (1772–1823) was a British economist and financier who is acknowledged as the founder of classical economic theory during the early 19th century. Ricardo's works contributed to the development of modern monetary theory during the early 19th century and his embrace of free-market capitalism had a significant influence on the economic thought of Victorian-era Britain.

Ricardo was born ca. April 18, 1772, in London, son of a Jewish banker and financier with Sephardic Jewish origins who had emigrated to London from the Netherlands. After a brief period of schooling, Ricardo went to work for his father as a stockbroker on the London Stock Exchange at the early age of 14. When he was 21, Ricardo fell in love with a Quaker, Priscilla Anne Wilkinson, whom he married after converting and renouncing Judaism, an event that led to permanent estrangement from his parents. Through his own business connections, however, Ricardo established a successful brokerage firm and swiftly accumulated a vast fortune and admiration from his colleagues for his savvy approach to financial matters. He and his wife had five daughters and two sons. By his early forties, he himself acknowledged in his personal correspondence that he felt he had acquired sufficient wealth to retire. In 1814, he purchased an estate at Gatcombe Park in Gloucestershire and began the life of a country gentleman and man of letters. At the suggestion of his close friend James Mill, father of John Stuart Mill, Ricardo decided to seek a Parliament seat in the House of Commons in 1819.

Ricardo's interest in economic theory evolved out of his initial reading of Adam Smith's *The Wealth of Nations* (1776) in 1799. In particular, Ricardo read Smith's work in light of the Bank Restriction Act of 1797, an Act of Parliament that relieved the Bank of England of its obligation to convert paper currency into gold out of concerns over the depreciation of the national currency and a shortage of **gold reserves** resulting from the financial pressures of war with France. As the currency situation worsened, Ricardo entered the public debate through a series of letters published in the *Morning Chronicle* during 1809–1810 in which he argued vigorously for a return to the convertibility of bank notes to gold. Ricardo's letters are credited with encouraging Parliament to appoint a committee to research the question and make its recommendations to the government. During February–May 1810, the committee selected to "enquire into the cause of the high price of gold bullion and to take into consideration the state of the circulating medium and of the exchanges between Great Britain and foreign parts" submitted its recommendations to the House of Commons (quoted material is actual subtitle of the committee). On his part, Ricardo addressed what would come to be known as the

Bullionist Controversy with a pamphlet published in 1810, "On the High Price of Bullion," in which he made a strong case for gold, arguing:

> The precious metals employed for circulating the commodities of the world, previously to the establishment of banks, have been supposed by the most approved writers on political economy to have been divided into certain proportions among the different civilized nations of the earth, according to the state of their commerce and wealth, and therefore according to the number and frequency of the payments which they had to perform. While so divided they preserved every where the same value, and as each country had an equal necessity for the quantity actually in use, there could be no temptation offered to either for their importation or exportation. Gold and silver, like other commodities, have an intrinsic value, which is not arbitrary, but is dependent on their scarcity, the quantity of labour bestowed in procuring them, and the value of the capital employed in the mines which produce them. (Ricardo 1810)

During the parliamentary proceedings, Bank of England directors refuted the suggestion that an increase in the availability of credit had any relationship to the depreciation of the pound and onset of inflation. Yet Ricardo's adamant position and theorization of the connections between price levels, money supply, foreign **exchange rates**, and **gold reserves** ultimately represented the foundations upon which a systematized theory of political economy and modern understandings of the function of central banking were constructed. The **Bullion Committee of 1810** largely validated Ricardo's recommendation that the convertibility of the currency to gold be reinstated.

Ricardo continued to expand on his theories in subsequent works such as *An Essay on the Influence of a Low Price of Corn on the Profits of Stock showing the inexpediency of Restrictions on Importation; with remarks on Mr Malthus' two last Publications* (1815), in which he engaged in debate with economist Thomas Malthus and presented a version of one of the most significant theories in the history of economics: the "law of diminishing returns," which introduced the theory of "natural wages" and identified correlations between wages, the cost of production, and prices. In the face of criticisms that this theory could be applied only to a single commodity, however, Ricardo later published *On the Principles of Political Economy and Taxation* (1817), which analyzed laws governing economic conditions across different sectors of the economy and proposed a theory of value that accounted for the ties between prices and the relative cost of labor used to produce diverse commodities. Ricardo himself acknowledged that his theory of value was significantly flawed due to its failure to account for the differential of capital levels across industries and regions. As an extension of these theories, he also presented a detailed case study to illustrate the benefits of free trade over protectionism.

Ricardo died at a relatively young age on September 11, 1823, at his estate at Gatcombe Park due to complications resulting from an ear infection.

The economic theory articulated by David Ricardo during his short life would continue to influence the thought of scholars such as Thomas Malthus, John Stuart Mill, and Karl Marx during the mid-19th century. After a period of decline in the late 19th century and early 20th century, "Ricardian economics" experienced a resurgence following Italian émigré economist Piero Sraffa's discovery of an equation for solving the dilemma in measuring value based on variable factors, an innovation that sparked what is known as the "Classical Revival" in contemporary economic theory.

See also: Banking and Credit; Bullion; Bullion Committee of 1810; Exchange Rates; Gold Reserves; Smith, Adam.

Further Reading

Bonar, J. 1923. "Ricardo's Ingot Plan." *Economic Journal* 33:281–304.
Heilbroner, Robert L. 1953. *The Worldly Philosophers*. New York: Simon and Schuster.
Hollander, Jacob. 1904. "The Development of Ricardo's Theory of Value." *Quarterly Journal of Economics* 18:455–491.
Hollander, Jacob. 1910–1911. "The Development of the Theory of Money from Adam Smith to David Ricardo." *Quarterly Journal of Economics* 25:429–470. http://socserv2.socsci.mcmaster.ca/~econ/ugcm/3ll3/hollander/money.html.
Hollander, Samuel. 1986. "Sraffa and the Interpretation of Value." *History of Political Economy* 32, no. 2:187–232.
McCulloch, J. R. 1888. *The Works of David Ricardo: A Notice of the Life and Writings of the Author*. London: John Murray. http://oll.libertyfund.org/?option=com_staticxt&staticfile=show.php%3Ftitle=1395&Itemid=27.
Ricardo, David. 1810. *The High Price of Bullion, a Proof of the Depreciation of Bank Notes*. Pamphlet printed in London by John Murray. http://socserv2.mcmaster.ca/~econ/ugcm/3ll3/ricardo/bullion.
Ricardo, David. 1815. *An Essay on the Influence of a Low Price of Corn on the Profits of Stock showing the inexpediency of Restrictions on Importation; with remarks on Mr Malthus' two last Publications*. http://socserv.mcmaster.ca/~econ/ugcm/3ll3/ricardo/profits.txt.
Ricardo, David. 1817. *On the Principles of Political Economy and Taxation*. Pamphlet printed in London.
Ricardo, David. 1951–1973. *The Works and Correspondence of David Ricardo*, ed. Piero Sraffa with help of M. H. Dobb, 11 vol. Cambridge: Cambridge Univ. Press.

RUMPELSTILTSKIN

First appearing in print in the 1812 collection by the Brothers Grimm, *Children's and Household Tales*, "Rumpelstiltskin" is a German fairy tale about a strange little man who could spin straw into gold. The title refers to the tale's main character and is a

Illustration of Rumpelstiltskin from a 1906 edition of *Household Tales by the Brothers Grimm*. (*Household Tales by the Brothers Grimm*. London: J. M. Dent & Co., 1906)

German word for a type of mischievous household goblin. Although this version has gained prominence in the modern West, similar tales have appeared in other cultures since the late 16th century. The Grimm story recounts the plot as follows:

> Once there was a miller who was poor, but who had a beautiful daughter. Now it happened that he had to go and speak to the King, and in order to make himself appear important he said to him, "I have a daughter who can spin straw into gold." The King said to the miller, "That is an art which pleases me well; if your daughter is as clever as you say, bring her tomorrow to my palace, and I will try what she can do."
>
> And when the girl was brought to him he took her into a room which was quite full of straw, gave her a spinning-wheel and a reel, and said, "Now set to work, and if by tomorrow morning early you have not spun this straw into gold during the night, you must die." Thereupon he himself locked up the room, and left her in it alone. So there sat the poor miller's daughter, and for the life of her could not tell what to do; she had no idea how straw could be spun into gold, and she grew more and more miserable, until at last she began to weep.
>
> But all at once the door opened, and in came a little man, and said, "Good evening, Mistress Miller; why are you crying so?" "Alas!" answered the girl, "I have to spin straw into gold, and I do not know how to do it." "What will you give me," said the manikin, "if I do it for you?" "My necklace," said the girl. The little man took the necklace, seated himself in front of the wheel, and "whirr, whirr, whirr," three turns, and the reel was full; then he put another on, and whirr, whirr, whirr, three times round, and the second was full too. And so it went on until the morning, when all the straw was spun, and all the reels were full of gold. By daybreak the King was already there, and when he saw the gold he was astonished and delighted, but his heart became only more greedy. He had the miller's daughter taken into another room full of straw, which was much larger, and commanded her to spin that also in one night if she valued her life. The girl knew not how to help herself, and was crying, when the door again opened, and the little man appeared, and said, "What will you give me if I spin that straw into gold for you?" "The ring on my finger," answered the girl. The little man took the ring, again began to turn the wheel, and by morning had spun all the straw into glittering gold.
>
> The King rejoiced beyond measure at the sight, but still he had not gold enough; and he had the miller's daughter taken into a still larger room full of straw, and said, "You must spin this, too, in the course of this night; but if you succeed, you shall be my wife." "Even if she be a miller's daughter," thought he, "I could not find a richer wife in the whole world."
>
> When the girl was alone the manikin came again for the third time, and said, "What will you give me if I spin the straw for you this time also?" "I have nothing left that I could give," answered the girl. "Then promise me, if you should become Queen, your first child." "Who knows whether that will ever happen?" thought the miller's daughter; and, not knowing how else to help herself in this strait, she promised the manikin what he wanted, and for that he once more span the straw into gold.
>
> And when the King came in the morning, and found all as he had wished, he took her in marriage, and the pretty miller's daughter became a Queen.
>
> A year after, she had a beautiful child, and she never gave a thought to the manikin. But suddenly he came into her room, and said, "Now give me what you promised." The Queen was horror-struck, and offered the manikin all the riches of the kingdom if he would leave her the child. But the manikin said, "No, something that is living is dearer to me than all the treasures in the world." Then the Queen began to weep and cry, so that the manikin pitied

her. "I will give you three days' time," said he, "if by that time you find out my name, then shall you keep your child."

So the Queen thought the whole night of all the names that she had ever heard, and she sent a messenger over the country to inquire, far and wide, for any other names that there might be. When the manikin came the next day, she began with Caspar, Melchior, Balthazar, and said all the names she knew, one after another; but to every one the little man said, "That is not my name." On the second day she had inquiries made in the neighborhood as to the names of the people there, and she repeated to the manikin the most uncommon and curious. "Perhaps your name is Shortribs, or Sheepshanks, or Laceleg?" but he always answered, "That is not my name."

On the third day the messenger came back again, and said, "I have not been able to find a single new name, but as I came to a high mountain at the end of the forest, where the fox and the hare bid each other good night, there I saw a little house, and before the house a fire was burning, and round about the fire quite a ridiculous little man was jumping: he hopped upon one leg, and shouted—

"To-day I bake, to-morrow brew, The next I'll have the young Queen's child. Ha! glad am I that no one knew That Rumpelstiltskin I am styled."

You may think how glad the Queen was when she heard the name! And when soon afterwards the little man came in, and asked, "Now, Mistress Queen, what is my name?" at first she said, "Is your name Conrad?" "No." "Is your name Harry?" "No."

"Perhaps your name is Rumpelstiltskin?"

"The devil has told you that! The devil has told you that!" cried the little man, and in his anger he plunged his right foot so deep into the earth that his whole leg went in; and then in rage he pulled at his left leg so hard with both hands that he tore himself in two.

When interpreted in the context of the folklore tradition in which the story was told, the moral of this tale is commonly believed to be a warning against the downside of arrogance, greed, and deception, with Rumpelstiltskin as the personification of evil, to whom one is vulnerable when engaged in dishonest deeds. Such themes also reflect fears of the unknown often associated with the practice of **alchemy**. The miller's insistence that his daughter can spin straw into gold likely relates to ties between gold and royalty. By virtue of her ability to produce gold, her father is implying she is fit to be queen.

Typical of other Brothers Grimm tales, the story plays to humanity's deepest fears through the threat of the loss of an infant child. Furthermore, the moral extends beyond a warning against greed when Rumpelstiltskin refuses to return the infant even after the queen offers him all the treasures of the kingdom. The goblin's reply that something living is more valuable than gold is meant as a potent reminder of the superficiality of material wealth.

See also: Alchemy.

Further Reading

Grimm, Jacob and Wilhelm. 1812. *Children's and Household Tales*, trans. Margaret Hunt. http://www.gutenberg.org/dirs/etext04/grimm10a.txt downloaded 1/17/2010.

Rand, Harry. 2000. "Who Was Rumpelstiltskin?" *International Journal of Psychoanalysis* 81:943–962.

SAN ESTEBAN

The *San Esteban* was a Spanish cargo galleon that was shipwrecked off the coast of Texas in 1554 along with two other ships, the *Espíritu Santo* and *Santa María de Yciar*, in a fleet comprised of four ships altogether carrying nearly 400 passengers and a wealth of cargo from the New World back to Spain. The wreckage of the *San Esteban* was discovered in the 1960s by Platoro, Inc., a private salvaging company.

The fleet arrived at the port of San Juan de Ulúa in Veracruz, Mexico, in March 1553 after a treacherous journey from Spain that had destroyed many of the other ships in the convoy, including the captain ship. The *San Esteban* and its fellow ships sat in port as its cargo was unloaded and repairs were made, forced to remain in New Spain for almost a year, having missed the opportunity to meet up with the armed convoy in Havana before heading back to Spain for the return voyage, an essential precaution due to the threat of pirates on the route home.

On April 9, 1554, the small fleet set sail without the convoy under the leadership of Captain General Antonio Corzo, carrying an estimated 2 million pesos in cargo and a likewise precious group of passengers comprised of conquistadors, rich merchants, laborers, and crew members. Having completed half of their route to Havana, on April 29 a storm forced the ships off course just south of the Bay of Corpus Christi, where the *San Esteban*, *Espíritu Santo*, and *Santa María de Yciar*, were wrecked on the sandbar surrounding Padre Island. The fourth ship, the *San Andrés*, narrowly survived to complete the journey to Havana in terrible condition. An estimated 300 passengers and crew members perished as a result of the wreck, which later assumed the name "The Wreck of the Three Hundred." Many victims drowned, yet a significant number managed to get to the beach, and a small group of seamen organized by *San Esteban* ship master Francisco del Huerto headed to Veracruz in a recovered dingy to get help. The survivors remaining on land mistakenly set out in a southward direction in an attempt to walk back to Mexico, running afoul of the Karankawa Indians, who attacked them along their route, or falling prey to the unforgiving conditions of the terrain.

Huerto reached Veracruz successfully, and in early June Spanish authorities sent out an immediate expedition to rescue any survivors. In July, a salvage expedition attempted to recover the lost wealth of cargo, recovering over 35,000 pounds of cargo, but unable to locate the remaining known 51,000 pounds of treasure made up of precious metals, coins, **jewelry**, and religious artifacts.

During the 1960s, a private American salvage company, Platoro, Inc., began to excavate treasures from the wreck of the *Espíritu Santo* when the Texas General

Land Office alerted state officials of the ship's discovery and the company's unpermitted recovery operation. Based on U.S. Supreme Court Legislation decision dating to 1960 that all underwater territory off the coast of Texas in the range of 10.35 miles remained the property of the state, Platoro was sued by the state attorney general to assert the state's jurisdiction and halt the private firm's excavation on the basis of both the court ruling and the concern that such a historic finding should be uncovered and preserved scientifically with an appreciation of the historical and archeological significance of the shipwreck—acknowledged to the present day as the oldest archeological excavation of a shipwreck site in the Western Hemisphere—and its cargo for Texas state history. The case remained tied up in the courts for several years over a controversy regarding state versus federal jurisdiction with respect to the preservation of the artifacts and the question of compensating the original private salvage firm. In the interim, the artifacts already recovered were taken to the Texas Archeological Research Laboratory at the University of Texas, Austin, to be catalogued and studied. In 1984 the state attorney general offered Platoro a cash settlement of $313,000 and acknowledged the state of Texas as the custodian of the artifacts, which were subsequently conserved at the Corpus Christi Museum. Among the items recovered from the *San Esteban* were silver and gold **bullion** and coins, gold religious artifacts, personal items belonging to the passengers, silver disks, cross bows, cannons, and three astrolabes, one of which is the oldest known extant astrolabe in the world, along with a wide range of other items of interest, such as rare native artifacts. The wreck site and Padre Island recovery and rescue sites remain preserved as part of the Padre Island National Seashore preserve and the Corpus Christi Museum of Science and History.

See also: Nuestra Señora de Atocha; Spanish Armada; SS *Republic*.

Further Reading

Arnold, J. Barto, III. 1979. *Documentary Sources for the Wreck of the New Spain Fleet of 1554*, trans. David McDonald. Austin: Texas Antiquities Committee Publication.

Arnold, J. Barto, III, and Melinda Arceneaux Wickman. "Padre Island Shipwrecks of 1554." *Handbook of Texas Online*. http://www.tshaonline.org/handbook/online/articles/PP/etpfe.html.

Arnold, J. Barto, III, and Robert Weddle. 1978. *The Nautical Archeology of Padre Island*. New York: Academic Press.

Olds, Doris. 1976. *Texas Legacy from the Gulf: A Report on Sixteenth Century Shipwreck Materials from the Texas Tidelands*. Austin: Texas Memorial Museum Publication 2, Misc. Papers 5, and the Texas Antiquities Committee.

SHEKEL

The term "shekel" is found in Old Testament and other ancient Semitic sources in reference simply to a standard measure of weight, the equivalent of 9 pennyweight troy. The word derives from the Hebrew verb *shaqal* ("to weigh"). The first evidence for the use of the shekel as a coin in the form of gold, silver, bronze, or iron is often considered to have derived from Persian sources, although Herodotus credits **Croesus**, King of Lydia, as the first to mint coins. Prior to the use of coins, precious metals were simply kept in a money pouch and weighed.

The earliest written sources documenting the shekel in use as a coin refer to the **minting** of a shekel coin of silver during the First Jewish Revolt in AD 66–67, when the Temple in Jerusalem was briefly reclaimed from Roman control. During this period, the shekel was minted as a symbol of the Jews' independence from Rome.

In 1984, the state of Israel changed its currency from the lira (pound) to the New Israeli Shekel in an effort to combat the devaluation of the original currency.

See also: Croesus.

Further Reading

Doty, Richard G. 1982. *Macmillan Encyclopedic Dictionary of Numismatics.* New York: Macmillan.

Hobson, Burton, and Robert Obojski. 1970. *Illustrated Encyclopedia of World Coins.* Garden City, NY: Doubleday.

Room, Adrian. 1987. *Dictionary of Coin Names.* London & New York: Routledge & Kegan Paul.

SILLA, ROYAL TOMBS OF

The Royal Tombs of Silla consist of vast burial mounds uncovered in the modern-day city of Gyeongju on the southeastern coast of South Korea, originally the site of the capital of the kingdoms of Old Silla (57 BC–AD 665) and Unified Silla (668–935), where members of the local monarchy and nobility were buried along with their personal treasures. The tombs were found to be in excellent condition during modern-day excavations as they were constructed by lowering wooden covered tombs into a pit that was sealed with clay and covered with mounds of earth and large boulders. The tombs also lacked any passageways to prevent the pillage of the treasures concealed within.

Excavations of the tombs revealed a vast wealth of gold artifacts that indicated a sophisticated

---

Vatican Bank Lawsuit

In 1999 a class action lawsuit was filed representing plaintiffs who were Holocaust survivors from Croatia, the Ukraine, and Yugoslavia against the Vatican Bank and the Franciscan Order alleging that the defendant organizations received large quantities of gold from exiled leaders of the Nazi collaborationist Croatian Fascist group Ustasha while they took refuge at the Pontifical Croatian College of St. Jerome in Rome after World War II. All of this gold was alleged to have been looted during the persecutions of Jews and other groups and then laundered by the Vatican's and the Franciscans' financial institutions to be distributed back to Ustasha and Nazi fugitives. The defendants hoped to obtain a complete accounting of the postwar dissemination of the Ustasha Treasury and restitution of gold transferred illicitly to the Vatican and the Franciscan Order. The U.S. State Department also investigated the claim in its 1998 report on the recovery and restoration of Nazi assets obtained during the war.

Both the Vatican Bank and the Franciscan Order filed motions to dismiss the case on various legal grounds. The Vatican Bank did not deny receiving the shipments of gold but disputed the existence of any evidence that the gold in question was tied to the plaintiffs' losses, claiming they were not subject to the lawsuit under the provisions of the 1976 Foreign Sovereign Immunities Act. The lawsuit was originally filed in U.S. District Court in San Francisco, which ruled in 2003 that the United States did not have jurisdiction in the case due to the Vatican's status as a sovereign entity. The case was reinstated in the Ninth Circuit Court of Appeals in 2005, yet the original ruling was upheld on the same grounds. Such a ruling allowed the courts to dismiss the case on political and jurisdictional grounds without having to address the question of the veracity of the plaintiffs' allegations.

The case against the Franciscan Order, however, has continued to the present day.

Source: "U.S. Court Rejects Vatican Bank Lawsuit." 2009. *CBS News*, December 29. http://www.cbsnews.com/stories/2009/12/29/world/main6036103.shtml.

U.S. State Department. 1998. "June 1998 Supplement to *Preliminary Study on U.S. and Allied Efforts To Recover and Restore Gold and other Assets Stolen or Hidden by Germany During World War II*." http://www.state.gov/www/regions/eur/rpt_9806_ng_links.html.

level of **goldsmithing** techniques for this era, leading scholars to believe that craftsmen were likely exposed to Chinese practices or even perhaps the work of the Etruscans and ancient Greeks due to examples of advanced metalwork, notably **granulation** and **filigree**. The tombs contained a wide array of pure gold treasures, including ornaments and accessories such as headpieces, earrings, necklaces, bracelets, belts, rings, and swords. Among the most famous discoveries within the tombs were six delicate crowns of pure gold that are now designated as national treasures in Korea.

The precious items buried along with kings, queens, and the high nobility not only provide examples of a high level of craftsmanship but also offer insight into social and gender relations as the manner in which they were buried and the adornments themselves provide important clues about how gold heirlooms and jewelry were used in ancient Korean culture as displays of social status, power, and wealth in the kingdom. Quality and design of the gold objects varied notably among degrees of royalty and nobility in the various tombs, an indication that the level of the goldsmith's craft itself was regarded as a reflection of the owner's status.

There is evidence that gold was originally imported to Silla by way of China. Gold ornaments were used among Silla elites and royals to convey their social status and political rank. As this practice flourished, gold was eventually mined and crafted locally in response to an increase in demand. With the conversion of the Silla monarchs to Buddhism in 528, however, the

Golden crown found in the tomb of Hwangnam-Dong at Silla, Three Kingdoms period, fifth to sixth century. Conserved at the National Museum in Gyeongju, South Korea. (DeA Picture Library / Art Resource, NY)

practice of burying these gold ornaments along with the deceased eventually died out in the later sixth century.

See also: Crowns, Royal; Goldsmithing; Granulation; Filigree; Jewelry; Ur, Royal Tombs of.

Further Reading

Adams, Edward B. 1991. *Korea's Golden Age: Cultural Spirit of Silla in Kyongju.* Seoul: Seoul International Pub. House.

Lee, Soyoung. 2003, October. "Golden Treasures: The Royal Tombs of Silla." In *Heilbrunn Timeline of Art History*. New York: Metropolitan Museum of Art. http://www.metmuseum.org/toah/hd/sila/hd_sila.htm.

Pak, Youngsook. 1988. "The Origins of Silla Metalwork." *Orientations* 19, no. 9:44–54.

SMITH, ADAM

Adam Smith (1723–1790) was a Scottish philosopher and political economist who is considered to be responsible for the development of modern theories of free market capitalism and is widely hailed as one of the founding thinkers of the field of economics for the theories he developed in his multivolume work *An Inquiry into the Nature and Causes of the Wealth of Nations* (1776).

Smith's exact date of birth is unknown, but his baptism occurred in early June 1723 in the small town of Kirkcaldy, near Edinburgh. Smith's father, Adam, who died months before his son was born, was a local customs comptroller. His mother, Margaret Douglas, was the daughter of a local landowning family of significance. The young Adam Smith received his elementary education in Kirkcaldy and was sent in 1737 to attend the University of Glasgow, where he was influenced by several of the early Scottish Enlightenment philosophers such as Philosophy Department Chair Francis Hutcheson. In 1740, Smith received a scholarship to attend Oxford, where he found the intellectual atmosphere to be restrictive and dull.

After having returned to Scotland, Smith accepted a position as professor of logic at the University of Glasgow and was promoted in his second year of teaching, 1752, to professor of moral philosophy, a field covering a wide range of subjects such as ethics, law, political economy, and natural philosophy. In 1758, he was promoted to dean of the faculty. During his tenure at the University of Glasgow, he formed a friendship with the Scottish philosopher **David Hume**, whose antimercantilist viewpoints and theories of the relationship between money and international trade had a significant influence on Smith's intellectual development as a political economist.

> "The quality of utility, beauty, and scarcity are the original foundation of the high price of those metals or of the great quantity of other goods for which they can be exchanged. This value was antecedent to, and independent of their being deployed as coin, and was the quality which fitted them for that employment."
>
> —ADAM SMITH

Smith published his first book, *The Theory of Moral Sentiments*, in 1759, a work that considered the interaction between essential facets of human nature and the larger organization and functioning of greater society. During the early 1760s, Smith accepted a job as tutor to the stepson of British Chancellor of the Exchequer Charles Townshend, whom he instructed over a period of several years in France. During intervals of free time, Smith traveled for extended periods to Paris and Geneva, where he established relationships with the great thinkers of the French Enlightenment. Smith subsequently returned to London under the employ of Townshend, after which he returned to Kirkcaldy, where he proceeded to write *An Inquiry into the Nature and Causes of the Wealth of Nations*, which was published in 1776 and met with wide acclaim.

In *The Wealth of Nations*, Smith explored the source of prosperity for sovereign nations as economies evolved historically from tribal hunting-based societies to nomadic agricultural communities to feudal societies based on local farming to, lastly, modern commercial-based market economies. Assessing the dynamic between individualistic aspects of human nature and the organizations and institutions governing human society, Smith developed a theory he called "system of perfect liberty," later known as free market capitalism, which posited man's natural inclination to self-interest and self-preservation should be viewed as a force of nature in which a healthy level of competition would in and of itself guide a steady progression toward social and economic well-being. In elaborating his theory, Smith explored the historical and legal foundations for private property; price levels and **inflation**; definitions and analysis of capital and monetary specie, taxation, trade policy, and tariffs; the relationship between wages, labor, and profits; patterns of consumption; and the function of government with respect to maintenance of the social order and economic policy. Smith was a vigorous antimercantilist who argued that protective tariffs created an artificial transfer of wealth that could not be sustained due to the inevitability of market forces with respect to supply and demand.

Smith's work was seen as having an almost immediate influence on government policies and the development of the work of subsequent classical economic theory, notably economist David Ricardo's theories of capital and his contributions to the Bullion Committee debate over prices, the circulation of money, and **gold reserves**. Despite the fact that Smith is not generally considered to have had a particular monetarist focus, his theories of the relationship of price levels to the availability of capital in the form of stock and currency and the productive capacity of the economy continue to inform policymakers in capitalist economies that adhere to many of the dictates of what has become known as laissez-faire capitalism.

Smith died in Edinburgh on July 17, 1790. He never married or had children.

See also: Bullion Committee of 1810; Hume, David; Ricardo, David.

Further Reading

Buchanan, James. 2006. *The Authentic Adam Smith: His Life and Ideas*. New York: W. W. Norton.

Heilbroner, Robert L. 1953. *The Worldly Philosophers*. New York: Simon and Schuster.

Hollander, Jacob. 1910–1911. "The Development of the Theory of Money from Adam Smith to David Ricardo." *Quarterly Journal of Economics* 25:429–470. http://socserv2.socsci.mcmaster.ca/~econ/ugcm/3ll3/hollander/money.html.

Laidler, David. 1981. "Adam Smith as a Monetary Economist." *Canadian Journal of Economics* 14, no. 2 (May):185–200.

Ross, Ian Simpson. 1995. *The Life of Adam Smith*. Oxford: Clarendon Press.

Smith, Adam. 1991. *The Wealth of Nations*. Amherst, NY: Prometheus Press. (Orig. pub. 1776.)

SPANISH ARMADA

The Spanish Armada was a large naval fleet built and assembled by King of Spain Philip II, formerly co-regent of England as the husband of Elizabeth's predecessor, Catholic Queen Mary I, during 1554–1558, to invade England and overthrow the monarchy of Protestant Queen Elizabeth I to reestablish a Catholic crown in England and thwart support for the Dutch Revolt against Spanish Habsburg rule in the Netherlands. Philip's plot to invade England was also motivated by a desire to retaliate against Elizabeth's implicit support of English privateers such as **Sir Francis Drake** who were pillaging Spanish treasure fleets transporting gold and other valuables from the New World and attacking Spanish settlements in the Americas.

The Spanish Armada of 1588. The armada of 124 ships sailed from Carunna, Spain, on July 12 for the invasion of England. The ships were manned by 8,500 seamen and galley slaves and carried 19,000 troops. Painting from the late 16th century. (Photos.com)

After nearly two years of preparation and setbacks due to weather once they were ready to launch, the armada set sail from the port at Lisbon with the promise of financial support from the papacy and a plan to meet up with the Spanish regent of the territories in the Netherlands, the Duke of Parma, who had provisioned a fleet with 30,000 of his men off the coast of Flanders in preparation for the Spanish king's plans to invade England on its southeastern coastline. The armada sailed for England in May 1588 with the Duke of Medina-Sidonia at its command as a lesser-experienced replacement for the original commander Alvaro de Bazan with the intention of joining Parma to provide a naval escort of his army to England for the invasion. Medina-Sidonia led a fleet comprised of a total of 130 ships, with approximately 22 armed galleons and the remaining ships serving as transports. The armada carried an estimated 8,000 sailors and 19,000 soldiers, largely infantry for the purpose of boarding the ships in the English fleet and invading the coast to wage a land offensive.

The English fleet provisioned to ward off the Spanish attack was commanded by Lord High Admiral Charles Howard, with **Sir Francis Drake** as the second command. Although the initial fleet prepared to meet the approaching Spanish was smaller and carried fewer soldiers than that of the Spanish, it had the advantage of more maneuverable ships with more effective and stronger heavy artillery. Yet when the armada was viewed approaching the English Channel off the coast of Cornwall on July 29, fear and uncertainty about England's vulnerability to potential attack from the supporting forces rumored to be coming from the Duke of Parma resounded among written testimonies and speeches of Queen Elizabeth and her advisors. In the face of such fears, Elizabeth gave perhaps her most iconic speech to reassure her subjects assembled to fend off the Duke of Parma at the Thames estuary in Essex of her steadfast commitment to their protection. Meanwhile, the sight of the approaching armada from the west was commemorated in contemporary memoirs as awe inspiring, with soldiers and military leaders outfitted in sumptuous golden trousers, bright cloaks, and extravagant lace.

The armada reached the Straits of Dover on August 6 and anchored in Calais, France, where the Spanish planned to await the Duke of Parma's army, whose trip would take a minimum of six days. This loose plan to coordinate with the Duke of Parma's army would prove to be a devastating tactical oversight, as the Spanish did not have access to a safe port where they could take refuge while they waited. This mistake would be remembered as one of the greatest blunders in the history of military strategy. To add insult to injury, at this early stage of the invasion, the prevailing winds remained in directions that were to the strong advantage of the English and thus the next night at midnight the English sent eight burning ships toward the Spanish fleet, which sat in the harbor fully exposed. To avoid catching on fire, the Spanish fleet was forced to exit the bay and sit idle in a position off of Gravelines, where the English then swiftly attacked. In the face of these early defeats in battle and persistent prevailing winds that made the possibility of the arrival of reinforcements from the Duke of Parma unlikely, the Spanish were forced to surrender and were escorted out of the English Channel to the northeast. The Spanish Armada was thus forced to return home to Spain by way of the northern coast of Scotland.

The long journey home through this disadvantageous route essentially destroyed the Spanish Armada after the initial losses from the naval battle, as over two dozen ships crashed on the West Coast of Ireland, with an estimated 30 more wrecking in the rough Atlantic passage home, and less than half of the original fleet arriving back in Spain in ruined condition. In addition to the hundreds of casualties sustained during battle, around 15,000 sailors and soldiers perished in the passage home, with many others being taken prisoner when they shipwrecked in Ireland. The quantity of ships lost at sea and on the shores of Ireland has led to centuries of speculation on the whereabouts of reported gold and silver ducats and gold **jewelry** and decorative items contained on the ships.

Almost immediately after the failed invasion, the Spanish Armada story assumed legendary proportions in the history of the Spanish Empire as a watershed event in the alleged imperial downfall of Spain due to administrative and military incompetence. Primary contributing factors to the campaign's defeat, among them poor planning and the unknown variable of fated winds that blew in England's favor, have been debated by historians through to the present day. The legacy of the Spanish Armada ultimately emphasized how the English victory preserved the English crown and the autonomy of the Dutch provinces. The story of the defeat of the Spanish Armada remains a source of national pride in the historiography of England.

See also: Drake, Sir Francis; Ducat.

Further Reading

Green, Wm. Spotswod. 1906. "The Wrecks of the Spanish Armada on the Coast of Ireland." *Geographical Journal* 27, no. 5 (May):429–448.
Knerr, Douglas. 1989. "Through the 'Golden Mist': A Brief Overview of Armada Historiography." *American Neptune* 49, no. 1:5–13.
Martin, Colin. 1975. *Full Fathom Five: Wrecks of the Spanish Armada*. New York: Viking.
Martin, Colin, and Geoffrey Parker. 1992. *The Spanish Armada*. New York: W. W. Norton.
Williams, Patrick. 2009. "The 'Chief Business': The Spanish Armada, 1588." *History Review* 65 (December):8–13.

SPANISH CONQUEST

The discovery of the Americas in 1492 by Genoese explorer **Christopher Columbus** on behalf of Spanish monarchs **Ferdinand II of Aragon** and **Isabella I of Castile** launched what would be come known as the Spanish Conquest, in reference to the establishment of a powerful colonial empire in Latin America undertaken initially as an extension of the crusading zeal of the final Reconquest of Moorish territories on the Iberian Peninsula by virtue of the Treaty of Granada in 1491. Columbus's voyage of discovery, charted in pursuit of a faster route to the gold-rich lands of the East, ultimately resulted in a steady flow of gold-filled treasure fleets returning to the ports of Spain from Central and South America with riches significant enough to propel the rise of Spain as a powerful global empire in the 16th century.

Following a relatively brief period of further discovery and exploration of regions in South and Central America and along the Pacific Coast of North America, conducted largely to locate the sources of the region's reputed wealth in gold, the

Spanish Crown claimed the territories for their own and set up a sophisticated governing bureaucracy to facilitate the further exploitation of the New World's abundant natural resources. After Columbus's establishment of the first Spanish settlement on the island of Hispaniola in the Caribbean in 1493, subsequent explorers accompanied by Catholic missionaries and representatives of the Spanish Crown claimed additional territories in the region "in the name of Christ and in the quest for gold" (Parry 1990, 25), beginning with **Vasco Nuñez de Balboa**, who, upon seeing the Pacific Ocean for the first time in 1513 after crossing the Isthmus of Panama claimed for the king and queen the newly discovered oceans and all lands whose shores it touched. Shortly after, **Hernán Cortés** conquered the Aztec empires centered in the capital of **Tenochtitlan** during 1519–1521 in present-day Mexico. The fall of the Mayan civilizations inhabiting the Yucatan Peninsula followed soon after. In 1532, conquistador **Francisco Pizarro** toppled the **Inca** Empire in Peru after he captured and executed Incan Emperor **Atahualpa**, subdued the capital in Cusco, and established the Spanish capital in Lima in 1535.

> "Much have I travell'd in the realms of gold, / And many goodly states and kingdoms seen"
>
> —JOHN KEATS, "ON FIRST LOOKING INTO CHAPMAN'S HOMER"

These swift and somewhat haphazard conquests were quickly formalized through the establishment of a strong centralized colonial administration that reported directly to the king through the offices of the Viceroyalty of New Spain in Central America and the Viceroyalty of Peru in Lima, as well as later viceroyal jurisdictions, all under the control of the Council of the Indies in Seville. These political offices were supported by the creation of Catholic missions throughout the Spanish territories to convert the native peoples to Christianity and in doing so socialize them to adapt to Western ways. Yet the brutal conditions of the conquest, in which torture, enslavement, and cruel massacre was inflicted on native peoples throughout Latin America as their lands were confiscated and exploited, drew criticism from among certain witnesses, most notably Bartolomé de Las Casas, a Dominican friar who documented the mistreatment of Indian populations in the New World and appealed tirelessly to Spanish emperor Charles V for reforms that would prohibit these inhumane practices. In an excoriating 1552 treatise dedicated to Spanish prince Philip II titled "A Brief Account of the Destruction of the Indies," with the dramatic subtitle "Or, a faithful NARRATIVE OF THE Horrid and Unexampled Massacres, Butcheries, and all manner of Cruelties, that Hell and Malice could invent, committed by the Popish Spanish Party on the inhabitants of West-India, TOGETHER With the Devastations of several Kingdoms in America by Fire and Sword, for the space of Forty and Two Years, from the time of its first Discovery by them," Las Casas lists the many ways in which Spanish conqueror behaved with "detestation, ignominy, and disgrace," inflicting in the name of Christianity and the Crown such tortures as, for example, when they bound "them hand and foot, laid them on the ground, and then pouring melted Gold down their Throats, cried out and called to them aloud in derision, yield, throw up thy Gold O Christian! Vomit and spew out the Mettal which hath so inqinated and invenom'd both Body and Soul."

The Spanish Conquest was a watershed event in the colonization of the New World and the 16th-century expansion of the Spanish Empire, which became one of the first and, for a time, the greatest seafaring empire of western Europe amid a time when several other nations sent out competing voyages of discovery and exploration. Imports of gold and silver from the New World escalated to an estimated 35 million ducats by the mid-16th century, injecting a level of wealth in gold into European economies that was unprecedented. Yet Spain's mismanagement of this newfound wealth, as the inflow of gold from New Spain declined precipitously during the latter half of the 16th century, would have significant consequences for the struggle over the balance of power on the European continent among the great powers emerging in an era of state building and increasingly centralized imperial governments.

Despite the shiploads of gold **bullion** transported on Spanish treasure fleets to the Royal Treasury in Seville during the 16th and 17th centuries, the imperial administration of Spain failed to convert this windfall into tangible economic development, and the gold left Spain almost as fast as it arrived as increased spending on imports spurred a disadvantageous trade balance and ensuing rampant inflation, exacerbated by a general tendency in Europe toward rising prices in the period known as the **Price Revolution**. During his reign, Emperor Charles V accumulated approximately 37 million ducats in foreign debt, nearly 2 million ducats more than the total of all the gold imported from New Spain to date. Within almost a century of its conquest and the most dramatic influx of gold and silver imports in the history of the West, Spain was on the verge of bankruptcy and experiencing a swift decline in political prominence. By the late 18th and early 19th centuries, the Spanish dominions in Latin America came unraveled and were eventually lost to waves of independence movements and fragmentation.

See also: Atahualpa; Balboa, Vasco Nuñez de; Bullion; Columbus, Christopher; Cortés, Hernán; Ferdinand II of Aragon; Inca; Isabella I of Castile; Pizarro, Francisco; Price Revolution.

Further Reading

Casas, Bartolomé de Las. 1552. "A Brief Account of the Destruction of the Indies." *Project Gutenberg*. http://www.gutenberg.org/etext/20321.
Hamilton, Earl J. 1929. "Imports of American Gold and Silver into Spain, 1503–1660." *Quarterly Journal of Economics* 43, no. 3 (May):436–472.
Hamilton, Earl J. 1970. *American Treasure and the Price Revolution in Spain, 1501–1650.* New York: Octagon Books. (Orig. pub. 1934.)
Haring, C. H. 1927. "The Genesis of Royal Government in the Spanish Indies." *Hispanic American Historical Review* 7, no. 2 (May):141–191.
Hemming, John. 1970. *The Conquest of the Incas.* London: Macmillan.
Pagden, Anthony. 1998. *Spanish Imperialism and the Political Imagination: Studies in European and Spanish-American Social and Political Theory 1513–1830.* New Haven, CT: Yale Univ. Press.
Parry, John H. 1990. *The Spanish Seaborne Empire.* Berkeley: Univ. of California Press.

Stern, Steve J. 1993. *Peru's Indian Peoples and the Challenge of Spanish Conquest: Huamanga to 1640*, 2nd ed. Madison: Univ. of Wisconsin Press.

Thomas, Hugh. 2004. *Rivers of Gold: The Rise of the Spanish Empire from Columbus to Magellan*. New York: Random House.

Vilar, Pierre. 1976. *A History of Gold and Money, 1450–1920*. London: New Left Books.

SS REPUBLIC

The SS *Republic* was a Civil War–era steamship that wrecked off the coast of Savannah, Georgia, in October 1865, while en route from New York to New Orleans to deliver goods and money crucial to the southern port city's reconstruction efforts.

The paddlewheel steamship was built in 1853 in the same northeastern shipyard as the infamous *Amistad* schooner and originally christened as the SS *Tennessee*. Intended to meet the needs of the rapidly expanding shipping trade stimulated by the gold rushes of the mid-19th century, the steamship operated as a merchant transport ship, shuttling cargo along the eastern seaboard and on regular routes to South America, on one occasion delivering mercenaries to an operation aimed at overthrowing the Nicaraguan government.

During the Civil War, it was commandeered for use as a blockade runner in the Confederate Navy, then captured by the Union Navy in New Orleans. At the end of the war, it was sold to a New York shipping company, renamed the SS *Republic*, and equipped as a merchant ship to serve ports from New York to the southeast.

On October 18, 1865, the ship left New York bound for New Orleans carrying a wide range of everyday goods for trade toward the Reconstruction effort, along with a privately consigned freight of 50,000 gold and silver coins at a contemporary estimated value of over $400,000. Due to the devaluation and lack of trust of the respective currencies of the North and South during the instability of the Civil War, trade transactions were largely conducted in such silver and gold currency during the Reconstruction era and were essential to the rebuilding of devastated cities such as New Orleans. Nearly a week later, the ship encountered a hurricane and wrecked about 100 miles off the coast of Savannah, Georgia, sinking in deep water down nearly 1,700 feet to the ocean floor. Astoundingly, all but 12 passengers and crew members were successfully rescued on lifeboats as the ship sank slowly.

On August 2, 2003, the ship was successfully identified by treasure salvage firm Odyssey Marine Exploration, after a year and a half of searching under the guidance of the firm's founders Greg Stemm and John Morris, who took great care to organize their expedition with respect for the archeological significance of the site, employing marine archeologists and adhering to state protocols for the recovery and distribution of artifacts. The location of the ship on the deep-sea floor was identified through the use of scarce surviving documents and side-scan sonar of a wide area of the Gulf Stream position where the ship was said to have met its end.

Upon their discovery of the ship, Stemm and Morris brought select artifacts with them to establish their claim to salvage rights in the federal courts, demonstrating their expertise in preservation and analysis of the items of significant historical value. Their efforts were initially thwarted when the ship's original insurer also laid claim to the salvaged wealth; however, Odyssey succeeded in working out a settlement and obtaining full legal ownership of the wrecked ship in March 2004. For the remainder of that year, the team recovered over 13,000 artifacts that shed significant light on everyday life during the mid-19th century and on the contributions of northern merchant enterprises to the reconstruction of war-torn cities in the South. Also recovered were 50,000 gold and silver coins, largely comprised of $10 and $20 gold pieces, along with extremely rare minted coins from the era, with a total estimated value of $75 million.

See also: Nuestra Señora de Atocha; San Esteban; Spanish Armada.

Further Reading

Vesilind, Priit J. 2005. *Lost Gold of the Republic: The Remarkable Quest for the Greatest Shipwreck Treasure of the Civil War Era.* Las Vegas, NV: Shipwreck Heritage Press.

SUMPTUARY LEGISLATION

Sumptuary legislation is any form of law that seeks to limit or regulate practices of consumption related to food, dress, and ritual, particularly those forms of consumption perceived to be overly lavish or involving a wasteful level of expenditure. Such laws have existed since antiquity in a variety of forms and contexts, yet, the creators of the laws have been motivated by certain essential, often ethical or religious rationales that have undertones of social control or may also seek to fulfill a protectionist or monetary economic purpose. Sumptuary laws often targeted women, minorities, or marginal social groups and were universally difficult to enforce.

The earliest known written records for sumptuary laws, dated to seventh-century-BC Greece, are found in a law code that forbade a range of lavish practices and included a prohibition against women wearing gold jewels and men wearing gold rings. Aristotle and his successors promoted moderation as an emulation of the supreme level of virtue for women. The men and women of ancient Sparta were forbidden to own gold or silver. Roman laws under both the republic and the empire also sought to limit lavish displays, feasting, and ostentatious funeral rituals. In the Law of the Twelve Tables, the earliest written example of Roman law, from ca. fifth century BC, it was forbidden to use gold clamps to affix false teeth. Later Roman sumptuary laws, such as the Lex Oppia of 215 BC, which limited the amount of gold jewelry women were permitted to wear to one-half ounce, were a product of both the social change of the increasing wealth of the Roman world as well as idealizations of the purity and simplicity of the heritage of the Roman people. This vision of the ideal Roman is evident in the writings of the poet Ovid, who praised the strength of women's Sabine ancestors and criticized the "gold-embroidered gowns" and ornate jewelry in fashion among Roman women, which he disdained as unfitting.

Early Christian texts also encouraged state regulation of sumptuous displays. In his "Letter to Timothy," Saint Paul advises that the females of the Christian congregation must be dressed modestly, without gold ornaments women in their hair. In a letter to Peter, Paul writes that married women should wear a modest hairstyle, and refrain from wearing gold bracelets and finery on their clothes and should adorn themselves only with "the ornament of a sweet and gentle disposition," caring only about what is "precious in the sight of God" (1 Peter 3:3–5). Saint Augustine echoed Paul's appeals when he wrote that men and women should seek to decorate themselves with "good character" (*Confessions*) rather than the false presentation of gold or lavish clothing. Such writings would reinforce later medieval sumptuary laws that cited religious precedents for modest dress.

Sumptuary laws also existed in the Islamic world, in certain passages of the Koran and in the Hadith. In the early years of the Ottoman Empire, the sultans attempted to restrict rising tendencies toward luxurious clothing and jewelry with a range of sumptuary laws. Sumptuary legislation also existed in China, starting in the Qin dynasty in 221 BC; and in feudal Japan, which exhibited the most widespread and detailed legal tradition for promulgating sumptuary laws, a tradition that saw a steep rise during the Tokugawa period during 1603–1867. In Central America during the mid-15th century, Aztec ruler Montezuma I promulgated a series of specific sumptuary laws restricting the wearing of **jewelry** and dress as a means of reinforcing the boundaries of social order in **Tenochtitlan**.

Patterns of urbanization and economic growth, with attendant social change, likely contributed to an increasing emphasis on sumptuary legislation in Europe and Britain during the 14th century. Whereas earlier legislation existed in Europe from the 12th and 13th centuries, the level of detail and targeted scope of the laws expanded in the course of the 1300s with a greater emphasis on distinguishing social groups than on ostentation in general categories. In 1362, English king Edward III enacted a law allowing only lords to wear cloth of gold and sable furs. Laws restricting gold **jewelry**, marriage and funeral banquets, and certain fabrics or types of clothing were common throughout Italian city-states such as Florence, Lucca, and Venice. A 1474 law in Bologna decreed that only the wives and daughters of knights were permitted to wear gold dresses, and only the wives of notaries and bankers might wear gold sleeves. A Bergamo law forbade its citizens to decorate their interiors with gold, silver, or ultramarine, and a 1476 Venetian law similarly forbade wall hangings of gold cloth and prohibited the use of gold or silver embroidery on fabric or the production of gold lace in the Point de Venice pattern. Late medieval kings in France and Spain limited the use of gold and silver fabrics, lace, and embroidery to certain noble or royal classes.

Along with the wealth and extravagance of the Renaissance came the rise of an increasingly wealthy merchant class that could afford to adorn themselves in the manner typically the reserve of titled elites. This period thus saw an expansion of sumptuary law with a greater emphasis on who could wear what when and limiting what could be served at banquets. In his first parliament, English king **Henry VIII** passed a law titled "An Act against wearing of costly Apparrell" in 1510, with three subsequent revisions of the law during his reign to restrict certain excesses in dress outside of the immediate nobility, such as fabrics of gold cloth and imported silks and furs. Elizabeth I also enacted detailed sumptuary laws with language justifying the law as a means of curbing excess that was potentially harmful to the fabric of British society. Indeed, during the 17th century, perhaps due to the rise of powerful states and a greater awareness of political affiliation and economic prosperity, sumptuary laws were often rationalized as a means of restricting foreign imports that were perceived as harmful to the domestic economy, or on moral and religious grounds for the benefit of the civic state. Such ideas were also exported to the American Colonies, for example, at the Massachusetts Bay Colony, which passed a law decreeing that only persons with a net worth of over 200 pounds were permitted to wear lace, metallic threads, and certain lavish accessories.

Democratization, industrialization, mass consumption, and shifting perceptions of social class rendered sumptuary legislation largely obsolete, although remnants of the tradition are echoed in contemporary campaigns to curb excessive dress or lavish expenditures amid periods of political or economic crisis.

See also: Henry VIII; Jewelry; Lace, Gold; Tenochtitlan.

Further Reading

Baldwin, Frances Elizabeth. 1926. *Sumptuary Legislation and Personal Regulation in England*. Baltimore: Johns Hopkins Press.
Brundage, James. 1987. "Sumptuary Laws and Prostitution in Late Medieval Italy." *Journal of Medieval History* 13:343–357.
Frick, Caroline Collier. 2002. *Dressing Renaissance Florence: Families, Fortunes, and Fine Clothing*. Baltimore: Johns Hopkins Univ. Press.
Hooper, Wilfrid. 1915. "The Tudor Sumptuary Laws." *English Historical Review* 30, no. 119 (July):433–449.
Hunt, Alan. 1996. *Governance of the Consuming Passions: A History of Sumptuary Law*. London: Macmillan.
Killerby, Catherine Kovesi. 2002. *Sumptuary Law in Italy, 1200–1500*. Oxford: Oxford Univ. Press.
Owen Hughes, Diane. 1983. "Sumptuary Legislation and Social Relations in Renaissance Italy." In *Disputes and Settlements: Law and Human Relations in the West*, ed. John Bossy, 69–99. New York: Cambridge Univ. Press.
Raffield, Paul. 2002. "Reformation, Regulation and the Image: Sumptuary Legislation and the Subject of Law" *Law and Critique* 13, no. 2 (May):127–150.
Rounds, J. 1979. "Lineage, Class, and Power in the Aztec State." *American Ethnologist* 6, no. 1 (February):73–86.
Shively, Donald H. (1964–1965). "Sumptuary Regulation and Status in Early Tokugawa Japan." *Harvard Journal of Asiatic Studies* 25:123–164.

SUTTER'S MILL

Sutter's Mill was a sawmill along the American River near present-day Sacramento at which carpenter James Wilson Marshall discovered **gold nuggets** in January 1848, a find that would ultimately launch the **California Gold Rush**.

The mill was part of the settlement of New Helvetia, a land claim staked by German-born Swiss immigrant John Augustus Sutter with the Mexican government in August 1839. Sutter was born as Johan August Suter in the German town of Kandern in Baden on February 15, 1803, son of a German papermaker who worked at a local paper mill. Raised in Switzerland for most of his early life while pursuing a career as a merchant, Sutter left his wife and children behind and

James Marshall discovered gold at Sutter's Mill, ca. 1850. (Library of Congress)

immigrated to the United States in 1834 to escape his creditors and avoid serving time in debtor's prison.

Upon his arrival in the United States, Sutter passed through New York, Saint Louis, Santa Fe, Oregon, Vancouver, Hawaii, and the Aleutians in search of land to settle before finally arriving in the town of Yerba Buena, present-day San Francisco, California. Having heard reports along his journey of the rich interior of California and its fertile river valleys, Sutter came into port at Monterey Bay to request permission for a land grant in the interior from the Spanish governor, then sailed back to Yerba Buena, where he exchanged his ocean vessel for a smaller boat suitable for river navigation. After thoroughly exploring the area, he returned to Yerba Buena once again to collect supplies and provisions for establishing a settlement in the region.

In August 1839, Sutter staked his claim with the Mexican government, after acquiring Mexican citizenship, for an area of land in the Sacramento River Valley, naming his settlement New Helvetia, or New Switzerland,

> "And a river went out of Eden to water the garden; and from thence it was parted, and became into four heads. The Name of the first is Pison: that is it which compasseth the whole land of Havilah, where there is gold; And the gold of that land is good."
> —GENESIS 2:10–12

which would later become Sacramento, the California state capital. Sutter had ambitious plans for his new settlement, which after additional land acquisitions ultimately comprised over 150,000 fertile acres in California's central valley, where he cultivated a variety of crops such as wheat and grapes and raised cattle, sheep, horses, and mules. By 1846, New Helvetia consisted of 60 buildings and had a bakery, military barracks, tannery, and cloth manufacture. In a strategic position at the convergence of the American and Sacramento rivers, the settlement served traders and farmers and soon grew to be an essential outpost for the interior of California.

Because the valley lacked adequate resources in timber to expand his development, Sutter built a timber mill in the area of Coloma on the American River about 50 miles east of New Helvetia. On January 24, 1848, head carpenter at the mill James Wilson Marshall discovered **gold nuggets** in the river at the building site and reported his discovery discretely to Sutter. Aware of the potential repercussions of this news on his development plans for New Helvetia, Sutter made every effort to keep the discovery a secret at least until the mill could be completed. Yet by the spring, word began to spread in other California settlements that gold had been found in the American River.

The rush of prospectors and the frenzy over the pursuit of gold would prove disastrous to Sutter's enterprises as his laborers left their jobs and homes to seek their fortune, and New Helvetia became deluged in 1849 with lawless gold prospectors who squatted on his lands and took his livestock and crops. This same year, his wife and four children finally came from Switzerland to reunite the family. Tragically, however, only a few years later, what was once a thriving trading post and farming community was destroyed and abandoned as a result of the gold rush. He moved his family to another farm near the Feather River shortly after.

Despite his best efforts to recover his investments and reestablish his land claims amid the chaos by deeding his lands to his son and appealing to the local courts, Sutter had essentially lost everything after his farm on the Feather River burnt down due to arson committed by an employee in 1865. In his own memoirs, Sutter exclaimed:

> By this sudden discovery of the gold, all my great plans were destroyed. Had I succeeded for a few years before the gold was discovered, I would have been the richest citizen on the Pacific shore; but it had to be different. Instead of being rich, I am ruined, and the cause of it is the long delay of the United States Land Commission of the United States Courts, through the great influence of the squatter lawyers. (Sutter 1857)

After an appeal to the U.S. Congress in Washington, D.C., for restitutions over some of his losses from his original land grant failed, Sutter and his wife moved to the small town of Lititz in Pennsylvania. On June 16, 1880, Congress adjourned its session without passing a proposed bill that would have granted Sutter $50,000 in restitution. He died of heart failure two days later on June 18 while in Washington, D.C.

Despite its tragic end, Sutter's Mill would thenceforth be equated with the onset of the California Gold Rush, a watershed event in California state history. The original building that served as Sutter's Fort is preserved today as a California state park in Sacramento.

See also: California Gold Rush; Gold Nuggets; Mining.

Further Reading

Brands, H. W. 2003. *The Age of Gold: The California Gold Rush and the New American Dream.* New York: Anchor Books.

Gudde, Erwin G. 1936. *Sutter's Own Story: The Life of General John Augustus Sutter and the History of New Helvetia in the Sacramento Valley.* New York: G. P. Putnam's Sons.

Sutter, John A. 1857. "The Discovery of Gold in California." *Hutchings' California Magazine* (November). http://www.sfmuseum.org/hist2/gold.html.

White, Richard. 1998. "The Gold Rush: Consequences and Contingencies." *California History* 77, no. 1 (Spring):42–55.

"THE TALE OF THE GOLDEN COCKEREL"

Alexander Pushkin's "The Tale of the Golden Cockerel" (1834) begins with the typical Russian fairy tale version of once upon a time in the opening verse, "Once in the Thrice-Ninth Kingdom, in the Thrice-Tenth Empire, lived the famous Tsar Dadon." Inspired by Washington Irving's short story "Legend of the Arabian Astrologer" from *Tales of the Alhambra* (1832), Pushkin's narrative verse tells the story of Tsar Dadon, a ruthless ruler who has spent a lifetime waging war and now seeks a "serene and festive" life for his later years. To achieve this he turns to a eunuch astrologer and sorcerer for help with keeping the belligerent armies of neighboring kingdoms at bay. The astrologer gives the tsar a Golden Cockerel to place on the palace's highest spire as a "faithful sentinel." When the kingdom faces a threat, the Cockerel will spin and crow, pointing to the direction of the danger to alert the tsar. When there is peace in the kingdom and no danger, the Golden Cockerel will remain at rest. In gratitude, the tsar offers to fulfill the astronomer's first wish as if it were his very own.

After two years of peace, the Golden Cockerel alerts the tsar of the threat of attack from the east. The tsar sends his army out with his eldest son at the head to thwart the offensive. Seven days later, with no word from the army, the Golden Cockerel shrieks and ruffles to warn of danger once more, and the tsar sends out two more sons to discover the fate of his eldest. With no news yet a week later, the Golden Cockerel crows again, and Dadon himself leads an army eastward. The tsar soon finds himself in unknown territory, amid the highest peaks of a mountain range where the terrain seems to have changed mysteriously. Here he encounters a silken tent encamped, around which is strewn a bloody battle scene in a meadow, where his three princes lie dead. Amid his grief, the curtains of the tent suddenly part and the enchantingly beautiful Queen of Shamakhan emerges to greet the tsar.

Forgetting his grief in an instant, the tsar joins the queen in her tent for a week of merriment before finally setting out to return home to his kingdom with the queen at his side. Upon his return he his greeted by the astrologer, who reminds him of his promise to grant him his first wish. He then asks for the tsar to give him the young radiant queen. The tsar refuses, but the astrologer persists with his request, angering Dadon, who then strikes the old man dead with his scepter as the queen stands by laughing. The Golden Cockerel then swoops down from the spire, pecks the tsar on his head, and the tsar slumps off the carriage, dead. In an instant, the queen also vanishes as if she had never existed. The tale concludes with a warning in riddle form, "Though my story is not true, 'tis a lesson, lads, to you."

Written in 1834 during a sojourn at his family's estate in Boldino as "*Skazka o Zolotom Petushke,*" Pushkin's tale is generally read as a critique of the autocratic Tsar Nicholas I during a period of political and financial pressure for the Russian writer. The Golden Cockerel was likely a symbol of the astrologer-sorcerer's magical powers, which were tied implicitly in the story to his abilities to preserve or take life, powers of creation often also associated with gold and perhaps **alchemy** as sorcery.

In 1907, Russian composer Nikolai Rimsky-Korsakov adapted the story into an opera, *The Golden Cockerel*, which premiered in 1909 on the stage in Moscow.

See also: Alchemy.

Further Reading

Bloom, Harold, ed. 1987. *Alexander Pushkin.* New York: Chelsea House.
Bethea, David M. 2005. *The Pushkin Handbook.* Madison: Univ. of Wisconsin Press.
Pushkin, Alexander. 1929. "The Golden Cock." Trans. Oliver Elton. *Slavonic and East European Review* 7, no. 21 (March):513–518
Vickery, Walter N. 1970. *Alexander Pushkin.* New York: Twayne.
Weir, Justin. 1998. "The Golden Cockerel between Realism and Modernism." In *Intersections and Transportations: Russian Music, Literature, and Society*, ed. and intro. Andrew Baruch Wachtel, 73–89. Evanston, IL: Northwestern Univ. Press.

TENOCHTITLAN

Tenochtitlan was the capital city of the Aztec empire in the central highlands of Mexico founded in 1325 on an island in the middle of Lake Texcoco. It was a location identified by the Mexica, a migrant people, as the place in which the Aztec god Huitzilopochtli foresaw in a vision that an eagle would land on a rock with a cactus growing out of it to indicate the site for the new capital that would serve as a type of promised land. Thus, beginning in 1325, the Mexica began to settle this marshy territory, unlikely as it was a choice for settlement, by reclaiming the land and building extensive causeways, canals, aqueducts, and floating gardens. By the time of Spanish contact in 1519, this legendary urban center was estimated to house a population of 200,000, with a sophisticated infrastructure and city plan compared to great cities the likes of Constantinople, Venice, and ancient Babylon. The city was laid out on an elaborate grid with sophisticated public sanitary and transportation networks. At the center of the city lay a network of religious temples and an adjacent complex of royal palaces with lush gardens.

The city flourished to its zenith during the rule of Aztec Emperor Montezuma II, who had inherited and further consolidated a vast empire consisting of vassal states in conquered regions required to maintain allegiance and submit tribute to the emperor. Upon reaching the Gulf of Mexico, Spanish conquistador **Hernán Cortés** inquired among the native people he encountered on the coast about the availability of gold in the territory and was told on numerous occasions that a great emperor ruled a city of vast riches inland. Cortés thus set out to travel inland and conquer and claim the riches of the famed city for the Spanish Crown. As he traveled toward the central highlands region to reach Tenochtitlan, he soon realized that he could use the resentments of Montezuma's conquered subject territories to his advantage, and along the way he succeeded in establishing important native allies who furnished him with essential strategic information and assistance regarding the Aztec capital.

In response to news of the approach of Cortés, Montezuma sent out a party of emissaries with gifts of tribute and a promise of more riches in an attempt to bribe him not to visit Tenochtitlan. Undeterred, Cortés continued along toward the Aztec capital, encountering garrisons of soldiers sent by Montezuma to thwart his approach, to no avail—Cortés arrived in Tenochtitlan on November 8, 1591. A series of extant firsthand accounts, including letters written by Cortés himself to the Emperor Charles V, provide essential details about the city and the impression that it made on the Spanish. In his second letter to the emperor, Cortés himself remarked that it was "more marvelous and richer than all the others" that he had encountered on his journey inland thus far. Cortés recounts that upon entering the city, he was greeted with many elaborate gifts, among them decorative gold disks; jewelry; headpieces and figurines in gold and silver with precious gems; strands of gold-beaded necklaces; ornate, sophisticated decorative objects of gold, silver, and **mosaic**; along with other precious items such as seashells.

According to the accounts of Cortés and other chroniclers that accompanied the Spanish, Montezuma interpreted the arrival of the Spanish as the fulfillment of the Aztec prophecy that the fair-skinned god Quetzalcoatl would one day return from a distant land to rule the Aztec people, although scholars still analyze and dispute competing claims about this first crucial encounter. Cortés and his men initially stayed in the city as honored guests, yet his repeated demands for gold as tribute and his disrespect of Aztec religion and culture, along with the natives' vulnerability to European-borne illness, particularly the transmission of smallpox, bred distrust and antipathy toward the newcomers. On his return journey to Tenochtitlan from a military expedition to Veracruz, Cortés learned that the Aztecs had turned on the Spanish troops after the captain in charge, Pedro de Alvarado, had murdered several Aztec nobles. Without heeding his peers' warnings about the dangers of returning, Cortés reentered the city to find armies of hostile Aztec troops awaiting him. Following the death of the emperor in the ensuing skirmish, the Spanish feared the worst faced with fierce Aztec troops greatly outnumbering their own. Gathering as much treasure in silver and gold as they could, they attempted to escape the city undetected during the night of July 1, 1520, but were caught and assaulted in a battle that killed a significant percentage of Cortés's troops and supporters. Remarkably, the next day the small remaining troop of Spanish conquistadors defeated the vast Aztec army through the use of skilled strategy by swiftly eliminating the military leaders. Over the next year, Cortés assembled his allies, sought reinforcements, and marched on Tenochtitlan in May 1521, conquering the city successfully for the Spanish and establishing it as the new capital of the Viceroyalty of New Spain.

Tragically, Cortés and his men destroyed the once great city almost entirely and used many of the original materials to construct the new Spanish city that would become present-day Mexico City. The majority of the little extant remains had lain entirely buried until significant Aztec sculptures were discovered in 1790, which prompted some small-scale excavations. During the 1970s, utility workers discovered important artifacts on the site of the original temple complex, prompting an extensive excavation of the area designated as the Templo Mayor.

See also: Cortés, Hernán; Mosaic; Spanish Conquest.

Further Reading

Jones, Julie. 2002. "Gold of the Indies." *Metropolitan Museum of Art Bulletin* New Series 59, no. 4 (Spring):3–4.

King, Heidi. 2004, October. "Tenochtitlan." In *Heilbrunn Timeline of Art History*. New York: The Metropolitan Museum of Art. http://www.metmuseum.org/toah/hd/teno_1/hd_teno_1.htm.

Matos Moctezuma, Eduardo. 1988. *The Great Temple of the Aztecs of Tenochtitlan*. New York: Thames & Hudson.

Mundy, Barbara E. 1998. "Mapping the Aztec Capital: The 1524 Nuremberg Map of Tenochtitlan, Its Sources and Meanings." *Imago Mundi* 50:11–33.

Rounds, J. 1979. "Lineage, Class, and Power in the Aztec State." *American Ethnologist*. 6, no. 1 (February):73–86.

Schwartz, Stuart B. 2000. *Victors and Vanquished: Spanish and Nahua Views of the Conquest of Mexico*. New York: Palgrave & Macmillan.

Serrato-Combe, Antonio. 2001. *The Aztec Templo Mayor: A Visualization*. Salt Lake City: Univ. of Utah Press.

Townsend, Richard F. 2000. *The Aztecs*. New York: Thames & Hudson.

TOUCHSTONE

Touchstones are an assaying tool used since the later Bronze Age to identify purity levels of metals, predominantly gold. Typically a stone of black siliceous schist, a category of medium-grained metamorphic rocks such as lydite, a touchstone is rubbed against an object and then the rubbings are compared to known pure-gold samples either by eye or by applying a treatment of acids to enhance the visual effects of the various alloys. In this way, one can determine the purity level of the unknown object.

The earliest confirmed archeological evidence for touchstones exists from Bronze Age sites in Europe dating to the late 8th century BC. Excavations at a Bronze Age settlement in Choisy-au-Bac in northern France during 1975–1982 uncovered a black Phtanite stone with visible traces of gold that appear under examination with a microscope to be linear markings on the stone. Lab tests reveal the markings to be composed of almost pure gold, with small traces of silver alloy. This discovery was found in the company of a small gold ingot and gold ring. The position and condition of these artifacts suggests that they were part of an early goldsmith's workshop. Similar assemblies of tools, reference metals, and touchstones have been identified for later periods at sites such as the **Royal Tombs of Ur** in ancient Mesopotamia and among the Asante people in Ghana. Scholars also have identified tools from earlier sites that may have potentially been used as a touchstone. Examples of earlier sites include: the archeological excavation in the Hisar district of Haryana, India, at Banawali; the Harappah period of the Indus Valley Civilization ca. 2600–1900 BC; an Iron Age site in Hampstead, England; and ancient Egyptian sites dating as far back as the 12th century BC. While the earliest

Touchstones are black, silica-containing stones that were first mentioned by Theophrastus (372–287 BC) in his treatise *On Stones* and are still in use today as a quick method of making an approximate estimate of the proportion of gold or silver in a sample of metal. The touchstone and needles shown here were made up according to instructions given in Lazarus Ercker's *Treatise on Ores and Assaying*, published in Prague in 1574. They lie on an open page of *A New Touchstone for Gold and Silver Wares*, a small treatise on assaying published in 1679, explaining the law relating to gold and silver. (Science & Society Picture Library/Getty Images)

known reference to touchstones in China was in AD 1387, the levels of accuracy reported at that time indicate the method had likely been in use for centuries.

The earliest known written reference to a touchstone is in the ancient Sanskrit treatise *Arthasastra* by Kautilya, dating to the fourth century BC. References to touchstones also appeared in early Greek texts such as in the work of sixth-century-BC poet Theognis and in fourth-century-BC philosopher Theophrastus's treatise *On Stones*. Roman natural philosopher Pliny the Elder included a section in his *Natural History* (ca. 77–79) on "Touchstones for Testing Gold," in which he describes the ancient method and praises its accuracy:

> A description of gold and silver is necessarily accompanied by that of the stone known as "coticula." In former times, according to Theophrastus, this stone was nowhere to be found, except in the river Tmolus, but at the present day it is found in numerous places. By some persons it is known us the "Heraclian," and by others as the "Lydian" stone. It is found in pieces of moderate size, and never exceeding four inches in length by two in breadth. The side that has lain facing the sun is superior to that which has lain next to the ground. Persons of experience in these matters, when they have scraped a particle off the ore with this stone, as with a file, can tell

in a moment the proportion of gold there is in it, how much silver, or how much copper; and this to a scruple, their accuracy being so marvellous that they are never mistaken.

With the increase in production of gold coinage in Europe during the Middle Ages and the Renaissance, goldsmiths and mint masters used touchstones to ensure the purity levels of the gold currency and to test **jewelry** and handicrafts. In 1300, King of England Edward I cited the mandatory regulation of the level of gold alloy by assaying with the "touch of Paris," a likely reference to the touchstone method. An early-13th-century Ayyubid treatise on the Cairo mint also makes reference to the use of a touchstone. By the 16th century, diagrams and descriptions of touchstones and reference tools were common in printed books and demonstrated certain technological advancements in terms of accuracy and range of assaying.

The technique of assaying by comparing traces of alloys by means of a chemical reaction, in existence for almost 4,000 years, represents the foundations of the processes of chemical analysis used in modern assaying methods. Techniques similar to the processes used since antiquity to assess purity levels of gold samples are still in use today.

See also: Asante Golden Stool; Assaying; Ur, Royal Tombs of.

Further Reading

Badcock, William. 1971. *A New Touchstone for Gold and Silver Wares*. New York: Praeger Publishers. (Orig. publ. 1679.)
Eluère, Christiane. 1986. "A Prehistoric Touchstone from France." *Gold Bulletin* 19, no. 2:58–61.
Morteani, Gioiulio, and Jeremy Peter Northover. 1995. *Prehistoric Gold in Europe: Mines, Metallurgy, and Manufacture*. New York: Springer.
Needham, Joseph, and Lu Gwei-Djenl. 1974. *Science and Civilization in China: Chemistry and Chemical Technology*. Cambridge: Cambridge Univ. Press.
Oddy, A. 1983. "Assaying in Antiquity." *Gold Bulletin* 15, no. 2:52–59.
Oddy, A. 1985. "Touchstones: Some Aspects of Their Nomenclature, Petrography and Provenance." *Journal of Archaeological Science* 12:59–80.
Philips, J. P.1965. "16th-Century Texts on Assaying," *Chem. Educ.* 12, no. 7:393–394.
Pliny the Elder. 1893. *The Natural History of Pliny*, vol. 6, ed. and trans. John Bostock and Henry Thomas Riley. London: George Bell & Sons.
Stockdale, D. 1924. "Historical Notes on the Assay of Gold." *Sci. Prog.* 18:476–479.
Wälchli, W. 1981. "Touching Precious Metals." *Gold Bulletin* 14, no. 4:154–158.

TOWER MINT

The original primary mint for the national coinage of England was located within the walls of the legendary Tower of London, a complex of fortifications and royal buildings with origins in the Roman period that was continually altered, remodeled, and refurbished through the modern era to accommodate shifting needs.

The building that housed the mint stands at the end of Water Lane, a road adjacent to the western Outer Wall that runs north. The mint was situated amid a series of functional buildings built along the wall on a strip of the street still known as Mint Street. The original building was completed at a cost of £729 17s 8 1/2 d during the reign of King Edward I (1272–1307), who ordered the construction of this row of buildings as part of his project to construct a new Outer Wall.

Because the fate of currency played a crucial role in the successes and failures of England and the British Empire, the Tower Mint and its keepers fulfilled a strategic function in governance and political intrigue throughout history. One of the first enactments of the newly crowned King Henry VII called for a reform of the currency and the production of a higher standard of gold coinage that would rival that being produced in Italian merchant centers such as Venice and Florence. Undertaken for both economic motives and as part of a larger effort to reinforce the monarchy's power and stability, Henry's reforms have been regarded by historians as a significant transition from medieval to modern concepts of currency in terms of standardization, quality, and value. As part of these reforms, Henry VII closed a series of ancillary mints in England and Ireland, appointed Sir Giles Danbeney and Bartholomew Reed as joint holders of the offices of Master and Worker of Monies and Keeper of the Exchange of the Tower, entrusting them with quality control at the mint and the ferreting out of illegitimate practices in coin production.

The economic pressures facing England during the reign of King Henry VIII (1509–1547) strained the treasury to its limits and accelerated the demand for production at the Tower Mint to a degree that required its remodel and expansion. In 1526, **Henry VIII** spent £480 clearing the building of military ordnance stored there, adding on to the building, and fixing the road to facilitate the increased cart traffic to and from the mint. With the treasury on the verge of bankruptcy, Henry ordered a debasement of the currency, a decision that undermined confidence in the monarchy both at home and abroad. The mint was further expanded in 1560 with the construction of the Irish Mint, also known as the Upper Mint, on the eastern section of the Outer Wall.

During the English Civil War of 1641–1651, Parliament assumed control of the Tower Mint in 1642, and Protector Oliver Cromwell monitored the continued production with great care as it was critical to the legitimacy of the government. However, select Royalist workers at the mint left, taking their tools and die casts with them, to seek King Charles in exile and thus enabled the king to continue to produce coinage sanctioned by the monarchy using gold donated by sympathizers and temporary mints set up outside of London. After the king's execution in 1649, the mint produced a new coinage free of symbols or images of the monarchy.

Upon his return to claim his right to the English throne in 1660, King Charles II brought a famous engraver from Antwerp, John Roettier, back with him to help reestablish the quality coinage of his earlier predecessors. During the Restoration period, many new techniques were employed at the mint, most significantly the conversion to mechanized processes of minting coins using horse-driven rolling mills, mechanized presses and edge markers, and blank-cutting machines to produce a milled coinage that would replace the earlier hammered coins. This involved another costly expansion of the mint, at a total cost of £400, during which Mint Street was again cleared of shops and tenants not directly affiliated with the mint. These updates also required the treasury to confiscate the old hammered coins as part of a recoinage effort to melt them down and mint them using mechanized techniques. Although costly, this process reduced the widespread problem of **coin clipping**, which steadily debased the old hammered currency over the course of its circulation.

One of the most famous subjects ever appointed to the office of Warden of the Royal Mint, and later Master of the Mint, was **Sir Isaac Newton** (1642–1727), who lived at the tower and oversaw the mint's production from the implementation of the **Great Recoinage of 1696** to his death. With the expansion of trade in the British Empire during this period, Newton held these posts during a critical time for additional expansion and quality control. Newton's scientific expertise in mathematics and early chemistry contributed greatly to the successful recoinage, elimination of corruption, and installation of mechanized mills.

As the British Empire grew and expanded throughout the 18th century and the scale of production in the mint rapidly increased, scrutiny of the use of space in the tower complex and the need to install steam-powered machines led to a decision to relocate the mint to a property directly outside of the Tower of London on Little Tower Hill, where the new mint was built and relocated in 1810. The old mint building was subsequently used as a military garrison and infirmary.

See also: Coin Clipping; Great Recoinage Act; Henry VIII; Newton, Sir Isaac.

Further Reading

Challis, Christopher. 1973. "A Contemporary Estimate of the Production of Silver and Gold Coinage in England, 1542–1556." *English Historical Review* 88, no. 346 (October):821–835.

Challis, Christopher, ed. 1992. *A New History of the Royal Mint.* Cambridge: Cambridge Univ. Press.

Craig, Sir John. 1953. *The Mint: A History of the London Mint from AD 287 to 1948.* Cambridge: Cambridge Univ. Press.

Trial of the Pyx

The Trial of the Pyx was a random sampling inspection protocol established at the Royal Tower Mint in London in 1279 under King Edward I to ensure quality control in production of Britain's gold and silver coinage. Versions of the trial may have existed as early as the reign of King Henry II in 1154. In the course of production at the mint, a plate of the king's gold was reserved and stored in the Chapel of the Pyx at Westminster Abbey to be used later as the standard for quality control. As new coins were minted, a random selection was placed in the "Pyx," a wooden box, the term of which derives from the Greek word for box. Annually or semi-annually, a Trial of the Pyx would be ordered to test the coinage. The trial was presided over by a jury with representatives from the independent goldsmiths guild. The goldsmiths would examine and assay the sample coins from the box and document their findings, comparing them to the specifications established by the Crown. If the results indicated that too little gold was used to mint the coinage, the Mint Master was obligated to reimburse the Royal Treasury for the difference, a policy intended to curtail any chance of debasement of the coinage. The Trial of the Pyx has continued in approximately the same format through the present day, although its function since the mid-19th century has largely been one of maintaining tradition and public confidence in the national currency.

Source: Stigler, Stephen M. 1977. "Eight Centuries of Inspection: The Trial of the Pyx." *Journal of the American Statistical Association* 72, no. 359:493–500.

Gould, J. D. 1952–1953. "The Royal Mint in the Early 17th Century." *Economic History Review* 5:242.

Li, Ming-Hsun. 1963. *The Great Recoinage of 1696 to 1699.* London: Weidenfeld and Nicolson.

Milnes, Nora. 1917. "Mint Records in the Reign of Henry VIII." *English Historical Review* 32, no. 126 (April):270–273.

Minney, R. J. 1970. *The Tower of London.* Englewood Cliffs, NJ: Prentice-Hall.

Rowse, A. L. 1972. *The Tower of London in the History of England.* New York: G. P. Putnam's Sons.

Wilson, Derek. 1978. *The Tower, 1078–1978.* London: Hamish Hamilton.

TRANS-SAHARAN GOLD TRADE

During the Middle Ages, the development of various trade routes from Mediterranean cities in North Africa and Egypt across the Sahara Desert to sub-Saharan kingdoms rich with gold and other precious commodities made a significant contribution to the gold supply to mints in the Mediterranean and played a key role in the development of merchant trading cities along the routes, cross-cultural exchange, and the spread of Islam in Africa.

During the Neolithic period, the barren desert environment of the Sahara region of the present-day southern Sudan was a fertile pastoral plain with lakes that filled seasonally on a savannah terrain that hosted a significant population. Out of these settlements developed the Kushite kingdom around 2400 BC, which established profitable trading relations selling luxury commodities such as gold to the Egyptian pharaohs and thus grew to develop a highly centralized state based in Kerma that would eventually be part of the Kingdom of **Nubia**, which became legendary as a source of gold for the New Kingdom Egyptian pharaohs such as **Tutankhamun**, who undertook military campaigns against the Nubians.

During the Roman era, the roads and trade routes to the Sudan were populated with Roman forts, and trade in gold from the area was essential to the production of Roman coinage. Despite the existence of descriptions of overland routes to modern-day Libya in early sources by fifth-century-BC Greek historian Herodotus, historians still debate the degree that trans-Saharan trade routes existed across the desert to West Africa in Greek and Roman antiquity.

The rise of trade routes across the Sahara desert from various coastal cities of the Mediterranean began to develop more systematically following the introduction of camels as pack animals around the fifth century BC. The use of camels as pack animals equipped Berber nomadic traders from North Africa with a swifter, more efficient means to establish annual, seasonal trade caravans along specific routes dealing primarily in gold and salt between West African kingdoms and Mediterranean cities. Empires in the region extending from the Niger River to the coast (such as the Soninke in the 7th–13th centuries, the Mali in the 13th–17th centuries, and the Asante in the 17th–19th centuries) expanded with the growing trade passing through legendary merchant trading centers such as Timbuktu and Gao.

The period during the 7th–12th centuries represented the peak of commercial activity along the Trans-Saharan trade routes and a level of gold production and intercultural exchange that had a significant effect on the expansion of Islam in

Africa and the economic expansion of various Mediterranean economies such as that of Byzantium due to the availability of gold for Western coinage. Contemporary accounts by Andalusian geographer Abu Ubaydallah al-Bakri attest to perceptions in western Europe and the Muslim world of West Africa as a "land of gold." Al-Bakri describes his impressions of the Soninke court in Kumbi Saleh as rich with lavish gold adornments beyond the expected, such as gold saddles and dog collars. A 1375 map drafted in Spain represents the region's legendary wealth in gold by depicting the king of Mali superimposed on the map holding up a large gold nugget and a gold scepter, and wearing an oversized gold crown. Indeed the pilgrimage of the Muslim king of Mali, Mansa Kankan Musa I, during 1324–1325 brought so much gold as tribute to Mecca and Egypt during his pilgrimage, or *haj*, that the price of gold plummeted in response to the windfall supply.

The level of trade along these overland routes began to trail off in the 14th century as coastal trade increasingly flourished, particularly among Dutch, Portuguese, and North African merchant trading fleets. In the modern era, efforts to reestablish these ancient routes by way of railroad or roads were stymied by the complexities of foreign relations among newly independent African nations in the postcolonial period; however, Nomadic Berber caravans exist along the traditional routes to the present day.

See also: Asante Golden Stool; Nubia; Tutankhamun.

Further Reading

Al-Bakri, Abu Ubaydallah. 1068. *The Book of Routes and Kingdoms.*
Bovill, Edward William. 1958. *The Golden Trade of the Moors.* London: Oxford Univ. Press.
Department of Arts of Africa, Oceania, and the Americas. 2000, October. "The Trans-Saharan Gold Trade (7th–14th century)." In *Heilbrunn Timeline of Art History.* New York: The Metropolitan Museum of Art. http://www.metmuseum.org/toah/hd/gold/hd_gold.htm.
Garrard, Timothy F. 1982. "Myth and Metrology: The Early Trans-Saharan Gold Trade." *Journal of African History* 234:443–461.
Liverani, Mario. 2000. "The Libyan Caravan Road in Herodotus IV.181–185." *Journal of the Economic and Social History of the Orient* 43, no. 4:496–520.
Vilar, Pierre. 1976. *A History of Gold and Money, 1450–1920.* London, New Left Books.

TUTANKHAMUN

Tutankhamun was an Egyptian pharaoh who ruled as a boy king and then young adult during the New Kingdom from 1333 to 1323 BC. Also known as King Tut, Tutankhamun remains one of the most famous pharaohs from this era due to the discovery of his funerary tomb, which was found almost completely intact in the Valley of the Kings in 1922 by British Egyptologist Howard Carter. Tutankhamun's tomb contained astounding quantities of treasures in gold, silver, and precious gems that demonstrate the sophisticated levels of artistry and metalwork flourishing in ancient Egypt during what has come to be known as its Golden Age.

Tutankhamun was born Tutankhaten, which translates as "Living Image of the Aten," a reference to his father Akhenaton's controversial religious revolution converting from the traditional pharaonic pantheon to monotheistic worship of the

Outer coffin of Egyptian king Tutankhamen. Tutankhamen became pharaoh over Egypt in 1333 BC as a member of the Eighteenth Dynasty and ruled until his death in 1323 BC. Known as the "boy king," he was approximately 18 years old when he died. (Corel)

sun god Aton. Tutankhamun was married to his half-sister Ankhesenamun and inherited the throne around the age of nine, ruling under the tutelage of powerful advisors who likely influenced the pharaoh's swift decision to reverse his father's unpopular religious reforms and reopen traditional temples. Little is known about the events of Tutankhamun's short-lived reign beyond details about the political and religious struggles he faced and military campaigns he waged in the face of border threats from the Hittites and the Nubians, and advancements in the sophistication, craftsmanship, and scope of art, architecture, and literature. King Tut and Ankhesenamun left no surviving heirs, and DNA samples from a 2010 study of two fetuses buried in royal Egyptian tombs confirmed that the two female mummy fetuses were their children. Tutankhamun died at the age of 20. Scholars assume his death came unexpectedly as he was buried in a small tomb not characteristic of royal Egyptian burial chambers. A forensic examination of his remains in 2010 indicated the cause of death may have been malaria and complications from a degenerative bone disease affecting his foot. His remains also had been thought to exhibit signs of a possible head injury by earlier scholars who posited that he may have been murdered; however, modern examinations indicate this was likely simply an aspect of the embalming process in which the pharaoh's brain was removed.

Despite his relatively insignificant reign, Tutankhamun remains one of the most widely recognized figures of ancient Egypt due to the discovery of his tomb complex by British archeologist Howard Carter, who had been working for Lord Carnarvon excavating sites in the Valley of the Kings since 1914. In his memoirs of the excavation of Tut's tomb, Carter recounts his first sight of the treasures within the burial chambers:

> At first I could see nothing, the hot air escaping from the chamber causing the candle flame to flicker, but presently, as my eyes grew accustomed to the light, details of the room within emerged slowly from the mist, strange animals, statues and gold—everywhere the glint of gold ... I was struck dumb with amazement, and when Lord Carnarvon, unable to stand the suspense any longer, inquired anxiously, "Can you

see anything?" It was all I could do to get out the words, "Yes, wonderful things." (Carter 1972, 95)

He had, indeed, discovered "wonderful things" within a series of burial chambers that had been undisturbed for three millennia. There was evidence that tomb raiders may have penetrated the front chambers soon after Tut's burial, yet the items were recovered and the tomb resealed, laying undiscovered until Carter's excavation. Amid a series of rooms crammed full of sacred and everyday objects, including sophisticated examples of New Kingdom **jewelry** and furnishings, intended to protect and accompany the pharaoh as he passed on to the afterlife, the mummy was preserved in four gilded shrines, an ornate sarcophagus, and three coffins, one constructed of solid gold. Many gold and silver artifacts were contained throughout the tomb, and those adorning the king's mummy are so famous they have since become iconic symbols of the corpus of ancient Egyptian culture itself. When the linen of the mummy's shroud was unwrapped, it revealed the body adorned with a gold pectoral piece in the shape of a falcon, symbolic of the protective god Horus, a gold dagger attached to his thigh for defense, and the royal crown, or diadem, on his head with Tut's personal protective totem symbols of a cobra and a vulture. These animal symbols were also present on his death mask, one of the most famous objects uncovered from the tombs, which was placed over the linens covering the mummy's face in the innermost coffin and crafted in the image of Tutankhamun out of over 24 pounds of solid gold inlaid with precious stones and colored glass.

The metalwork uncovered in Tutankhamun's tomb indicates a higher level of artistry and technological skill in metalwork during the second half of the 18th Dynasty, a development that also coincided with linguistic changes as new vocabulary emerged for the first time to describe gold specifically as a separate entity from other precious metals in the same category, such as electrum, as well as descriptive words to indicate with more precision the purity levels and quality of gold. The Egyptians associated gold with the sun god Amon Re and, like other cultures of antiquity, rebirth. It was thus an additional symbolic element connecting the semidivine pharaoh to this singular god as he proceeded on his transformational journey to eternal life.

Tutankhamun's mummy remains in his original tomb in the Valley of Kings, and the treasures that once surrounded him are

---

Egyptian Red Gold

Gold artifacts excavated from ancient Egyptian tombs display a reddish-purple tone often referred to as Egyptian red gold. Scientists have concluded that this discoloration is a tarnishing effect resulting from exposure to significant levels of sulfide ions at high temperatures for sustained periods in the sealed tombs, which caused the gold to tarnish and interact with other metals in the alloy based on natural electrum from the region.

Source: Frantz, Tony, and Deborah Schorsch. 2007, March. "Egyptian Red Gold." In *Heilbrunn Timeline of Art History*. New York: The Metropolitan Museum of Art. http://www.metmuseum.org/toah/hd/rgod/hd_rgod.htm.

conserved at the Museum of Egyptian Antiquities in Cairo. Since the 1970s, various exhibits displaying the treasures of Tutankhamun's tomb have toured as part of temporary shows at museums and other venues throughout the world. A 2005 study sponsored by the National Geographic Society in the United States created a reconstruction of Tut's likely appearance based on three-dimensional CT scans of the mummy with separate, autonomous contributions from scientists in Egypt, France, and the United States.

See also: Crowns, Royal; Jewelry; Nubia.

Further Reading

Carter, Howard. 1972. *The Tomb of Tutankhamen*. New York: E. P. Dutton.
Edwards, I. E. S. 1976. *The Treasures of Tutankhamun*. New York: Ballantine.
Hawass, Zahi. 2005. *Tutankhamun and the Golden Age of Pharaohs*. Washington, DC: National Geographic Society.
James, T. G. H. 1972. "Gold Technology in Ancient Egypt," *Gold Bulletin* 5, no. 2:38–42.
James, T. G. H., and Araldo DeLuca. 2002. *Tutankhamun*. New York: Metro Books.
Metropolitan Museum of Art. 1976–1977. "Tutankhamun." *Metropolitan Museum of Art Bulletin* 34, no. 3 (Winter):1–48.
Radford, Tim. 2005. "Revealed after 3,300 years: The Face of King Tut." *Guardian*, May 11. http://www.guardian.co.uk/uk/2005/may/11/science.research.
Reeves, Nicholas. 1990. *The Complete Tutankhamun: The King, His Tomb, the Royal Treasure*. London: Thames and Hudson.
Schorsch, Deborah. 2001. "Precious-Metal Polychromy in Egypt in the Time of Tutankhamun." *Journal of Egyptian Archaeology* 87:55–71.

UNITED STATES BULLION DEPOSITORY

The United States Bullion Depository, commonly referred to as Fort Knox, is a fortified vault complex adjacent to the United States Army Armor Center in Fort Knox, Kentucky, constructed in 1936 to store the national supply of gold reserves in an era when the value of the U.S. dollar was based on the gold standard.

Following New Deal legislation enacted in 1933 that prohibited the exchange of gold as currency by individuals, the government collected gold **bullion** and bullion coins in exchange for minted U.S. dollars at market value, thereby amassing a larger supply of this precious metal than ever before. To house these gold reserves securely, the U.S. Treasury Department arranged a land transfer with the Fort Knox military complex and began construction of the vault in 1936 at a total cost of $560,000.

The depository consists of a subterranean vault constructed of steel with an outer layer of granite that is further encased in concrete. The 105 x 121 foot vault is sealed with a 22-ton blast-proof door surrounded by a fortified outer structure. The combination to the vault's lock is not known in its entirety by any one staff member; rather, opening the vault requires 10 employees who must each enter individual combinations that are classified and only known by each employee. The complex is further protected by additional security measures such as a network of alarms, video cameras, and armed guards, along with the full military support of the adjacent fort. No visitors are allowed at the site, which is a fully classified facility.

The vault officially opened in 1937 when the first shipments of gold arrived by rail. In addition to gold bullion and coins, the vault also contains several valuable government documents, including the Declaration of Independence, the U.S. Constitution, the Articles of Confederation, and Abraham Lincoln's Gettysburg address. Additionally, the vault has been a safe haven for other nations' treasures during times of strife, housing the Magna Carta for Great Britain during World War II and the Crown of St. Stephen, along with the royal orb, scepter, and cape, for Hungary during the Cold War.

The value of the present-day holdings at the depository is approximately 147.3 million ounces of gold recorded as an asset of the government at a rate of $42.22 per ounce. The standard weight of bars of gold bullion contained in the vault is 400 ounces. Since its construction, the complex's reputation for high security has led to the use of the term "Fort Knox" as a euphemism for impenetrable security.

See also: Banking and Credit; Bullion; Crowns, Royal; Gold Standard.

Further Reading

Bagby, Milton. 2001. "Uncle Sam's House of Gold." *Military History* 18, no. 2 (June).
U.S. Mint. "Fun Fact Sheet." http://www.usmint.gov/about_the_mint/fun_facts/?action=fun_facts13.

UR, ROYAL TOMBS OF

The Royal Tombs of Ur represent a portion of a much larger burial complex adjacent to the Great Ziggurat of Ur, once the locus of the ancient Sumerian civilization centered around the city known in antiquity as Ulrim, also commonly called Ur, that flourished in the Early Dynastic Period ca. 2900–2300 BC. In ancient Mesopotamia, Ur was a strategic coastal city settled at the mouth of the Euphrates River where it met the Persian Gulf. Due to centuries of deposited sediment and the river's shifting course, the site of the ancient city is now located in the modern-day city of Tell el-Mukayyar, in the southern Iraqi province of Dhi Qar.

In 1922, British archaeologist C. Leonard Woolley undertook a joint excavation of the site of ancient Ur sponsored by the British Museum, London, and the University of Pennsylvania Museum, Philadelphia. Although the project originally focused on uncovering the ziggurat and adjacent buildings surrounding the area of a central complex of sacred temples constructed of mud brick, Woolley discovered an enormous field of burials, containing over 1,800 tombs, 660 of which dated from the Early Dynastic Period. Concerned that he lacked the expertise to excavate the burial sites at that time, Woolley continued to concentrate on the building sites and reassumed the trenching of the burial sites in 1926.

Ram Caught in a Thicket, a gold, silver, lapis lazuli, copper, shell, red limestone, and bitumen sculpture, found in the royal tombs at Ur. (AP/Wide World Photos)

From 1926 to 1934, Woolley and his team excavated the burial sites to reveal thousands of private tombs, in most of which a single body was buried in a coffin or wrapped in cloth of river reeds and accompanied by various personal items, most commonly a cup or other vessel and in some cases jewelry and ornaments made of gold and precious stones. Lying among these tombs were 16 more elaborate grave complexes, distinct from the others in their structure and the wealth of the personal items within. These burial chambers were identified by Woolley as the "Royal Tombs" on the basis of the cuneiform inscriptions found inside, the sumptuous personal adornments of the deceased, and the evidence of ritual sacrifice. These latter tombs consisted of passageways connecting a series of rooms, sometimes at different levels, and containing many bodies.

A majority of the royal tombs had been pillaged over the millennia, and identification of the contents and the royal status of the deceased remains open to debate among scholars of ancient Mesopotamia, particularly regarding one tomb that contained sumptuous artifacts and a male body adorned with a golden helmet, yet contained neither explicit written evidence of royalty nor certain elements of the ritual evidence of the other royal tombs. The only tomb excavated intact is identified as that of Queen Puabi based on the cuneiform inscription on a cylinder seal discovered near the body. Puabi's tomb consisted of a ramp extending down to a network of rooms, with the queen's body buried at the very far end of the pit. Male bodies clutching copper daggers, presumably soldiers, lay just inside the door of the tomb, and 10 female bodies considered to be the queen's personal attendants lay in the chamber before her, sumptuously adorned and surrounded by instruments such as the lyre and flute. The body of the queen was decorated with an exquisite array of precious beadwork covering her torso consisting of hundreds of rows of gold, carnelian, and lapis lazuli beads. Her head was adorned with a gold headdress with carnelian beads and a gold and lapis lazuli hair pin, an extravagant example of the typical headdress worn by women of ancient Sumeria. Along with some of the other royal tombs, the inner chambers of Puabi's tomb contained a wheeled type of cart likely to have been pulled by oxen. Whereas the presence

---

Bactrian Hoard

In 1978, Russian archeologist Viktor Sarianidi led a Soviet-Afghan team that discovered a series of burial sites from among the nobility of the ancient Bronze Age kingdom of Bactria containing over 20,000 gold objects dating from the first century AD while excavating a mound in northern Afghanistan referred to by locals as Tillya Tepe "Hill of Gold." The vast quantities of objects included jewelry, armor, and other luxury items demonstrating the highly sophisticated, refined nomadic culture in the region. Thought to have been made by local craftsmen, the artifacts were created in the typical styles of the Eurasian steppes with evidence of other international influences they likely encountered due to their position on the Silk Road, used for centuries by Hellenistic, Chinese, Indian, and Indo-European cultures.

The completion of the excavations was threatened by the onset of the Soviet war in Afghanistan in 1979. Sarianidi succeeded in excavating around 20,000 objects, which he presented to the National Museum of Afghanistan in Kabul. Soon after, however, the treasure known as the "Bactrian hoard" went missing. During the war, the Taliban melted down many precious objects, and the treasure was thought to have been either looted or destroyed during the chaos of the war. In 2003, however, Afghan president Hamid Karzai announced he had identified a secret vault located in the presidential palace in Kabul. Sarianidi was invited with other scholars and dignitaries to attend the opening of the vault, which dramatically revealed that the treasures of the Bactrian hoard remained intact.

of a multitude of other bodies in the royal tombs has been cited as evidence of human sacrifice in Mesopotamian culture, scholars largely consider this as an example of self-immolation, in which the loyal retinue of the deceased king or queen decided to accompany their master on their journey to the afterlife by taking their own life, likely by drinking poison.

The goldwork discovered among the Royal Tombs of Ur provided an unprecedented glimpse of early Mesopotamian artistic techniques and style. Ornate gold beads were decorated with an early method of filigree along with cloisonné of lapis lazuli. A large lyre featuring a bull's head and elaborate inlaid panel, known as the "Great Lyre," found in the grave of a king at Ur, was constructed of a base of wood and bitumen and decorated with hammered gold and silver, along with lapis lazuli and shell inlay. The majority of the artifacts recovered in these tombs are preserved at the British Museum, London; the University of Pennsylvania Museum, Philadelphia; and the National Museum of Iraq in Baghdad, which was tragically looted during the U.S. offensive of the Iraq War in April 2003. Thousands of precious objects and manuscripts from ancient Mesopotamia were stolen, many of which have since been recovered, although an international team of experts with the support of the U.S. Marines continues to seek the recovery of the remaining looted artifacts.

See also: Crowns, Royal; Filigree, Gold; Gilding; Jewelry.

Further Reading

Department of Ancient Near Eastern Art. 2003, October. "Ur: The Royal Graves." In *Heilbrunn Timeline of Art History.* New York: The Metropolitan Museum of Art. http://www.metmuseum.org/toah/hd/urrg/hd_urrg.htm.

Moorey, P. R. S. 1977. "What Do We Know About the People Buried in the Royal Cemetery?" *Expedition* 20, no. 1:24–40.

Pollock, Susan. 2007. "The Royal Cemetery of Ur: Ritual, Tradition, and the Creation of Subjects." In *Representations of Political Power: Case Histories from Times of Change and Dissolving Order in the Ancient Near East*, ed. Marlies Heinz and Marian H. Feldman, 89–110. Winona Lake, IN: Eisenbrauns.

Woolley, C. Leonard, et al. 1934. *The Royal Cemetery: A Report on the Predynastic and Sargonid Graves Excavated between 1926 and 1931. Ur Excavations, vol. 2.* London and Philadelphia: Joint Expedition of the British Museum and of the University Museum, University of Pennsylvania.

Zettler, Richard, and Horne, Lee, eds. 1998. *Treasures from the Royal Tombs of Ur.* Philadelphia: Univ. of Pennsylvania Press.

VERSAILLES, PALACE OF

The Palace of Versailles (Château de Versailles) lies about 12 miles southwest of Paris in the Île-de-France region of northern France. It is the former residence of the French monarchy in the late 17th and 18th centuries and the present-day UNESCO World Heritage site in Versailles. Versailles' opulent interior décor and surrounding gardens with outside statuary, gates, and fountains were designed to reflect the highest grandeur of the French kings who lived there, and thus gold-leaf accents, gold-threaded fabrics, solid-gold household items, and gilt bronze along with pure silver features conveyed the splendor of the rooms and gardens.

> "Every passion in Paris resolves into two terms: gold and pleasure."
> —HONORÉ DE BALZAC, THE GIRL WITH THE GOLDEN EYES (1833)

The château in Versailles was originally constructed during the reign of Louis XIII (1610–1643) in 1631–1634 to serve as a royal hunting lodge. Under his successor, Louis XIV (1643–1715), and amid the rise and consolidation of absolutism for the French monarchy, the palace emerged as the focal point of the elaborate French court and was renovated and expanded accordingly during 1661–1710 by a team of architects and artists who created one of the most stunning and extravagant royal complexes in history. Designed principally by architects Jules Hardouin-Mansart, Robert da Cotte, and Louis Le Vau, the renovation expanded the original structure by adding extensive private apartments for the royal family members as well as stately common rooms, with court painter Charles Le Brun orchestrating the elaborate baroque interiors and landscape architect André Le Nôtre designing extravagant, stylized formal gardens. Almost every element of the interior and exterior design of the château included symbolic and allegorical glorifications of the power of the king and his family.

Following the series of renovations, estimated to have cost the modern equivalent of roughly 2 billion dollars, the Palace of Versailles became the official royal residence of the king and home to his court in 1682, the beginning of a period in which the famous buildings became the stage for rituals of pomp and circumstance intended to revere the king and reinforce his supremacy at court and as the head of France. Louis XIV himself acknowledged the importance of the physical space in the construction of his role and legacy as king and, therefore, oversaw certain details with a shrewd eye to maintaining perfection and opulence befitting the "Sun King" and his likewise heralded queen, Maria Theresa of Spain. Adorned in

A gilder works on the restoration of the windows of the Hall of Mirrors at the Palace of Versailles. Thousands of slips of gold leaf have been meticulously adhered to the windows surrounding the Halls of Mirrors. (Marc Deville/Corbis)

gold themselves, and depicted wearing elaborate, sumptuous clothing and jewelry in paintings, panels, and engravings throughout the décor of the palace, the king and queen surrounded themselves with gold and are even pictured wearing shoes of silver and gold. The palace is indeed draped in gold with the enormous iron entrance gates gilded in gold leaf; gilded windows and cartouches; gold-threaded tapestries, upholstery, and linens; solid gold tableware and decorative ornaments; gilded iron balconies; carved wood furnishings accented with gilt-bronze to give the effect of gold; and candles and chandeliers in abundance to reflect the golden hues of the surroundings. Perhaps one of the most famous rooms, the Hall of Mirrors, consists of an arched hallway with mirrors in each archway reflecting a row of windows; marble pilasters on either side and gilded capitals with various symbols of France carved into them; gilded sculptures lining the hall; and gold leaf on the paneled paintings adorning the ceiling. In contemporary accounts of court life at Versailles, it is evident that the king utilized precious items made out of gold and silver in ritual and pageantry intended to reinforce his supremacy: for example, in distinguishing between the royal tableware or by using a gold communion chalice exclusively for his own sacrament.

With the death of Louis XIV, Versailles saw a brief reprieve as the royal residence from 1715 to 1722 when his successor, Louis XV, age five, and his regent Philippe d'Orléans lived in Paris. After returning to the palace in 1622, Louis XV undertook additional renovations during the remainder of his reign through 1774. His successor, Louis XVI (r. 1774–1791), would continue with the completion of various renovations undertaken by his father and grandfather and attempt to update the gardens. With the outbreak of the French Revolution, however, Louis XVI and his queen, Marie Antoinette, were forced to relocate to Paris, and the fate of Versailles remained uncertain as the citizens of Paris debated how to handle its wealth of art and furniture. Following debates over the possibility of liquidating certain categories of items within, it was decided to attempt to preserve the palace as the property of the revolutionary Republic of France.

Following the restoration of the French monarchy in 1830, Versailles was returned to the crown and established as a museum by King Louis-Philippe (r. 1830–1848). In the modern era, Versailles has continued to serve certain government and political functions as the site for significant meetings of state or diplomacy. In 1919, it served as the base for the Paris Peace Conference at the end of World War I and the site for the signing of the Treaty of Versailles.

In 1979, UNESCO designated the Palace of Versailles and its gardens a World Heritage Site. Continued restoration efforts to maintain this famous landmark have proven to be more costly than its actual construction. Just the restoration project to reconstruct the gold leaf gates at the entrance to the palace complex in 2008 required an estimated 100,000 gold leaves, an effort for which "private donors contributed £4 million to rebuild the 15-ton work, and a plethora of historians and top craftsmen—sculptors, gilders, wrought iron craftsmen and ornament makers—were drafted in to ensure an exact replica of the original built by Jules Hardouin-Mansart in the 1680s" (Samuel 2008)—an effort befitting the glory of one of France's most famous tourist attractions today.

See also: Gilding.

Further Reading

Arizzoli-Clémentel, Pierre. 2002. *Versailles: Furniture of the Royal Palace, 17th and 18th Centuries*, vol. 2. Dijon: Faton.
Berger, Robert W. 1986. *Versailles: The Château of Louis XIV.* University Park: The College Arts Association.
Kisluk-Grosheide, Daniëlle O. "French Royal Furniture in the Metropolitan Museum." *Metropolitan Museum of Art Bulletin* 63, no. 3 (Winter):4–48.
Parker, James. 1953. "Louis XVI Gilt Bronze Ornaments." *Metropolitan Museum of Art Bulletin* Vol. 12 No. 3 (Nov. 1953):68–73.
Raggio, Olga. 2000. "French Decorative Arts during the Reign of Louis XIV (1654–1715)." *Heilbrunn Timeline of Art History.* New York: Metropolitan Museum of Art. http://www.metmuseum.org/toah/hd/frde/hd_frde.htm.
Samuel, Henry. 2008. "Palace of Versailles Golden Gate Restored." *Telegraph* United Kingdom, July 1. http://www.telegraph.co.uk/news/worldnews/europe/france/2228807/Palace-of-Versailles-golden-gate-restored.html.
Spawforth, Antony. 2008. *Versailles: A Biography of a Palace.* New York: St. Martin's.
Walton, Guy. 1986. *Louis XIV's Versailles.* Chicago: Univ. of Chicago Press.

WAT TRAIMIT

Known as the Temple of the Golden Buddha, Wat Traimit is a Buddhist temple in the Chinatown district of Samphanthawong of Bangkok, Thailand, that houses a large solid gold statue of Buddha, called Phra Phuttha Maha Suwan Patimakon (in the Thai language). The statue stands at 15 feet 9 inches high, including the base; weighs around 5 tons; and is recorded as the largest pure gold statue of Buddha in the 1991 edition of the *Guinness Book of World Records*.

Although its precise origins are unknown, because of the style of the statue, it is thought to have been constructed during the Sukhothai period under the reign of King Ramkhamhaeng the Great in 1279–1299 and housed originally in a temple in the ancient city of Ayutthaya, where it was covered in plaster at some point during the late 18th century to conceal the solid gold beneath during an attack on the city by the Burmese. The statue was later moved to Bangkok during the rule of King Rama III (1824–1851), where it was on display in the Wat Phrayakrai, a temple that was demolished due to disrepair in the 1930s, at which time the statue, still covered in plaster, was relocated to a crude tin shelter near Wat Traimit. During the mid-1950s, a new temple was erected at Wat Traimit, and during an attempt to install the statue in the new structure, a cable on the crane snapped and the statue dropped to the ground and cracked. The solid gold statue beneath was discovered when a Buddhist monk inspected the dropped statue

Solid gold Buddha statue and a praying woman in Wat Traimit temple, Bangkok, Thailand. (Alexey Stiop/Dreamstime.com)

where it lay. It was thus restored to its original glory and remains on display as one of the richest treasures of Buddhist traditions in Thailand, considered by many to be priceless.

The Buddha is depicted in a typical pose known as the Bhumisparsha Mudrā, in which the Buddha is seated cross-legged with his right palm resting facedown on his right knee, a gesture signifying the earthbound enlightenment of Buddha underneath the Bodhi Tree at Bodh Gaya, India, and his left hand resting flat on his lap.

The present-day temple housing the giant Buddha also contains a gallery outlining its history and a Chinese cultural museum. The complex and the statue function as a strong symbol of Chinese heritage and history in Thailand.

Further Reading

Aasen, Clarence. 1998. *Architecture of Siam: A Cultural History and Interpretation.* Oxford: Oxford Univ. Press.
Tripathi, Manote. 2010. "Sealed in Solid Gold." *Nation*, January 24.
Van Beek, Steve, and Luca Invernizzi. 1999. *The Arts of Thailand.* Boston: Periplus Editions.

WITWATERSRAND GOLD RUSH

In 1886, the political and social history of South Africa charted a significant course as a result of the discovery of gold by diamond speculator George Harrison while digging up stone to clear the land for building on Widow Oosthuizen's farm near Witwatersrand, also known as Rand, in the Transvaal. Based on his experiences with previous speculation efforts in South Africa, Harrison recognized the stone as the type that more than likely contained gold when crushed and processed. He had discovered what would come to be acknowledged as the largest gold-bearing reef ever discovered. The discovery of gold in South Africa would launch a different type of gold rush, however, due to the fact that gold supplies in the region existed in low-grade ore found in outcrops or deep underground veins that required significant capital and technology to mine. This would lead to the swift growth of Johannesburg as a vibrant, industrial city.

The discovery precipitated a flood of immigration from the United Kingdom to the settlement that was established at Johannesburg, with an estimated 75,500 British citizens moving to South Africa during 1895–1898; however, as profits and yields from the outcrops in the gold fields diminished within three years of their discovery, and the relatively small gold yield from processing large amounts of ore discouraged corporate speculators, the South African gold rush began to decline, with many contemporaries predicting that Johannesburg would also fade away.

In 1889, however, Scottish representatives of the African Gold Extracting Company introduced the technology patented as the MacArthur-Forrest Process for extracting a higher percentage of pure gold from the ore, up to 96 percent, using **cyanide** to dissolve and filter the gold from the ore. Gold extraction utilizing these methods required a significant level of capital, industrial development, and technology to be profitable. The gold-mining industry thus developed in the region as a powerful consortium of large companies with extensive political and financial influence in the Transvaal. The emergent industrial capitalism and settlement of Johannesburg and surrounding areas accelerated the displacement of native African chiefdoms and the exploitation of Africans as laborers. Gold **mining** is also credited with fueling the enmity between the ruling Afrikaaners, or Boers, and the British business leaders and settlers who predominated the **mining** industry. The tense relationship between these two groups resulted in the Jameson Raid of 1895–1896, an attempt by British governor of Cape Town Cecil Rhodes to incite insurrection in Johannesburg and establish official British control of the commercial **mining** industry.

By 1899, foreigners known as *uitlanders* represented the majority of the population in Johannesburg and the area of the Rand as a result of the gold rush. Intent on curbing the political and economic influence of the British and other foreign residents, President of the South African Republic Paul Kruger enacted legislation restricting their civil rights and imposing heavy taxes and tariffs on foreign industrial operations and individuals. The escalation of tensions that had been growing between the respective sides since the Jameson Raid provoked the Second Boer War of 1899–1902 between Great Britain and the Boer republics of Transvaal (South African Republic) and the Orange Free State. The British emerged victorious from this brutal war to assert control over the riches of the gold mines and enter into a peace that is attributed as a harbinger of the future apartheid movement by creating a system of self-rule for the new South African Republic that excluded black Africans from the vote.

Today, South Africa remains one of the world's largest producers of gold, along with other valuable stones and minerals such as diamonds, platinum, and coal.

Krugerrand

The krugerrand is a gold bullion coin introduced by the Republic of South Africa in 1967 in coordination with the Chamber of Mines of South Africa, a representative of the national gold mining industry, and the South Africa Reserve Bank as a means of promoting the holding of South African gold as an investment in international markets. The original krugerrands were minted from 22-karat-gold bullion blanks produced by the Rand Refinery weighing one troy ounce and engraved with a portrait of Transvaal president Stephanus Johannes Paulus Kruger (1825–1904). Hence, the name "krugerrand" was taken from president's last name (Kruger) and the Witwatersrand mining region of the Transvaal (Rand). Although the South African Treasury acknowledged the krugerrand as legal tender, it has largely served as an investment vehicle. During the 1970s and 1980s, import of krugerrands to many Western countries was severely restricted due to the economic sanctions against apartheid. In 1994, when apartheid was abolished, the sanctions were lifted. After 1980, additional denominations of one-half, one-quarter, and one-tenth of a troy ounce of gold were minted. Today, the krugerrand remains the most widely held and exchanged gold coin in international markets.

See also: California Gold Rush; Cyanide; Klondike Gold Rush; Mining.

Further Reading

Firaz, C. E. 1988. "How the MacArthur-Forrest Cyanidation Process Ensured South Africa's Golden Future." *Journal of the South African Institute of Mining and Metallurgy* (September):309–318.

Richardson, Peter, and Jean Jacques van Helten. 1984. "The Development of the South African Gold Mining Industry, 1895–1918." *Economic History Review* 37, no. 3:319–340.

Van Onselen, C. 1982. *Studies in the Social and Economic History of the Witwatersrand. Volume 1: New Babylon.* New York: Raven Press.

YELLOW BRICK ROAD

In literary criticism since the mid-20th century, Frank L. Baum's classic fantasy novel *The Wizard of Oz* (1900) has been widely considered to be in many instances a political allegory of one the most prominent issues of the late 19th century, namely, the populist advocacy of the adoption of a bimetallic currency standard to replace the **gold standard** as a solution to contemporary farmers' struggles to combat deflation.

Although Baum himself never indicated that he intended to convey any political associations in the story, beginning with the 1964 publication of Henry Littlefield's "*The Wizard of Oz*: Parable on Populism" scholars have suggested that the tale is a product of the era in which it was produced, and, whether intentional or not, it includes a symbolic cast of characters that reflected various aspects of the radical monetary reforms advocated by Populist Democrat William Jennings Bryan.

A variety of theories have been put forth about what exactly these representations are, some heavily debated, but the prevailing theory is founded on Littlefield's suggestion that the Yellow Brick Road signifies the **gold standard**; that Dorothy's shoes, which were silver in the original book, represent the solution silver offered to the monetary situation; the Emerald City stands for the ineffective political system or perhaps even the devalued dollar (greenback); the scarecrow is inspired by the plight of farmers; the Tin Man represents dehumanized industrial workers; the Lion is perhaps William Jennings Bryan himself, who had been ridiculed by the press as a coward for his opposition to the 1898 Spanish American War; the Wicked Witch of the East represents Wall Street and big business; the Wicked Witch of the West, the unpredictable destructiveness of Mother Nature; and Glinda the Good Witch of the North represents the hope offered by the potential votes from Bryan's most likely allies in the Democratic Southern states.

In the story, Dorothy's house lands on the Wicked Witch of the East and kills her, which frees the Munchkins, a scene which has been viewed as an allegory for the liberation of the common people from the machinations of the barons of big business.

> **Dorothy:** [*in the Wizard's Throne Room with the three others, having returned from the Witch's castle*] Please, sir. We've done what you told us. We brought you the broomstick of the Wicked Witch of the West. We melted her!
>
> **Wizard of Oz:** Oh, you liquidated her, eh? Very resourceful!
>
> —*Wizard of Oz* (1939 film)

Dorothy's silver slippers have the ability to achieve what she desires, but only if she follows the Yellow Brick Road. The combination of the silver slippers and the journey down the Yellow Brick Road has been interpreted as signifying the path to a bimetallic currency standard.

Baum's true political affiliations still remain largely unknown. It has been widely believed that Baum was a Democrat due to the memoirs of his eldest son, which depict him as such; however, there is little evidence of his true position. Baum appears to have been largely apolitical, a fact that may have reflected a certain skepticism about the political system of both sides.

See also: Gold Standard.

Further Reading

Barnes, James A. 1937. "Myths of the Bryan Campaign." *Mississippi Historical Review* 34 (December):367–400.

Baum, Frank L. 1900. *The Wonderful Wizard of Oz*. Chicago: George M. Hill.

Dighe, Ranjit S., ed. 2002. *The Historian's "Wizard of Oz": Reading L. Frank Baum's Classic as a Political and Monetary Allegory*. Westport, CT: Praeger.

Littlefield, Henry. 1964. "The *Wizard of Oz*: Parable on Populism." *American Quarterly* 16, no. 3:47–58.

Rockoff, Hugh. 1990. "The *Wizard of Oz* as a Monetary Allegory." *Journal of Political Economy* 98:739–760.

Resource Guide

American Institute of Mining, Metallurgical, and Petroleum Engineers
P.O. Box 270728
Littleton, CO 80127-0013
(303) 948-4255
http://www.aimeny.org/

American Museum of Natural History
Gold Exhibit Online Catalogue
Central Park West at 79th St.
New York, NY 10024
(212) 769-5606
http://www.amnh.org/exhibitions/gold/

British Museum
Great Russell St.
London, UK WCIB3DG
+ 44 (0) 2073238195
http://www.britishmuseum.org/

Canadian Institute of Mining, Metallurgy, and Petroleum
3500 de Maisonneuve Blvd. W., Suite 1250
Westmount, QC H3Z 3C1
(514) 939-2710
http://www.cim.org

Federal Reserve
20th Street and Constitution Avenue NW
Washington, DC 20551
http://www.federalreserve.gov/

Gold Bulletin: Journal of Gold Science, Technology and Applications
Published by Springer
233 Spring Street
New York, NY 10013
(212) 460-1500
http://www.goldbulletin.org

Heilbrunn Timeline of Art History
The Metropolitan Museum of Art
1000 Fifth Avenue at 82nd Street
New York, NY 10028-0198
(212) 535-7710
http://www.metmuseum.org/toah/

International Council on Mining and Metals
35/38 Portman Square
London, UK
W1H 6LR
+44 (0) 20 7467 5070
http://www.icmm.com

International Precious Metals Institute
5101 North 12th Ave., Suite C
Pensacola, FL 32504
(850) 476-1156
http://www.ipmi.org/

Library of Economics and Liberty
Liberty Fund, Inc.
8335 Allison Pointe Trail, Suite 300
Indianapolis, IN 46250-1684
(800) 866-3520
http://www.econlib.org

Money Museum
Hadlaubstrasse 106
CH-8006 Zurich
+41 443507380
www.moneymuseum.com

National Mining Association
101 Constitution Ave. NW
Suite 500 East
Washington, D.C., 20001
(202) 463-2600
http://www.nma.org

Society for Mining, Metallurgy, and Exploration
830 Shaffer Parkway
Littleton, CO
(303) 948-4200
http://www.smenet.org/

United States Geological Survey
12201 Sunrise Valley Drive
913 National Center
Reston, VA 20192
(703) 648-6110
http://minerals.er.usgs.gov/

United States Treasury
1500 Pennsylvania Ave. NW
Washington, DC 20220
(202) 622-2000
http://web.archive.org/web/20080131183526/www.ustreas.gov/education/

World Gold Council
10 Old Bailey
London
EC4M 7NG
+44 (0)20.7826.4700
http://www.gold.org

Bibliography

Aasen, Clarence. 1998. *Architecture of Siam: A Cultural History and Interpretation.* Oxford: Oxford Univ. Press.

Adams, Edward B. 1991. *Korea's Golden Age: Cultural Spirit of Silla in Kyongju.* Seoul: Seoul International Pub. House.

Aldrich, Wilbur. 1903. *Money and Credit.* New York: Grafton Press.

Alexander, Jonathan J.G. 1992. *Medieval Illuminators and Their Methods of Work.* New Haven: Yale Univ. Press.

Altick, Richard. 1978. *The Shows of London.* Cambridge, MA: Harvard Univ. Press.

Arizzoli-Clémentel, Pierre. 2002. *Versailles: Furniture of the Royal Palace, 17th and 18th Centuries.* vol. 2. Dijon: Faton.

Arnold, J. Barto, III, and Robert Weddle. 1978. *The Nautical Archeology of Padre Island.* New York: Academic Press.

Bachman, H.G., and Robert B. Cook. 2003. *Gold: The Noble Metal.* Ann Arbor: Univ. of Michigan Press.

Bachman, H.G., Steven Lindberg, and Jorg Vollnagel. 2006. *The Lure of Gold: An Artistic and Cultural History.* New York: Abbeville Press.

Bain, Francis William. 1896. *The Bullion Report and the Foundation of the Gold Standard.* Oxford: James Parker.

Baldwin, Frances Elizabeth. 1926. *Sumptuary Legislation and Personal Regulation in England.* Baltimore: Johns Hopkins Press.

Baldwin, John W. 1959. "The Medieval Theories of Just Price." *Transactions of the American Philosophical Society* 49, no. 4:15–92.

Ball, Laurence M. 2008. *Money, Banking, and Financial Markets.* New York: Worth Publishers.

Bandelier, Adolph. 1893. *The Gilded Man: El Dorado and Other Pictures of the Spanish Occupancy of South America.* New York: D. Appleton.

Barsali, Isa Belli. 1969. *Medieval Goldsmiths Work.* London: Pauly Hamlyn.

Becker, Thomas W. 1970. *The Coin Makers.* New York: Doubleday.

Bellamy, Caroline. 2008. "Medicine Is Going For Gold as Experts Rediscover the Ancient Properties of this Precious Metal." *Daily Mail,* June 7.

Belliveau, Denis, and Francis O'Donnell. 2008. *In the Footsteps of Marco Polo.* Lanham: Rowman & Littlefield Publishing Group.

Berger, Robert W. 1986. *Versailles: The Château of Louis XIV.* University Park, MD: The College Arts Association.

Bergreen, Laurence. 2008. *Marco Polo: From Venice to Xanadu.* New York: Knopf.

Bernholz, Peter. 2003. *Monetary Regimes and Inflation.* Cheltenham, UK: Edward Elgar.

Bernstein, Peter L. 1968. *A Primer on Money, Banking, and Gold.* 2nd edition. New York: Random House.

Berton, Pierre. 2003. *The Life and Death of the Last Great Gold Rush.* New York: Carroll and Graf Publishers.

Beuster, Kirsten. 2007. *Gold: Gilding History and Techniques*. Atglen, PA: Schiffer Publishing.

Bigelow, Deborah, ed. 1980. *Gilded Wood, Conservation, and History*. Madison, CT: Sound View Press.

Bohlander, Richard E., ed. 1992. *World Explorers and Discoverers*. New York: Macmillan Publishing.

Bonar, J. 1923. "Ricardo's Ingot Plan." *Economic Journal* 33:281–304.

Bond, Geoffrey C., Catherine Louis, and David T. Thompson. 2006. *Catalysis by Gold*. London: World Scientific Publishing.

Bordo, M. D. and Eichengreen, B. eds. 1993. *A Retrospective on the Bretton Woods System: Lessons for International Monetary Reform*. Chicago: Univ. of Chicago Press.

Bovill, Edward William. 1958. *The Golden Trade of the Moors*. London: Oxford Univ. Press.

Bowersock, G. W. 2006. *Mosaics as History: The Near East from Late Antiquity to Islam*. Cambridge, MA: Belknap.

Boyle, Robert W. 1987. *Gold: History and Genesis of Deposits*. New York: Van Nostrand Reinhold.

Brands, H. W. 2003. *The Age of Gold: The California Gold Rush and the New American Dream*. New York: Anchor Books.

Braudel, Fernand, and Franck Spooner. 1967. "Prices in Europe from 1450 to 1570." In *The Cambridge Economic History of Europe*, vol. 4, eds. E. E. Rich and C. H. Wilson, 338–442. Cambridge: Cambridge Univ. Press.

Bray, Warwick. 1980. *The Gold of El Dorado*. New York: Abrams.

Bryer, Anthony, and Judith Herrin, eds. 1977. *Iconoclasm*. Birmingham, England: Univ. of Birmingham Centre for Byzantine Studies.

Buchan, James. 2008. "When Keynes Went to America." *New Statesman*, November 10.

Buenker, John D., and Joseph Buenker, eds. 2005. *Encyclopedia of the Gilded Age and Progressive Era*. Armonk, NY: M. E. Sharpe.

Bunker, Emma C. 1993. "Gold in the Ancient Chinese World: A Cultural Puzzle." *Artibus Asiae* 53, no. 1–2:27–50.

Bunt, Cyril G. E. 1926. *The Goldsmiths of Italy, Some Accounts of Their Guilds, Statutes, and Work Compiled from the Published Papers, Notes, and other Material Collected by Sidney J. A. Churchill*. London: Martin Hopkinson.

Burton, R. E. 1979. *The Gold-mines of Midian and the Ruined Midianite Cities (1878)*. New York: Courier Dover.

Calabi, Donatella. 2004. *The Market and the City: Square, Street, and Architecture in Early Modern Europe*. Burlington, VT: Ashgate Publishing.

Calkins, Robert G. 1983. *Illuminated Books of the Middle Ages*. Ithaca, NY: Cornell Univ. Press.

Campbell, Marion. 2009. *Medieval Jewellry in Europe, 1100–1500*. London: V & A Publications. Carter, Howard. 1972. *The Tomb of Tutankhamen*. New York: E. P. Dutton.

Casas, Bartolomé de Las. 1552. "A Brief Account of the Destruction of the Indies." *Project Gutenberg*. http://www.gutenberg.org/etext/20321.

Cellini, Benvenuto. 2006. *The Treatises of Benvenuto Cellini on Goldsmithing and Culture*. Translated by C. R. Ashbee. Whitefish, MT: Kessinger Publishing.

Cennini, Cennino d'Andrea. 1954. *The Craftsman's Handbook*. Translated by Daniel Varney Thompson. New York: Thompson Courier Publications.

Challis, Christopher. 1967. "The Debasement of the Coinage, 1542–51." *Economic History Review* New Series 20, no. 3 (December):441–466.

Challis, Christopher. 1973. "A Contemporary Estimate of the Production of Silver and Gold Coinage in England, 1542–56." *English Historical Review* 88, no. 346 (October):821–835.
Challis, Christopher. 1979. *The Tudor Coinage.* Manchester, UK: Manchester Univ. Press.
Challis, Christopher, ed. 1992. *A New History of the Royal Mint.* London: Cambridge Univ. Press.
Chalquist, Craig. 2009. "Sir Isaac Newton, Alchemist." *Psychological Perspectives* 52, no. 2:199–218.
Chavarria, Joaquin. 1999. *The Art of Mosaics: A Guide to the History, Material, Equipment, and Techniques.* New York: Watson-Guptill.
Chown, John F. 1994. *A History of Money: From AD 800.* London: Psychology Press.
Cirlot, Juan Eduardo. 2002. *A Dictionary of Symbols.* New York: Courier Dover.
Clarke, Peter. 2009. *Keynes: The Rise, Fall, and Return of the Twentieth Century's Most Influential Economist.* London: Bloomsbury Publishing.
Corti, Christopher W., Richard J. Holliday, and David T. Thompson. "Potential of Catalysis by Gold for Fuel Cell and Pollution Control Applications " *World Gold Council.* http://www.nacatsoc.org/18nam/Posters/P020-Potential%20of%20Catalysis%20by%20Gold%20for%20Fuel%20Cell.pdf.
Craig, John. 1963. "Isaac Newton and the Counterfeiters." *Notes and Records of the Royal Society of London* 18, no. 2 (December):136–145.
Craig, John Herbert, Sir. 1946. *Newton at the Mint.* London: Univ. Press.
Craig, Sir John. 1953. *The Mint: A History of the London Mint from AD 287 to 1948.* New York: Cambridge Univ. Press.
Crane, Walter Richard. 1908. *Gold and Silver: Comprising an Economic History of Mining in the United States.* New York: John Wiley and Sons.
Danaher, Kevin, ed. 1994. *50 Years Is Enough: The Case Against the World Bank and the International Monetary Fund.* Cambridge, MA: South End Press.
Davidson, Neiles H. 1997. *Columbus Then and Now: A Life Reexamined.* Norman: Univ. of Oklahoma Press.
Davies, Glyn. 1995. *A History of Money from Ancient Times to the Present Day.* Cardiff: Univ. of Wales Press.
Department of Ancient Near Eastern Art. 2003, October. "Ur: The Royal Graves." In *Heilbrunn Timeline of Art History.* New York: The Metropolitan Museum of Art. http://www.metmuseum.org/toah/hd/urrg/hd_urrg.htm.
Diaz Del Castillo, Bernal. 2004 ed. *The Discovery and Conquest of Mexico, 1817–21.* Cambridge, MA: Da Capo Press.
Dighe, Ranjit S., ed. 2002. *The Historian's "Wizard of Oz": Reading L. Frank Baum's Classic as a Political and Monetary Allegory.* Westport, CT: Praeger.
Dillard, Dudley D. 1948. *The Economics of John Maynard Keynes: The Theory of Monetary Economy.* Upper Saddle River, NJ: Prentice-Hall.
Dobbs, B. 1991. *The Foundations of Newton's Alchemy, or, "The Hunting of the Green Lyon."* Cambridge: Cambridge Univ. Press.
Dobbs, B. 1992. *The Janus Face of Genius: The Role of Alchemy in Newton's Thought.* London: Cambridge Univ. Press.
Dodd, Agnes F. 1911. *The History of Money in the British Empire and the United States.* London: Longmans, Green & Co.
Dodwell, C. R. 1993. *The Pictorial Arts of the West, 800–1200.* Pelican History of Art. New Haven, CT: Yale Univ. Press.

Doty, Richard G. 1982. *Macmillan Encyclopedic Dictionary of Numismatics.* New York: Macmillan.

Drost, E., and J. H. Hanbelt. 1992. "Uses of Gold Jewellry." *Interdisciplinary Science Review* 17:271–280.

Dunbabin, Katherine. 2001. *Mosaics of the Greek and Roman World.* London: Cambridge Univ. Press.

Dürer, Albrecht. 1973. *Albrecht Dürer: Diary of His Journey to the Netherlands.* Edited by J. A. Goris and G. Marlier. Greenwich, CT: New York Graphic Society.

Edwards, I. E. S. *The Treasures of Tutankhamun.* New York: Ballantine.

Edwards, John. 2004. *Ferdinand and Isabella.* New York: Longman.

Eichengreen, Barry. 1996. *Globalizing Capital: A History of the International Monetary System.* Princeton, NJ: Princeton Univ. Press.

Eichengreen, Barry J., Jorge Bragade Macedo, and Jaime Reis, eds. 1996. *Currency Convertibility: The Gold Standard and Beyond.* London: Routledge.

Eisen, Arri, and Gary Laderman, eds. 2007. *Science, Religion, and Society: An Encyclopedia of History, Culture, And Controversy.* Armonk, NY: M. E. Sharpe.

Emmerich, Andre. 1965. *Sweat of the Sun and Tears of the Moon: Gold and Silver in Pre-Columbian Art.* Seattle: Univ. of Washington Press.

Evans, J. A. S. 2005. *The Emperor Justinian and the Byzantine Empire.* Westport, CT: Greenwood.

Evans, Joan. 1982. *A History of Jewelry, 1100–1870.* London: British Museum Publications.

Faber, Toby. 2008. *Fabergé's Eggs: The Extraordinary Story of the Masterpieces that Outlived an Empire.* New York: Random House.

Feathering, George. 1997. *The Gold Crusades: A Social History of Gold Rushes, 1849–1929.* Toronto: Univ. of Toronto Press.

Feldman, Burton. 2001. *The Nobel Prize: A History of Genius, Controversy, and Prestige.* New York: Arcade Publishing.

Ferns, H. S. 1992. "The Baring Crisis Revisited." *Journal of Latin American Studies* 24 no. 2 (May):241–273.

Fischer, David Hacket. 1996. *The Great Wave: Price Revolution and the Rhythm of History.* London: Oxford Univ. Press.

Fivaz, C. E. 1998. "How the MacArthur-Forrest Cyanidation Process Ensured South Africa's Golden Future." *Journal of the South African Institute of Mining and Metallurgy* (September):309–318.

Flores, Juan Huitzi. 2006. "A Microeconomic Analysis of the Baring Crisis, 1880–90." Carlos III University, Economics History Department, Madrid. http://emlab.berkeley.edu/users/webfac/eichengreen/e211_fa05/211_baring.pdf.

Forbes, Robert James. 1964. *Studies in Ancient Technology.* Boston, MA: Brill.

Foss, C. 1990. *Roman Historical Coins.* London: Seaby.

Frick, Caroline Collier. 2002. *Dressing Renaissance Florence: Families, Fortunes, and Fine Clothing.* Baltimore: Johns Hopkins Univ. Press.

Friedman, Milton. 1953. "The Case for Flexible Exchange Rates." In *Essays in Positive Economics,* 157–203. Chicago: Univ. of Chicago Press.

Friedman, Milton. 1992. *Money Mischief: Episodes in Monetary History.* New York: Harcourt Brace Jovanovich.

Friedman, Milton, and Anna Schwartz. 1963. *A Monetary History of the United States, 1867–1960.* Princeton, NJ: National Bureau of Economic Research and Princeton Univ. Press.

Garrard, Timothy F. 1982. "Myth and Metrology: The Early Trans-Saharan Gold Trade." *Journal of African History* 234:443–461.
Gerdesmeier, Dieter, Francesco Paolo Mongelli, and Barbara Roffia. 2007. "The Eurosystem, the U.S. Federal Reserve, and the Bank of Japan: Similarities and Differences." *Journal of Money, Credit and Banking* 39 no. 7 (October):1785–1819.
Gibbs, Laura. 2002. *Aesop's Fables*. Oxford: Oxford Univ. Press.
Godwin, Malcolm. 1994. *The Holy Grail: Its Origins, Secrets, and Meaning Revealed*. London: Labyrinth Books.
Goldthwaite, Richard. 2009. *Economy of Renaissance Florence*. Johns Hopkins Univ. Press.
Goodman, David. 1994. *Gold Seeking: Victoria and California in the 1850s*. Stanford, CA: Stanford Univ. Press.
Gordon, John Steele. 1997. *Hamilton's Blessing: The Extraordinary Life and Times of Our National Debt*. New York: Walker.
Gough, Hugh, and John Horne, eds. 1994. *De Gaulle and Twentieth-Century France*. Oxford: A Hodder Arnold Publications.
Gould, J.D. 1952–1953. "The Royal Mint in the Early 17th Century." *Economic History Review* 5:242.
Gould, J.D. 1970. *The Great Debasement: Currency and the Economy in Mid-Tudor England*. Oxford: Clarendon Press.
Green, Timothy. 1993. *The World of Gold: The Inside Story of Who Mines, Who Markets, Who Buys Gold*. London: Rosendale Press. First published in 1968.
Green, Timothy. 2007. *The Ages of Gold*. London: GFMS Ltd.
Green, Wm. Spotswod. 1906. "The Wrecks of the Spanish Armada on the Coast of Ireland." *Geographical Journal* 27, no. 5 (May):429–448.
Gregory, Timothy E. 2010. *A History of Byzantium*. New York: John Wiley & Sons.
Grierson, P. 1991. *Coins of Medieval Europe*. London: Seaby.
Grice-Hutchinson, Marjorie. 1952. *The School of Salamanca: Readings in Spanish Monetary Theory*. London: Oxford Univ. Press.
Grierson, Philip. 1982. *Byzantine Coins*. Berkeley: Univ. of California Press.
Grimm, Jacob and Wilhelm. 1812. *Children's and Household Tales*. Translated by Margaret Hunt. http://www.gutenberg.org/dirs/etext04/grimm10a.txt.
Grossman, Richards. 2010. *Unsettled Account: The Evolution of Banking in the Industrialized World since 1800*. Princeton, NJ: Princeton Univ. Press.
Habashi, Fathi. 2009. *Gold, History, Metallurgy, Culture*. Quebec City: Metallurgie Extractive Quebec.
Hagen, Victor Wolfgang von. 1968. *The Gold of El Dorado: The Quest for the Golden Man*. Farnborough, UK: Saxon House.
Hahn, Emily. 1980. *Love of Gold*. New York: Lippincott and Crowell.
Hamilton, Earl J. 1929. "Imports of American Gold and Silver into Spain, 1503–1660." *Quarterly Journal of Economics* 43, no. 3 (May):436–472.
Hamilton, Earl J. 1970. *American Treasure and the Price Revolution in Spain, 1501–1650*. New York: Octagon Books. (Orig. pub. 1934.)
Hargraves, Edward Hammond. 1855. *Australia and Its Goldfields*. London: H. Ingram.
Haring, C. H. 1927. "The Genesis of Royal Government in the Spanish Indies." *Hispanic American Historical Review* 7, no. 2 (May):141–191.
Harmston, Steven. 1998. "Gold as a Store of Value." *World Gold Council* Research Study, no. 22 (November). http://www.goldbullion.com.au/pdf/research_study_22.pdf.

Hawass, Zahi. 2005. *Tutankhamun and the Golden Age of Pharaohs.* Washington, DC: National Geographic Society.

Hawthorne, J. G., and C. S. Smith. 1979. *Theophilius, On Divers Arts: The Foremost Medieval Treatise on Painting, Glassmaking, and Metalwork.* New York: Courier Dover Publications.

Hayward, J. F. 1976. *Virtuoso Goldsmiths and the Triumph of Mannerism.* New York: Rizzoli International.

Healy, J. E. 1978. *Mining and Metallurgy in the Greek and Roman World.* London: Thames & Hudson.

Healy, J. E. 1985. "The Processing of Gold Ores in the Ancient World." *Canadian Inst. Min. Metall. Bull.* 78, no. 874:84–88.

Heffernan, Shelagh. 2005. *Modern Banking.* Oxford: Wiley and Sons.

Heilbroner, Robert L. 1953. *The Worldly Philosophers.* New York: Simon and Schuster.

Hemming, John. 1970. *The Conquest of the Incas.* San Diego, CA: Harcourt and Brace.

Hendy, Michael F. 2008. *Studies in the Byzantine Monetary Economy c. 300–1450.* Cambridge: Cambridge Univ. Press.

Herrin, Judith. 2009. *Byzantium: The Surprising Life of a Medieval Empire.* Princeton, NJ: Princeton Univ. Press.

Hiebert, Fredrik, and Pierre Cambon, eds. 2008. *Afghanistan: Hidden Treasures from the National Museum, Kabul.* Washington, DC: National Geographic Society.

Hill, George F. 1977. *Ancient Methods of Coining.* New York: Attic Books.

Hill, Gerald. 2007. *Fabergé and the Russian Master Goldsmiths.* New York: Universe.

Hirshfeld, Alan. 2010. *Eureka Man: The Life and Legacy of Archimedes.* New York: Walker.

Hobson, Burton, and Robert Obojski. 1970. *Illustrated Encyclopedia of World Coins.* Garden City, NY: Doubleday.

Hollander, Jacob. 1910–1911. "The Development of the Theory of Money from Adam Smith to David Ricardo." *Quarterly Journal of Economics* 25:429–470. http://socserv2.socsci.mcmaster.ca/~econ/ugcm/3ll3/hollander/money.html.

Hollander, Jacob. 1904. "The Development of Ricardo's Theory of Value." *Quarterly Journal of Economics* 18:455–491.

Hooper, Wilfred. 1915. "The Tudor Sumptuary Laws." *English Historical Review* 30, no. 119 (July):433–445.

Hunt, Alan. 1996. *Governance of the Consuming Passions: A History of Sumptuary Law.* London: Macmillan.

Isard, Peter. 1995. *Exchange Rate Economics.* London: Cambridge Univ. Press

Jackson, A. V. Williams. 1905. "The Magi in Marco Polo and the Cities in Persia from Which They Came to Worship the Infant Christ." *Journal of the American Oriental Society* 26:79–83.

James, T. G. H., 1972. "Gold Technology in Ancient Egypt," *Gold Bulletin* 5, no. 2:38–42.

Jastram, Roy. 1977. *The Golden Constant.* New York: John Wiley and Sons.

Jones, Julie. 2002. "Gold of the Indies." *Metropolitan Museum of Art Bulletin* 59, no. 4 (Spring):3–4.

Keel, Trevor, Richard Holliday, and Tim Harper. 2010. "Gold for Good: Gold and Nanotechnology in the Age of Innovation." *World Gold Council Bulletin* (January).

Kenwood, A. G., and A. L. Longheed. 1999. *The Growth of the International Economy, 1820–2000.* London: Routledge.

Keynes, John Maynard. 1925. *The Economic Consequences of Mr. Churchill.* London: L. and V. Woolf.

Keynes, John Maynard. 1971–1989. *The Collected Writings of John Maynard Keynes.* 30 vol. Edited by Austin Robinson and Donald Moggridge. London: Macmillan.

Killorby, Catherine Kovesi. 2002. *Sumptuary Law in Italy, 1200–1500.* Oxford: Oxford Univ. Press.

Kindleberger, Charles. 1985. *Keynesianism versus Monetarism and Other Essays in Financial History.* London: George Allen and Unwin.

Kindleberger, Charles. 1993. *A Financial History of Western Europe.* New York: Oxford Univ. Press.

Kindleberger, Charles. 1990. *Spenders and Hoarders: The World Distribution of Spanish American Silver, 1550–1750.* Pasir Pajang, Singapore: ASEAN Economic Research Unit.

King, Heidi. 2002. "Gold in Ancient America." *Metropolitan Museum of Art Bulletin* New Series 59, no. 4 (Spring):5–55.

King, Heidi. 2004, October. "Tenochtitlan." In *Heilbrunn Timeline of Art History.* New York: The Metropolitan Museum of Art. http://www.metmuseum.org/toah/hd/teno_1/hd_teno_1.htm.

Kleinbauer, W. Eugene, Antony White, and Henry Matthews. 2004. *Hagia Sophia.* London: Scala Publishers.

Klenke, M. Amelia. 1955. "Chrétien's Symbolism and Cathedral Art." *PMLA* 70, no. 1 (March):223–243.

Koch, John T. 2006. *Celtic Culture: A Historical Encyclopedia.* Santa Barbara, CA: ABC-CLIO.

Lane, Frederic, and Reinhold Mueller. 1985. *Money and Banking in Medieval and Renaissance Venice, vol. 1: Coins and Money of Account.* Baltimore: Johns Hopkins Univ. Press.

LaNiece, Susan. 2009. *Gold.* London: British Museum.

Lee, Soyoung. 2003, October. "Golden Treasures: The Royal Tombs of Silla." In *Heilbrunn Timeline of Art History.* New York: Metropolitan Museum of Art. http://www.metmuseum.org/toah/hd/sila/hd_sila.htm.

Leon-Portillo, Miguel, and Lysander Kemp, trans. 2006. *The Broken Spears: The Aztec Account of the Conquest of Mexico.* Boston: Beacon Press.

Lewis, Bernard. 1958. "Some Reflections on the Decline of the Ottoman Empire." *Studia Islamica* no. 9:111–127.

Li, Ming-Hsun. 1963. *The Great Recoinage of 1696 to 1699.* London: Weidenfeld and Nicolson.

Lind, Michael. 1994. "Hamilton's Legacy" *Wilson Quarterly* 18, no. 3 (Summer):40–52.

Littlefield, Henry. 1964. "The *Wizard of Oz:* Parable on Populism." *American Quarterly* 16, no. 3:47–58.

Liverani, Mario. 2000. "The Libyan Caravan Road in Herodotus IV.181–185." *Journal of the Economic and Social History of the Orient* 43 no. 4:496–520.

Lleras, Roberto, Clara Isabel Botero, and Santiago Lodono. 2007. *The Art of Gold: The Legacy of Pre-Hispanic Colombia.* New York: Skira.

Lloyd, Seton. 1999. *Ancient Turkey: A Traveller's History.* New York: Longitude.

Loomis, Roger Sherman. 1959. *The Grail: From Celtic Myth to Christian Symbol.* Cardiff: Univ. of Wales Press.

Lordkipanidze, Otar. 2001. "The Golden Fleece: Myth, Euhemeristic Explanation and Archaeology." *Oxford Journal of Archaeology* 20:1–38.

MacDonald, George. 1916. *The Evolution of Coinage.* Cambridge: Cambridge Univ. Press.

Mackenzie, Donald Alexander. 1978. *Myths of Pre-Columbian America.* Boston: Longwood Press.

Mango, Cyril. 1962. *The Mosaics of St. Sophia at Istanbul.* Locust Valley, NY: J. J. Augustin.

Marichal, Carlos. 1989. *A Century of Debt Crises in Latin America: From Independence to the Great Depression, 1820–1930.* Princeton, NJ: Princeton Univ. Press.

Marsden, John and Iain House. 2006. *The Chemistry of Gold Extraction.* Littleton, CO: Society for Mining Metallurgy and Exploration.

Marshall, Peter. 2001. *The Philosopher's Stone: A Quest for the Secrets of Alchemy.* London: Pan.

Martin, Colin. 1975. *Full Fathom Five: Wrecks of the Spanish Armada.* New York: Viking.

Martin, Colin, and Geoffrey Parker. 1992. *The Spanish Armada.* New York: W. W. Norton.

Maryon, Herbert. 1954. *Metalwork and Enamelling: A Practical Treatise on Gold and Silversmiths' Work and Their Allied Craft.* London: Chapman & Hall.

Marx, Jennifer. 1978. *The Magic of Gold.* New York: Doubleday.

Matos Moctezuma, Eduardo. 1988. *The Great Temple of the Aztecs of Tenochtitlan.* New York: Thames & Hudson.

Matthewson, R. Duncan. 1986. *Treasure of the* Atocha. Hialeah, FL: Dutton.

Maxwell-Hyslop, Rachel K. 1972. "The Art of Granulation in Early Iranian Gold Jewelry." *Proceedings of the 5th International Congress of Iranian Art and Archaeology.* http://www.cais-soas.com/CAIS/Art/granulation.htm.

McCulloch, J. Huston. 1982. *Money and Inflation: A Monetarist Approach.* 2nd edition. New York: Academic Press.

McQuillan, Lawrence J., and Peter C. Montgomery, eds. 1999. *The International Monetary Fund—Financial Medic to the World?: A Primer on Mission, Operations, and Public Policy Issues.* Chicago: Hoover Institution Press.

Medema, Steven G., and Warren J. Samuels. 2003. *The History of Economic Thought: A Reader.* New York: Routledge.

Meltzer, Allan H. 2003. *A History of the Federal Reserve, 1913–51.* Chicago: Univ. of Chicago Press.

Milnes, Nora. 1917. "Mint Records in the Reign of Henry VIII." *The English Historical Review* 32, no. 126 (April):270–273.

Minney, R. J. 1970. *The Tower of London.* Englewood Cliffs, NJ: Prentice-Hall.

Moggridge, D. E. 1968. *The Return to Gold, 1925.* Cambridge: Cambridge Univ. Press.

Moggridge, D. E. 1972. *British Monetary Policy, 1924–31: The Norman Conquest of $4.86.* Cambridge: Cambridge Univ. Press.

Mohide, Thomas Patrick. 1981. *Gold.* Toronto: Ontario Minister of Natural Resources.

Molnar, Michael R. 1999. *The Star of Bethlehem: The Legacy of the Magi.* Piscataway, NJ: Rutgers Univ. Press.

Morteani, Gioiulio, and Jeremy Peter Northover. 1995. *Prehistoric Gold in Europe: Mines, Metallurgy, and Manufacture.* New York: Springer.

Mundy, Barbara E. 1998. "Mapping the Aztec Capital: The 1524 Nuremberg Map of Tenochtitlan, Its Sources and Meanings." *Imago Mundi* 50:11–33.

Needham, Joseph, and Lu Gwei-Djen. 1974. *Science and Civilization in China: Chemistry and Chemical Technology.* London: Cambridge Univ. Press.

Newman, William R. 2004. *Promethean Ambitions: Alchemy and the Quest to Perfect Nature.* Chicago: Chicago Univ. Press.

Oddy, A. 1983. "Assaying in Antiquity" *Gold Bulletin* 15, no. 2:52–59.

Oddy, A. 1985. "Touchstones: Some Aspects of Their Nomenclature, Petrography and Provenance." *Journal of Archaeological Science* 12:59–80.

Oddy, Andrew, and Susan La Niece. 1986. "Byzantine Gold Coins and Jewellery: A Study of Gold Contents." *Gold Bulletin* 19, no. 1:19–27.

Ogden, J. M. 1992. "Gold in Antiquity." *Interdisciplinary Science Reviews* 17:262–263.

Olds, Doris. 1976. *Texas Legacy from the Gulf: A Report on Sixteenth Century Shipwreck Materials from the Texas Tidelands.* Texas Memorial Museum Publication 2, Misc. Papers 5, and the Texas Antiquities Committee, Austin.

Oliver, Andrew, Jr. 1966. "Greek, Roman, and Etruscan Jewelry." *Metropolitan Museum of Art Bulletin* 24, no. 9 (May).

Owen Hughes, Diane. 1983. "Sumptuary Legislation and Social Relations in Renaissance Italy." In *Disputes and Settlements: Law and Human Relations in the West.* Edited by John Bossy, 69–99. New York: Cambridge Univ. Press.

Pächt, Otto. *Book Illumination in the Middle Ages.* Oxford: Oxford Univ. Press, 1986.

Pagden, Anthony. 1998. *Spanish Imperialism and the Political Imagination: Studies in European and Spanish-American Social and Political Theory 1513–1830.* New Haven, CT: Yale Univ. Press.

Parry, John H. 1990. *The Spanish Seaborne Empire.* Berkeley: Univ. of California Press.

Parsons, Wayne. 1997. *Keynes and the Quest for Moral Science: A Study of Economics and Alchemy.* Cheltenham, UK: Edward Elgar.

Patinkin, Don. 1976. *Keynes' Monetary Thought: A Study of Its Development.* Durham, NC: Duke Univ. Press.

Philips, Clare. 1996. *Jewelry: From Antiquity to the Present.* London: Thames and Hudson.

Philips, J. P. 1965. "16th-Century Texts on Assaying." *Chem. Educ.* 12, no. 7:393–394.

Pliny the Elder. 1893. *The Natural History of Pliny.* Vol. 6. Translated and Edited by John Bostock and Henry Thomas Riley. London: George Bell & Sons.

Pollen, John Hungerford. 1878. *Gold and Silversmiths' Work.* London: R. Clay and Sons.

Polo, Marco. 1903. *The Travels of Marco Polo.* Translated by Henry Yule. http://www.gutenberg.org/catalog/world/readfile?fk_files=44328.

Prescott, William H. 1847. *History of the Conquest of Peru.* Boston: Harper & Bros.

Prescott, William H. 2003. *History of the Reign of Ferdinand and Isabella the Catholic.* Honolulu, HI: Univ. Press of the Pacific.

Prescott, William H. 2010. *History of the Conquest of Mexico.* New York: The Modern Library. (Orig. pub. 1843.)

Pressnell, L. S. 1968. "Gold Reserves, Banking Reserves, and the Barings Crisis of 1890." In *Essays in Honor of R. S. Sayers.* Edited by C. R. Whittlesey and J. S. G. Wilson, 167–288. Oxford: Clarendon Press.

Principe, Lawrence M. 2007. *Chymists and Chymistry: Studies in the History of Alchemy and Early Modern Chemistry.* Sagamore Beach, MA: Watson Publishing International.

Procopius. 2011. *Of the Buildings of Justinian.* Translated by Aubrey Stuart. Port Chester, NY: Adegi Graphics, LLC.

Quinn, Stephen. "Gold, Silver, and the Glorious Revolution: International Bullion Arbitrage and the Origins of the English Gold Standard." *Economic History Association.* http://eh.net/Clio/Conferences/ASSA/Jan_95/Quinn.shtml.

Raffield, Paul. 2002. "Reformation, Regulation and the Image: Sumptuary Legislation and the Subject of Law." *Law and Critique* 13, no. 2 (May):127–150.

Raggio, Olga. 2000. "French Decorative Arts during the Reign of Louis XIV (1654–1715)." In *Heilbrunn Timeline of Art History.* New York: Metropolitan Museum of Art. http://www.metmuseum.org/toah/hd/frde/hd_frde.htm.

Ramage, Andrew, and Paul Craddock. 2000. *King Croesus' Gold: Excavations at Sardis and the History of Gold Refining.* London: British Museum Publications.

Ramsey, Peter H. 1971. *The Price Revolution in Sixteenth-Century England.* London: Metheun.

Reeves, Nicholas. 1990. *The Complete Tutankhamun: The King, His Tomb, the Royal Treasure.* London: Thames and Hudson.

Ricardo, David. 1810. *The High Price of Bullion, a Proof of the Depreciation of Bank Notes.* Pamphlet printed in London by John Murray. http://socserv2.mcmaster.ca/~econ/ugcm/3ll3/ricardo/bullion.

Ricardo, David. 1951–1973. *The Works and Correspondence of David Ricardo.* 11 vol. Edited by Piero Sraffa, with help of M. H. Dobb. Cambridge: Cambridge Univ. Press.

Richards, John F. 1983. *Precious Metals in the Later Medieval and Early Modern Worlds.* Durham, NC: Carolina Academic Press.

Richardson, Glenn. 2002. *Renaissance Monarchy: The Reigns of Henry VIII, Francis I, and Charles V.* London: Arnold.

Richardson, Peter, and Jean Jacques van Helten. 1984. "The Development of the South African Gold Mining Industry, 1895–1918." *Economic History Review* 37, no. 3:319–340.

Rickard, T. A. 1932. *Man and Metals.* 2 vols. New York: McGraw-Hill.

Rockoff, Hugh. 1990. "*The Wizard of Oz* as a Monetary Allegory." *Journal of Political Economy* 98 (August):739–760.

Rohrbough, Malcolm J. 1997. *Days of Gold: The California Gold Rush and the American Nation.* Berkeley: Univ. of California Press.

Room, Adrian. 1987. *Dictionary of Coin Names.* London & New York: Routledge & Kegan Paul.

Rose, Thomas. K. 2008. *The Metallurgy of Gold: Mining and Mineral Processing.* 6th ed. Palm Springs, CA: Wexford College Press.

Rosenthal, Eric. 1970. *Gold. Gold. Gold.* London and New York: Macmillan.

Rossabi, Morris. 1990. *Kubilai Khan: His Life and Times.* Berkeley: Univ. of California Press.

Rothbard, Murray N. 2002. *A History of Money and Banking in the United States: The Colonial Era to World War II.* Auburn, AL: Ludwig von Mises Institute.

Rounds, J. 1979. "Lineage, Class, and Power in the Aztec State." *American Ethnologist* 6, no. 1 (February):73–86.

Rowse, A. L. 1972. *The Tower of London in the History of England.* New York: G. P. Putnam's Sons.

Sale, Kirkpatrick. 1990. *The Conquest of Paradise: Christopher Columbus and the Columbian Legacy.* New York: Plume-Penguin.

Sarris, Peter. 2006. *Economy & Society in the Age of Justinian.* Cambridge: Cambridge Univ. Press.

Scarisbrick, Diana. 1993. *Rings: Symbols of Wealth, Power, and Affection.* New York: Abrams.

Scarisbrick, Diana, Christophe Vachaudez, and Jan Walgrave, eds. 2008. *Royal Jewels: From Charlemagne to the Romanovs.* New York: Vendome Press.

Schmidbaur, Hubert, ed. 1999. *Gold: Progress in Chemistry, Biochemistry, and Technology.* Chichester, UK: Wiley and Sons.

Schnabel, Carl. 1907. *A Handbook of Metallurgy.* New York: MacMillan.

Schwartz, Anna. 1973. "Secular Price Change in Historical Perspective." *Journal of Money, Credit, and Banking* 5:243–279.

Schwartz, Stuart B. 2000. *Victors and Vanquished: Spanish and Nahua Views of the Conquest of Mexico.* New York: Palgrave & Macmillan.

Serrato-Combe, Antonio. 2001. *The Aztec Templo Mayor: A Visualization.* Salt Lake City: Univ. of Utah Press.

Shively, Donald H. 1964–1965. "Sumptuary Regulation and Status in Early Tokugawa Japan." *Harvard Journal of Asiatic Studies* 25:123–164.

Solomon, R. 1982. *The International Monetary System, 1945–1981: An Insider's View.* New York: Harper and Row.

Spawforth, Antony. 2008. *Versailles: A Biography of a Palace.* New York: St. Martin's.

Stern, Steve J. 1993. *Peru's Indian Peoples and the Challenge of Spanish Conquest: Huamanga to 1640.* 2nd ed. Madison: Univ. of Wisconsin Press.

Stockdale, D. 1924. "Historical Notes on the Assay of Gold." *Sci. Prog.* 18:476–479.

Strange, S. 1976. *International Economic Relations of the Western World 1959–1971, 2: International Monetary Relations.* London and New York: Oxford Univ. Press.

Sutherland, C.H.V. 1969. *Gold: Its Beauty, Power, and Allure.* 2nd rev. ed. New York: McGraw Hill.

Swainson, Harold. 2005. *The Church of Sancta Sophia Constantinople: A Study of Byzantine Building.* Boston: Adamant Media Corp.

Tait, H. 1986. *Seven Thousand Years of Jewelry.* London: British Museum Publications.

Taylor, John H. 1991. *Egypt and Nubia.* London: British Museum Press.

Teichova, Alice, Ginette Kurgan van Hentenryk, and Dieter Ziegler. 1997. *Banking, Trade, and Industry: Europe, America, and Asia from the Thirteenth to the Twentieth Century.* London: Cambridge Univ. Press.

Temple, R. C. 1899. "Beginnings of Currency." *Journal of the Anthropological Institute of Great Britain and Ireland.* 29(1–2).

Thomas, Hugh. 1995. *Conquest: Montezuma, Cortés, and the Fall of Old Mexico.* New York: Simon and Schuster.

Thomas, Hugh. 2004. *Rivers of Gold: The Rise of the Spanish Empire from Columbus to Magellan.* New York: Random House.

Thompson, Daniel Varney. 1956. *The Materials and Techniques of Medieval Painting.* New York: Courier Dover.

Tóth, Sándor László. 1988. "The 'Price Revolution' in the Ottoman Empire at the End of the Sixteenth Century." *Acta Universitatis Szegediensis de Attila Jozsef Nominatae: Acta Historica* 87 (July):35–47.

Townsend, Richard F. 2000. *The Aztecs.* New York: Thames & Hudson.

Trexler, Richard. 1997. *The Journey of the Magi.* Princeton, NJ: Princeton Univ. Press.

Triffin, Robert. 1960. *Gold and the Dollar Crisis.* New Haven, CT: Yale Univ. Press.

Tyldesley, Joyce. 1998. *Hatchepsut: The Female Pharaoh.* New York: Penguin.

Tylecote, R. E. 1976. *A History of Metallurgy.* London: The Metals Society.

Van Beek, Steve, and Luca Invernizzi. 1999. *The Arts of Thailand.* Berkeley, CA: Periplus Editions.

Van Canwenberghe, Eddy, ed. 1989. *Precious Metals, Coinage, and the Changes of Monetary Structures in Latin America, Europe, and Asia (Late Middle Ages-Early Modern Times).* Leuven, Belgium: Leuven Univ. Press.

Van Onselen, C. 1982. *Studies in the Social and Economic History of the Witwatersrand. Volume 1: New Babylon.* New York: Raven Press.

Varon Gabai, Rafael. 1997. *Francisco Pizarro and His Brothers: The Illusion of Power in Sixteenth-Century Peru.* Norman: Univ. of Oklahoma Press.

Vermeule, Cornelius C. 1954. *Some Notes on Ancient Dies and Coining Methods.* London: Spink.

Vesilind, Priit J. 2005. *Lost Gold of the Republic: The Remarkable Quest for the Greatest Shipwreck Treasure of the Civil War Era.* Las Vegas, NV: Shipwreck Heritage Press.

Vickery, Walter N. 1970. *Alexander Pushkin.* New York: Twayne.
Vielstich, Wolf, et al., eds. 2009. *Handbook of Fuel Cells: Advances in Electrocatalysis, Materials, Diagnostics, and Durability.* 6 vols. Hoboken, NJ: Wiley.
Vilar, Pierre. 1976. *A History of Gold and Money, 1450–1920.* London: New Left Books.
Wardwell, Anne E. 1992. "Two Silk and Gold Textiles of the Early Mongol Period." *Bulletin of the Cleveland Museum of Art* 79, no. 10 (December):354–378.
Ware, Mike. 2006. *Gold in Photography. The History and Art of Chrysotype.* Brighton, UK: Cromwell Press.
Watt, Melinda. 2000. "Textile Production in Europe: Lace, 1600–1800." *Heilbrunn Timeline of Art History.* New York: Metropolitan Museum of Art. http://www.metmuseum.org/toah/hd/txt_l/hd_txt_l.htm.
Wells, Donald R. 2004. *The Federal Reserve System: A History.* Jefferson, NC: McFarland.
Welsby, D, and J. Anderson, eds. 2004. *Sudan Ancient Treasures.* London: British Museum Press.
White, Michael. 1977. *Isaac Newton: The Last Sorcerer.* Reading, MA: Addison-Wesley.
White, Richard D. 2000. "Political Economy and Statesmanship: Smith, Hamilton, and the Foundation of the Commercial Republic." *Public Administration Review* 60.
Whitley, Kathleen P. 2000. *The History and Technique of Manuscript Gilding.* New Castle, DE: Oak Knoll Press.
Wilks, Ivor, ed. 1993. *Forests of Gold: Essays on the Akan and the Kingdom of Asante.* Athens: Ohio Univ. Press.
Williams, J., ed. 1997. *Money: A History.* London: British Museum.
Wilson, Derek. 1978. *The Tower, 1078–1978.* London: Hamish Hamilton.
Winthrop, Samuel K., et al., eds. 1961. *Essays in Pre-Columbian Art and Archaeology.* Cambridge, MA: Harvard Univ. Press.
Wood, John Harold. 2005. *A History of Central Banking in Great Britain and the United States.* Cambridge: Cambridge Univ. Press.
Wood, Michael. 1968. *Conquistadors.* Berkeley: Univ. of California Press.
Woolley, C. Leonard, et al. 1934. *The Royal Cemetery: A Report on the Predynastic and Sargonid Graves Excavated between 1926 and 1931. Ur Excavations, vol. 2.* London and Philadelphia: Joint Expedition of the British Museum and of the University Museum, University of Pennsylvania.
World Gold Council. "The Caratage (Karatage) System for Gold Jewelry." http://www.utilisegold.com/jewellery_technology/caratage/.
World Gold Council. "Gold as a Reserve Asset." http://www.reserveasset.gold.org/.
World Gold Council. 2010, March. "World Gold Holdings—Volume and Value." http://www.reserveasset.gold.org/.
Wright, Louis B. 1970. *Gold, Glory and the Gospel: The Adventurous Lives and Times of the Renaissance Explorers.* New York: Atheneum.
Wright, Robert E. 2002. *Hamilton Unbound: Finance and the Creation of the American Republic.* Westport, CT: Greenwood.
Zettler, Richard, and Lee Horne, eds. 1998. *Treasures from the Royal Tombs of Ur.* Philadelphia: Univ. of Pennsylvania Press.

Index

Page numbers in **bold** indicate the main article on a subject.

Academy Awards, 220–221
adornment. *See* crowns, royal; jewelry; lace, gold; sumptuary legislation
Adzic, Radoslav, 111
aerospace industry, 1–2, 95, 111
   See also electroplating; fuel cells
Aesop, 140
African American pop culture, 36–37
Al-Bakri, Abu Ubaydallah, 270
Al-Tabari, 33
alchemy, 2–6
   Aquinas, Thomas, 8, 9
   gold deposit formation theories, 201–202
   Holy Grail, 157, 159
   Keynes, John Maynard, 188–189
   Newton, Sir Isaac, 210–212
   pre-Columbian gold, 233
   Rumpelstiltskin, 239–241
   "The Tale of the Golden Cockerel," 262
   See also alloying
alcoholic beverages containing gold, 136
Alexander III, 101–102
allergies, 6–7, 29, 86, 136
   See also alloying; biomedical research
alloying, 7–8
   allergies, 6
   Archimedes, 10, 11, 12
   Croesus, 77
   dental crowns, 86
   lace, gold, 196
   See also karat
Almagro, Diego de, 223–224, 225
American Academy of Motion Picture Arts and Sciences, 220–221
Anne Boleyn, 155–156
Apollonius of Rhodes, 128–129
Aquinas, Thomas, **8–10**

Archimedes, **10–12**, 80
   See also alloying
"Archimedes Principle," 10, 11–12
architecture. *See* art, artifacts, and architecture
Argentina, **12–14**, 26–27
   See also Baring Crisis (1890)
Argonautica (Apollonius of Rhodes), 128–129
art, artifacts, and architecture
   Asante Golden Stool, **14–16**
   Ca d'Oro, **47–48**
   chrysography, **60–61**, 141, 166, 167–168
   chrysotype, **61–62**
   Fabergé, House of, **101–102**
   Gospel Books of Saint Médard de Soissons, 60, **140–142**, 168
   illuminated manuscripts, 140–141, **166–169**
   lace, gold, **195–196**
   mosaic, 148, **204–207**, 263
   Ponte Vecchio, **228–230**
   pre-Columbian gold, 91–92, **230–233**
   Silla, Royal Tombs of, 79–80, **245–247**
   Versailles, Palace of, 139, **279–281**
   Wat Traimit, **283–284**
   See also Hagia Sophia; jewelry; Tenochtitlan; Ur, Royal Tombs of
arthritis treatment, 34, 35
artifacts. *See* art, artifacts, and architecture
Asante Golden Stool, **14–16**
assaying, **16–17**, 264–266
   See also mining
Aswân High Dam, 215
Atahualpa, **17–19**
   Pizarro, Francisco, and, 224–225
   pre-Columbian gold, 232

Spanish Conquest, 170, 252
  See also Inca; Pizarro, Francisco;
    Spanish Conquest
Atocha, 216–218
Au, **19–20**
Australian Gold Rush, 151–152
Avila, Pedro Arias de, 223
Aztec, 231, 233, 262–264
  See also Tenochtitlan

Bactrian hoard, 277
bagels with edible gold leaf, 136
Bakri, Abu Ubaydallah al-, 270
Balboa, Vasco Nuñez de, **21–23**, 223, 252
  See also Spanish Conquest
Bangkok, Thailand, 283–284
Bank Conservation Act of 1933, 96–97
Bank of England, 27–28, 44–45,
    121–122, 237–238
  See also Baring Crisis (1890); Bullion
    Committee of 1810
banking and credit, **23–26**
  Aquinas, Thomas, 9
  Argentina, 12–13
  Bullion Committee of 1810, 44–45
  Emergency Banking Relief Act of 1933,
    95–97
  Federal Reserve System, 103–104
  financial crisis, contemporary, 13–14, 26
  See also Emergency Banking Relief Act
    of 1933; gold standard
banking and finance
  Argentina, **12–14**, 26–27
  Baring Crisis, 13, **26–28**
  Bullion Committee of 1810, 24–25,
    43–45, 237–238
  Emergency Banking Relief Act of 1933,
    95–97
  Federal Reserve System, 97, **103–104**,
    120, 172
  Gold Standard Act of 1925, 119–120,
    123, **124–126**, 187
  Price Revolution, 172, **233–235**, 253
  United States Bullion Depository, 120,
    134–136, **275–276**
  See also banking and credit; Bretton
    Woods System; exchange rates; gold
    reserves; gold standard; inflation

Baring Crisis (1890), 13, **26–28**
  See also banking and credit; gold
    standard
Barris, George, 117
Bastidas, Rodrigo de, 21
Baum, Frank L., 67, 122, 287–288
beauty products, **28–30**
  See also nanotechnology
bezant, **30–31**, 163, 184
  See also Justinian I
biblical magi, **31–34**, 129, 227
biomedical research, 6–7, **34–36**, 209
  See also nanotechnology
bling, **36–37**, 182
  See also jewelry
Bobadilla, Francisco de, 71
Bond, James (fictional character),
    134–136
bonding wire, **37–38**, 94
Bouton, Matthew, 65–66
Bretton Woods System, **38–41**
  banking and credit, 25
  de Gaulle, Charles, 84–85
  exchange rates, 98
  gold standard, 123
  inflation, 172
  International Monetary Fund, 173–174
  Keynes, John Maynard, 188
  See also exchange rates; International
    Monetary Fund; Keynes, John
    Maynard
bridge, 228–230
"A Brief Account of the Destruction of the
    Indies" (Las Casas), 252–253
Briot, Nicholas, **41–42**, 63
  See also Tower Mint
Britain. *See* Great Britain
Bronze Age, 264, 277
Brookhaven Lab, 111
Brothers Grimm, 239–241
Brugnatelli, Luigi, 94
Bryan, William Jennings, 122, 287
bullion, **42–43**
  Emergency Banking Relief Act of 1933,
    96
  gold reserves, 118, 119, 120
  Goldfinger (film), 135
  Nuestra Señora de Atocha, 216, 218

Price Revolution, 234
San Esteban, 244
Spanish Conquest, 253
United States Bullion Depository, 275
See also Bullion Committee of 1810; Price Revolution; United States Bullion Depository
Bullion Committee of 1810, 24–25, **43–45**, 237–238
See also banking and credit; bullion; exchange rates; gold standard
Burton, James, 175
Byzantine Empire
    bezant, 30–31
    crowns, royal, 80
    Hagia Sophia, 147–148
    iconoclasm, 163–165
    mosaic, 206

Ca d'Oro, **47–48**
Cadillac, gold, 117
California Gold Rush, **48–52**, 151, 259–260
    See also gold rushes; mining
cancer treatment, 34–35
The Canterbury Tales (Chaucer), 5
Capac, Maco, 169
capitalism, laissez-faire, 248
carat. *See* karat
Carmack, George Washington, 190
Carter, Howard, 271–272
Cassioli, Giuseppe, 219–220
casting, **52–54**, 86, 182, 232
catalysis, **54–56**, 110–111, 209
Catherine of Aragon, 155
cell phone recycling, 38
Cellini, Benvenuto, **56–59**
    casting, 53
    goldsmithing, 139
    Ponte Vecchio statue, 229, 230
    See also Ponte Vecchio
Charlemagne, 165
Charles I, King of England, 42
Charles II, King of England, 143, 144–145
Charles III, Duke of Bourbon, 57
Charles V, Holy Roman Emperor
    Cortés, Hernán, and, 73, 74, 75
    Field of the Cloth of Gold, 107–108

Isabella I of Castile and, 177
    Pizarro, Francisco, and, 224
    Spanish Conquest, 252, 253
Charlie and the Chocolate Factory (book and film), 134
Chaucer, Geoffrey, 5
Chevalier, Michel, **59–60**
    See also gold standard
Chilkoot Pass, 191
China, 3, 4, 203
Choisy-au-Bac, France, 264
Chrétien de Troyes, 157–158
chrysography, **60–61**, 141, 166, 167–168
    See also gilding; Gospel Books of Saint Médard de Soissons; illuminated manuscripts
chrysotype, **61–62**
church. *See* Hagia Sophia
Churchill, Winston, 123, 125
City of Gold, 91–93
Clement VII, Pope, 57, 155
coin clipping, **62–64**
    coin stamping and, 66
    Great Recoinage Act of 1696, 143
    guinea, 145
    Newton, Sir Isaac, 210
    Tower Mint, 267–268
    See also Tower Mint
coin stamping, **64–66**
coinage. *See* money and coinage
Coinage Act of 1873, **66–68**
    See also gold standard
Columbus, Christopher, **68–73**
    biblical magi, 33
    Ferdinand II of Aragon and, 105, 106
    Isabella I of Castile and, 176, 177
    Polo, Marco, and, 226, 228
    pre-Columbian gold, 230, 233
    Spanish Conquest, 251, 252
    See also Ferdinand II of Aragon; Isabella I of Castile; Spanish Conquest
Comptroller of the Currency, 96–97
Constantine V, 164, 165
Contarini, Marino, 47–48
Cortés, Hernán, **73–75**, 231, 252, 262–263
    See also Spanish Conquest; Tenochtitlan

cost-push theory of inflation, 172
Cox, James, **75–77**
　See also goldsmithing
Cox's Timepiece, 76
credit. *See* banking and credit
Croesus, **77–78**, 200, 244
crowns, royal, **78–81**
　Archimedes, 10, 11, 12
　filigree, 109
　Silla, Royal Tombs of, 246
　Tutankhamun, 272
　See also Silla, Royal Tombs of; Tutankhamun; Ur, Royal Tombs of
culture
　bling, **36–37**, 182
　Field of the Cloth of Gold, **107–108**, 156
　Gold Album Certification, **116–117**
　Golden Ticket (*Charlie and the Chocolate Factory*), **134**
　Goldfinger (film), **134–136**
　Goldschläger, **136–137**
　"The Goose and the Golden Egg" (Aesop), **140**
　Nobel Prize medals, **212–214**
　Nubia, 202, **214–216**, 269
　Olympic gold medal, **219–220**
　Oscar statuette, **220–221**
　Yellow Brick Road, 67, 122, **287–288**
　See also Inca
cyanide, **81–82**, 94, 95, 203
Cyrus, King, 78

Dahl, Roald, 134
Dandolo, Doge Giovanni, 89
de Boron, Robert, 158
de Gaulle, Charles, **83–86**
　See also gold standard
debasement, 7
Deir al-Bahri complex, 153
dental crowns, **86**
developing countries, 175
Djeser Djeseru, 153
Drake, Sir Francis, **87–89**, 250
　See also Spanish Armada
ducat, **89–90**, 251
Dummling (fictional character), 130–132
Dürer, Albrecht, 231

Easter eggs, 101–102
The Economic Consequences of the Peace (Keynes), 186–187
Egypt, 4, 152–154, 179, 214–215
　See also Tutankhamun
Egyptian red gold, 272
El Dorado, **91–93**
　See also gilding; pre-Columbian gold; Spanish Conquest
"Eldorado" (Poe), 93
electronics manufacturing, 37–38
electroplating, 1–2, 37–38, **94–95**, 115
　See also bonding wire
elements, periodic table of, 19–20
Eligius, Saint, 130
Elizabeth I, 87, 88, 249, 250, 256
Elkington, George, 94
Elkington, Henry, 94
Elsner reaction, 82
The Emerald Tablet of Thoth, 4
Emergency Banking Relief Act of 1933, **95–97**
　See also banking and credit
Encisco, Martin Fernández de, 21, 22
England. *See* Great Britain
Epistles (Jerome, Saint), 167
Espiritu Santo, 243–244
An Essay on the Influence of a Low Price of Corn on the Profits of Stock (Ricardo), 238
Essays, Moral and Political (Hume), 161–162
ethic of reciprocity, 133
Etruscan civilization, 86, 180
exchange rates, **97–99**
　Bretton Woods System, 39, 40
　Bullion Committee of 1810, 43–45
　gold standard and, 123–124
　International Monetary Fund, 174
　See also Bretton Woods System; gold standard

Fabergé, House of, **101–102**
　See also jewelry
Fabergé, Peter Carl, 101–102
face lift, 20
Famous Thirteen, 224

Federal Agricultural Research Center (Germany), 55
Federal Open Market Committee, 103
Federal Reserve Act of 1913, 97, 103
Federal Reserve System, 97, **103–104**, 120, 172
    See also banking and credit
Ferdinand I, Duke, 229
Ferdinand II of Aragon, **104–107**
    Balboa, Vasco Nuñez de, and, 22
    Columbus, Christopher, and, 69, 70–71, 72
    Isabella I of Castile and, 176, 177
    Spanish Conquest, 251–252
    See also Isabella I of Castile; Spanish Conquest
Field of the Cloth of Gold, **107–108**, 156
    See also Henry VIII
filigree, **108–109**
    crowns, royal, 79, 80
    Gospel Books of Saint Médard de Soissons, 141
    granulation, 142
    pre-Columbian gold, 232
    Silla, Royal Tombs of, 246
    See also jewelry
finance. *See* banking and finance
financial crisis, contemporary, 13–14, 26, 189
Fisher, Mel, 218
fixed exchange rates, 97, 98, 123–124
Flamel, Nicolas, 2
    See also alchemy
floating exchange rates, 97, 98, 123–124
Florence, Italy, 24, 109–110, 228–230
florin, **109–110**
    See also gold standard
flu vaccine, 35
folklore. *See* myths, legends, and folklore
food additive, gold as, 136–137
fool's gold, as term, 203
Fort Knox. *See* United States Bullion Depository
France, 41–42, 83–85, 195–196, 264
    See also Versailles, Palace of
Franchetti, Giorgio, 48
Francis I, 57, 107–108
Franciscan Order lawsuit, 245–246

free market capitalism, 248
Free Silver Movement, 67
fuel cells, 1, **110–112**, 209
    See also catalysis; nanotechnology

Gates of Paradise (Ghiberti), 115
The General Theory of Employment, Interest, and Money (Keynes), 187
Genoa, Italy, 24
Germany, 55
Ghana, 14–16
Ghiberti, Lorenzo, 115
Gilded Age, **113–115**
Gilded Man, 91–93
gilding, **115–116**
    Ca d'Oro, 47, 48
    chrysography, 60
    El Dorado, 91–92
    illuminated manuscripts, 166, 167–168
    Oscar statuette, 221
    Versailles, Palace of, 280, 281
    See also electroplating; illuminated manuscripts
Godescalc Gospels, 60
Gold Album Certification, **116–117**
gold bread, 48
gold dermatitis. *See* allergies
gold mine, as term, 201
gold nuggets, **117–118**, 190, 259
    See also mining; Sutter's Mill
gold reserves, **118–121**
    exchange rates and, 98
    Federal Reserve System, 104
    gold standard and, 123
    Ricardo, David, 237–238
    United States Bullion Depository, 275
    See also gold standard; United States Bullion Depository
gold rushes
    Australian Gold Rush, 151–152
    California Gold Rush, **48–52**, 151, 259–260
    Hargraves, Edward Hammond, **151–152**
    Klondike Gold Rush, **190–192**
    Sutter's Mill, 49, **257–260**
    Witwatersrand Gold Rush, **284–286**

gold standard, **121–124**
   Argentina, 13
   banking and credit, 25
   Baring Crisis, 27–28
   Bullion Committee of 1810, 44–45
   Chevalier, Michel, 59–60
   Coinage Act of 1873, 66–67
   de Gaulle, Charles, 84–85
   exchange rates, 98
   Federal Reserve System, 103, 104
   florin, 109–110
   gold reserves and, 118–120
   Hamilton, Alexander, 150
   International Monetary Fund and, 173
   Keynes, John Maynard, 187–188
   Newton, Sir Isaac, 144, 145
   Ricardo, David, 237–238
   United States Bullion Depository, 275
   Yellow Brick Road, 67, 287–288
   See also banking and credit; gold reserves; Gold Standard Act of 1925
Gold Standard Act of 1925, 119–120, 123, **124–126**, 187
   See also gold standard
Gold to Go, 120
Golden Buddha, Temple of the, 283–284
Golden Calf, **126–128**, 163
Golden Cockerel, 261–262
Golden Crown, 10, 11, 12
Golden Fleece, **128–130**, 202
Golden Goose, **130–133**
Golden Man, 91–93
Golden Rule, **133**
Golden Ticket (*Charlie and the Chocolate Factory*), **134**
Goldfinger (film), **134–136**
Goldschläger, **136–137**
goldsmithing, **137–139**
   Cellini, Benvenuto, 56–58
   Cox, James, 75–76
   filigree, 108–109
   jewelry, 181–182
   pre-Columbian gold, 231, 232
   Silla, Royal Tombs of, 245–246
   See also alloying; assaying; casting; gilding
"The Goose and the Golden Egg" (Aesop), **140**

Gospel Books of Saint Médard de Soissons, 60, **140–142**, 168
granulation, 108–109, **142–143**, 246
   See also filigree; jewelry
Great Britain
   Asante Golden Stool, 14–15
   Bank of England, 27–28, 44–45, 121–122, 237–238
   Baring Crisis, 13, **26–28**
   Bullion Committee of 1810, 24–25, 43–45, 237–238
   coin clipping, 63–64
   crowns, royal, 81
   gold standard, 121–122, 123, 124–125, 144, 145
   Gold Standard Act of 1925, 119–120, 123, 124–125, 187
   guinea, 143, 144–146
   sumptuary legislation, 256
   See also Great Recoinage Act of 1696; Henry VIII; Tower Mint
Great Depression, 95–97, 103–104, 123, 187–188
Great Recoinage Act of 1696, **143–144**
   coin clipping, 63–64
   guinea, 145
   Newton, Sir Isaac, 210, 268
Greece
   alchemy, 4
   Golden Fleece, 128–129
   jewelry, 180
   mosaic, 204–205
   Olympic gold medal, 219–220
   sumptuary legislation, 255
Gregory II, Pope, 164
Grimm Brothers, 239–241
Guatavita, Lake, 91, 93
guinea, 143, **144–146**

Hagia Sophia, **147–148**
   iconoclasm, 163
   Justinian I, 184, 206
   mosaic, 206
Hall of Mirrors, 280
hallmarks, karat, 185
Hamilton, Alexander, **148–151**
Hargraves, Edward Hammond, **151–152**
Harrison, George, 284

Hatshepsut, **152–154**
Hawkins, Sir John, 87, 88
health and medicine
 allergies, **6–7**, 29, 86, 136
 beauty products, **28–30**
 biomedical research, 6–7, **34–36**, 209
 dental crowns, **86**
 See also nanotechnology
"Hen Egg," 101–102
Henry IV, 176–177
Henry VII, 267
Henry VIII, **154–157**
 Field of the Cloth of Gold, 107–108
 Price Revolution, 234
 sumptuary legislation, 181, 256
 Tower Mint, 267
 See also Tower Mint
Henwood, Doug, 114
Hermes the Magician, 4
Herod, King, 31–32
Herschel, John, 61–62
Hieron II, 10, 11, 12, 80
hip-hop culture, 36–37
Holocaust survivors, 245–246
Holy Grail, **157–160**
 See also alchemy
Huerto, Francisco del, 243
Hume, David, 150, **160–162**, 171, 247
hybrid exchange rates, 97–98
Hydrogen Fuel Initiative, 111

ice cream sundae, most expensive, 137
iconoclasm, 30, **163–166**, 167, 206
ideas and movements
 Gilded Age, **113–115**
 Golden Rule, **133**
 iconoclasm, 30, **163–166**, 167, 206
 trans-Saharan gold trade, 14, 214–215, 234, **269–270**
 See also alchemy; Spanish Conquest
illuminated manuscripts, 140–141, **166–169**
 See also chrysography; gilding; Gospel Books of Saint Médard de Soissons
IMF. *See* International Monetary Fund
Imperial State Crown, 81
Inca, **169–171**
 Atahualpa, 17–19
 goldsmithing, 138

 Pizarro, Francisco, and, 224–225
 pre-Columbian gold, 232–233
 Spanish Conquest, 252
 See also Atahualpa; Spanish Conquest
India, 4, 179–180, 182
inflation, **171–173**
 Argentina, 13
 Chevalier, Michel, 59–60
 coin clipping and, 62–63
 Hume, David, 162
 Price Revolution, 233–235, 253
 See also Federal Reserve System
An Inquiry into the Nature and Causes of the Wealth of Nations (Smith), 237, 248
International Monetary Fund (IMF), **173–176**
 Bretton Woods System, 39, 40–41
 de Gaulle, Charles, 85
 gold reserves, 121
 gold standard, 123
 Keynes, John Maynard, 188
 See also Bretton Woods System; exchange rates; gold standard
investment casting, 52–54, 86, 182, 232
Ireland, 196–197
Irene, Empress, 164–165
Iron Crown of Lombardy, 80
Isabella I of Castile, **176–177**
 Columbus, Christopher, and, 69, 70–71
 Ferdinand II of Aragon and, 104–106
 Spanish Conquest, 251–252
 See also Ferdinand II of Aragon; Spanish Conquest
Italy
 banking and credit, 24
 Florence, 24, 109–110, 228–230
 florin, 109–110
 Genoa, 24
 lace, gold, 195
 Venice, 24, 47–48, 89–90, 195

Japan, 38
Jason and the Argonauts, 128–129
Jerome, Saint, 167
Jesus Christ, 31–33
jewelry, **179–183**
 alloying, 7
 bling, 37

casting, 53
electroplating, 94
Fabergé, House of, 101–102
filigree, 108–109
goldsmithing, 137, 138, 139
granulation, 142, 143
Nuestra Señora de Atocha, 218
pre-Columbian gold, 232
San Esteban, 243
Spanish Armada, 251
sumptuary legislation, 255–256
touchstone, 266
Tutankhamun, 272
Ur, Royal Tombs of, 276, 277
See also bling; filigree; goldsmithing; sumptuary legislation
Johannesburg, South Africa, 284–285
John II, 69, 70
Justinian I, **183–184**
   bezant, 30
   Hagia Sophia, 147, 206
   iconoclasm, 163
   mosaic, 206
   See also bezant; Hagia Sophia

karat, **185**
   alloying, 7
   gold nuggets, 117
   Nobel Prize medals, 212
   Olympic gold medal, 219
   Oscar statuette, 221
Keynes, John Maynard, **186–190**
   Bretton Woods System, 40
   gold standard, 119, 120
   Gold Standard Act of 1925, 125
   inflation, 171–172
   International Monetary Fund, 173
   Newton, Sir Isaac, and, 211–212
   Price Revolution, 234
   See also Bretton Woods System; gold standard; Newton, Sir Isaac
The Kiss (Klimt), 115
Klimt, Gustav, 115
Klondike Gold Rush, **190–192**
   See also gold rushes
Koran, 127
Korea, 79–80, 245–247
Kruger, Paul, 285

krugerrand, 285
Kublai Khan, **192–194**, 226

lace, gold, **195–196**
laissez-faire capitalism, 248
Las Casas, Bartolomé de, 252–253
laws
   Bank Conservation Act of 1933, 96–97
   Coinage Act of 1873, **66–68**
   Emergency Banking Relief Act of 1933, **95–97**
   Federal Reserve Act of 1913, 97, 103
   Gold Standard Act of 1925, 119–120, 123, **124–126**, 187
   See also Great Recoinage Act of 1696
legends. *See* myths, legends, and folklore
Leo III, 163–164
Leo IV, 164
Leo V, 165
leprechauns, **196–197**
Lindeman, Henry R., 66–67
liqueurs, 136
Lister, John, 151
Locke, John, 143, 144
London Gold Delivery Bars, 43
lost wax casting, 52–54, 86, 182, 232
Louis XIV, 279–280
Ludderdale, William, 27–28
Lyon, Eugene, 218

MacArthur-Forrest Process, 82, 284–285
magi, biblical, **31–34**, 129, 227
Malthus, Thomas, 238
map, oldest surviving geological, 215
Marcellus, 12
Maria Fedorovna, Tsarina, 101–102
Mark, Saint, 89
Marshall, James Wilson, 49, 259
medals
   Nobel Prize medals, **212–214**
   Olympic gold medal, **219–220**
Medici family, 228–229, 230
Mehmet II, 148
Mendeleev, Dmitri, 19–20
Mesopotamia, 52–53
   See also Ur, Royal Tombs of
metallurgy. *See* technology and metallurgy
Metamorphoses (Ovid), 200

Midas, King, **199–201**
Middle Ages, 65, 138, 195, 269–270
mining, **201–204**
    assaying, 16
    California Gold Rush, 50–51
    cyanide, 81–82
    developing countries, 175
    gold nuggets, 117–118
    Klondike Gold Rush, 191–192
    nanotechnology, 209
    Nubia, 215, 216
    pre-Columbian gold, 232
    Witwatersrand Gold Rush, 284–285
    See also gold rushes
minting, 62–63, 143–144, 145, 245
    See also money and coinage
money and coinage
    bezant, **30–31**, 163, 184
    coin stamping, **64–66**
    Coinage Act of 1873, **66–68**
    ducat, **89–90**, 251
    florin, **109–110**
    guinea, 143, **144–146**
    shekel, **244–245**
    See also bullion; coin clipping; Great Recoinage Act of 1696; Tower Mint
Montagu, Charles, 143–144
Montezuma II, 74, 262–263
More, Sir Thomas, 155, 156
Morris, John, 254–255
mosaic, 148, **204–207**, 263
    See also Hagia Sophia; Justinian I
Mosely, Henry, 20
Moses (Biblical leader), 126–127
movements. *See* ideas and movements
myths, legends, and folklore
    biblical magi, **31–34**, 129, 227
    El Dorado, **91–93**
    Golden Calf, **126–128**, 163
    Golden Fleece, **128–130**, 202
    Golden Goose, **130–133**
    Golden Rule, **133**
    "The Goose and the Golden Egg" (Aesop), **140**
    Holy Grail, **157–160**
    leprechauns, **196–197**
    Rumpelstiltskin, **239–241**
    "The Tale of the Golden Cockerel," **261–262**

Nanostellar, Inc., 55
nanotechnology, **209–210**
    beauty products, 28–29
    biomedical research, 35
    catalysis, 54, 55
    fuel cells, 110, 111
NASA, 1–2
Natural History (Pliny the Elder), 265–266
Nazi gold, 245–246
Newton, Sir Isaac, **210–212**
    alchemy, 5
    coin clipping, 63
    gold standard, 121
    Great Recoinage Act of 1696, 143–144, 268
    guinea, 145
    Keynes, John Maynard, and, 188–189
    Tower Mint, 268
    See also alchemy; Tower Mint
"Newton, the Man" (Keynes), 212
Nicholas II, 102
Nippon Shokubai, 55
Nobel, Alfred Bernhard, 212, 213
Nobel Foundation, 212
Nobel Prize medals, **212–214**
Nubia, 202, **214–216**, 269
Nuestra Señora de Atocha, **216–218**
    See also Spanish Conquest

Odyssey Marine Exploration, 254–255
Olympic gold medal, **219–220**
On Stones (Theophratus), 265
"On the Balance of Trade" (Hume), 161–162
"On the High Price of Bullion" (Ricardo), 237–238
On the Principles of Political Economy and Taxation (Ricardo), 238
Open Market System, 103
Opticks (Newton), 210, 211
Order of the Golden Fleece, 129
Oscar statuette, **220–221**
Outkast, 36
Ovid, 200

pandoro, 48
papyrus, 215
Paracelsus, 34
Paris Mint, 41–42
Parma, Duke of, 250

Parzival (Wolfram von Eschenbach), 159
Paul the Apostle, Saint, 255–256
Peacock Clock, 76
Pedrarias, 22
Perceval, ou Le Conte du Graal (Chrétien de Troyes), 157–158
periodic table of elements, 19–20
Pétain, Philippe, 83–84
Philip I of Castile, 106, 177
Philip II, King of Spain, 249, 252
philosopher's stone, 2, 211
  See also alchemy
photography, alternative, 61–62
Pizarro, Francisco, **223–225**
  Atahualpa and, 17–19
  Inca and, 170
  pre-Columbian gold, 232
  Spanish Conquest, 252
  See also Atahualpa; Inca; Spanish Conquest
Platoto, Inc., 243
Pliny the Elder, 265–266
Poe, Edgar Allen, 93
Point de Venice, 195
Polk, James K., 50
Polo, Maffeo, 193, 226
Polo, Marco, **225–228**
  biblical magi and, 33
  Columbus, Christopher, influence on, 68–69
  Kublai Khan and, 192, 193–194
  See also Kublai Khan
Polo, Niccolò, 193, 226
Ponte Vecchio, **228–230**
  See also Cellini, Benvenuto
pre-Columbian gold, 91–92, **230–233**
Presley, Elvis, 116, 117
Price Revolution, 172, **233–235**, 253
  See also bullion; inflation
price-wage spiral, 172
prima matera, 3–4
Puabi, 52–53, 277–278
Pushkin, Alexander, 261–262
pyrite, 203
Pyx, Trial of the, 268

Raleigh, Sir Walter, 92
reciprocity, ethic of, 133
Record Industry Association of America, 116–117
records, gold, 116–117
recycling of precious metals, 38
religion and spirituality
  biblical magi, **31–34**, 129, 227
  Golden Calf, **126–128**, 163
  Golden Rule, **133**
  Holy Grail, **157–160**
  iconoclasm, 30, **163–166**, 167, 206
  Wat Traimit, **283–284**
  See also alchemy; Hagia Sophia
Renaissance, 138–139, 181, 202, 231
Republic, SS, **254–255**
Reynaud, Paul, 83–84
rheumatoid arthritis treatment, 34, 35
Ricardo, David, 44, **237–239**, 248
  See also Bullion Committee of 1810; gold standard
Roettier, John, 145
Roger II of Sicily, 89
Roman Empire
  crowns, royal, 80
  jewelry, 180–181
  mosaic, 205–206
  Nubia and, 215
  sumptuary legislation, 255
Roosevelt, Franklin D.
  Emergency Banking Relief Act of 1933, 95–96, 97
  Federal Reserve System, 103–104
  gold standard, 120, 123
Rueff, Jacques, 85
Rumpelstiltskin, **239–241**
Rustichello of Pisa, 227

Saliera (Cellini), 57–58
San Esteban, **243–244**
Santa Margarita, 217, 218
Sapa Inca, 169
Sarianidi, Viktor, 277
Second Council of Nicaea, 165
Serendipity 3 restaurant, 137
shekel, **244–245**
Sherman, John, 67
shipwrecks
  Espiritu Santo, 243–244
  Nuestra Señora de Atocha, **216–218**

San Esteban, **243–244**
Spanish Armada, 88, **249–251**
SS *Republic*, **254–255**
Silla, Royal Tombs of, 79–80, **245–247**
Smith, Adam, 150, 160, 234, 237, **247–249**
Solon, 77, 78
South Africa, 203, 284–285
Spanish Armada, 88, **249–251**
   See also Drake, Sir Francis
Spanish Conquest, **251–254**
   Atahualpa, 17–19
   Balboa, Vasco Nuñez de, 21–23
   Columbus, Christopher, 69–72
   Cortés, Hernán, 73–75
   El Dorado, 91–92
   Ferdinand II of Aragon, 106
   Hume, David, on, 162
   Inca and, 170–171
   Isabella I of Castile, 176, 177
   Pizarro, Francisco, 223–225
   pre-Columbian gold, 230–232
   Price Revolution and, 234
   Tenochtitlan, 262–263
   See also Columbus, Christopher; Cortés, Hernán; Ferdinand II of Aragon; Isabella I of Castile; *Nuestra Señora de Atocha*; Pizarro, Francisco; Tenochtitlan
spirituality. *See* religion and spirituality
SS *Republic*, **254–255**
St. Edward's Crown, 81
Stemm, Greg, 254–255
structural theory of inflation, 172
subprime mortgage crisis, 26
Summa Theologica (Aquinas), 8–10
sumptuary legislation, **255–257**
   Henry VIII, 155
   jewelry, 180, 181
   lace, gold, 195
   See also jewelry; lace, gold
Sutter, John Augustus, 49, 257–259
Sutter's Mill, 49, **257–260**
   See also California Gold Rush; gold nuggets

Tabari, al-, 33
Talbot, William Henry Fox, 61

"The Tale of the Golden Cockerel," 261–262
technology and metallurgy
   aerospace industry, **1–2**, 95, 111
   assaying, **16–17**, 264–266
   Au, **19–20**
   biomedical research, 6–7, **34–36**, 209
   bonding wire, **37–38**, 94
   casting, **52–54**, 86, 182, 232
   catalysis, **54–56**, 110–111, 209
   chrysotype, **61–62**
   cyanide, **81–82**, 94, 95, 203
   electroplating, 1–2, 37–38, **94–95**, 115
   fuel cells, 1, **110–112**, 209
   gold nuggets, **117–118**, 190, 259
   granulation, 108–109, **142–143**, 246
   touchstone, **264–266**
   See also alloying; filigree; gilding; goldsmithing; karat; mining; nanotechnology
Temple of the Golden Buddha, 283–284
Tenochtitlan, **262–264**
   Cortés, Hernán, 74–75
   pre-Columbian gold, 232
   Spanish Conquest, 252
   sumptuary legislation, 256
   See also Cortés, Hernán; Spanish Conquest
Thailand, 283–284
Theophratus, 265
Thornton, Henry, 44
Thutmose I, 152, 153–154
Thutmose III, 152–153
Tierra Firme treasure fleet, 216–218
Tom, James, 151
tombs, royal. *See* Silla, Royal Tombs of; Tutankhamun; Ur, Royal Tombs of
touchstone, **264–266**
   See also assaying
Tower Mint, **266–269**
   Briot, Nicholas, 42
   coin clipping, 63
   guinea, 144–145
   Henry VIII, 156
   Newton, Sir Isaac, 144, 210
   Price Revolution, 234
   See also coin clipping; Henry VIII; Newton, Sir Isaac

trans-Saharan gold trade, 14, 214–215, 234, **269–270**
The Travels of Marco Polo (Polo), 227–228
Très Riches Heures, 168
Trial of the Pyx, 268
troy ounce, 43
Turin papyrus, 215
Tutankhamun, 79, 138, 269, **270–273**
Twain, Mark, 113, 114

UMO (cosmetics company), 29
United Kingdom. *See* Great Britain
United States
   Bank Conservation Act of 1933, 96–97
   Coinage Act of 1873, **66–68**
   Constitution, 149–150
   Department of Energy's Brookhaven Lab, 111
   Emergency Banking Relief Act of 1933, **95–97**
   Federal Reserve Act of 1913, 97, 103
   gold standard, 120–121, 122, 123
   Keynesianism after global economic recession, 189
   leprechauns, 197
   NASA, 1–2
United States Bullion Depository, 120, 134–136, **275–276**
   See also bullion; gold reserves
Ur, Royal Tombs of, **276–278**
   casting, 52–53
   crowns, royal, 79
   filigree, 108
   goldsmithing, 138
   granulation, 142
   jewelry, 179
   touchstone, 264
Ustasha, 245–246

Valverde, Vicente de, 18, 19
Vasari Corridor, 229, 230
Vatican Bank lawsuit, 245
Velazquez, Diego, 73–74, 75
vending machines, gold, 120
Venice, Italy, 24, 47–48, 89–90, 195
Versailles, Palace of, 139, **279–281**
   See also gilding
Victor Meyer (jewelry company), 102

Wadi Allaqi, 214–215
Ware, Mike, 62
Washington, George, 149, 150
Wat Traimit, **283–284**
Watt, James, 65–66
The Wealth of Nations (Smith), 237, 248
"Welcome Stranger" gold nugget, 118
Western New York Hotel, 136
White, Harry Dexter, 40
Willy Wonka and the Chocolate Factory (film), 134
Witwatersrand Gold Rush, **284–286**
   See also gold rushes
The Wizard of Oz (Baum), 67, 122, 287–288
Wolfram von Eschenbach, 159
Wolsey, Thomas, 107, 155
Woolley, C. Leonard, 276–277
World Bank, 39, 41
World Gold Council, 175
World War I, 119, 123, 186–187
World War II, 83–84, 119, 221, 229

Yellow Brick Road, 67, 122, **287–288**
   See also gold standard

zecchino, 89–90, 251
Zuan di Franza, 48
Abolition, 22, 25, 26, 57, 68, 196–97, 212, 226, 256, 259.
See also Antislavery; Reform
Abolitionism, 27, 30, 32, 111–12, 176, 216, 240.
See also Animal rights and animal welfare
Activism and protests, **1–8**, 27–31, 33, 35, 36, 42, 43, 50, 75, 88, 90–91, 94, 96–97, 109, 114, 119, 147, 149–51, 158, 166, 167–68, 173, 175, 179, 181, 197, 256, 266.
See also Alcott, William A.; American Vegetarian Society; Baur, Gene; Clubb, Henry S.; Dinshah, H. Jay; Francione, Gary L.; Graham, Sylvester; Houston, Lorri; International Vegetarian Union; Lyman, Howard F.; Macfadden, Bernarr; Metcalfe, Rev. William; Newkirk, Ingrid; Pacelle, Wayne; Pacheco,

Alex; People for the Ethical Treatment of Animals; Reform; Salt, Henry S.; Vegetarians and vegans, celebrity; Vegetarians and vegans, noted; Vegetarian Society of America; Youth

Adams, Carol J., xx, 4, 74, 90, 158, 240, 251
Adventism.
See Seventh-day Adventists
African Americans, xiv, 12–14, 79, 93, 96, 178, 205, 244, 245, 252

About the Author

Shannon L Venable, veteran writer and educator, is an authority on the economic and social implications of luxury and wealth. Widely published, Shannon's research has encompassed a vast range of topics in her respective fields, covering, among other areas, political aspects of conversion to Christianity during the 4th century, the significance of sumptuary legislation in medieval and Renaissance Italian city-states, and the late-20th-century emergence of sophisticated debt instruments in global financial markets. A longtime editor and writer for *Historical Abstracts*, Venable presently leads annual history and culture programs in Europe as founder and director of Arte al Sole. As proprietor of an editorial and writing services agency specializing in history, lifestyle, and travel, Santa Barbara, California–based Shannon is a passionate advocate for the study of history and society.